MW01516906

WARFARE IN CULTURAL CONTEXT

Amerind Studies in Archaeology

series editor JOHN WARE

Volume 1

Trincheras Sites in Time, Space, and Society

Edited by Suzanne K. Fish, Paul R. Fish, and M. Elisa Villalpando

Volume 2

Collaborating at the Trowel's Edge: Teaching and Learning in Indigenous Archaeology

Edited by Stephen W. Silliman

Volume 3

Warfare in Cultural Context: Practice, Agency, and the Archaeology of Violence

Edited by Axel E. Nielsen and William H. Walker

Warfare in Cultural Context

Practice, Agency, and the Archaeology of Violence

EDITED BY **Axel E. Nielsen**
AND **William H. Walker**

The University of Arizona Press
Tucson

The University of Arizona Press
© 2009 The Arizona Board of Regents
All rights reserved

www.uapress.arizona.edu

Library of Congress Cataloging-in-Publication Data
Warfare in cultural context : practice, agency, and the archaeology of
violence / edited by Axel E. Nielsen and William H. Walker.
        p. cm. — (Amerind studies in archaeology ; v. 3)
    Includes bibliographical references and index.
    ISBN 978-0-8165-2707-6
    1. Indians—Warfare. 2. Indians—Rites and ceremonies. 3. Warfare,
Prehistoric—America. 4. Social archaeology—America. 5. Violence—
America. I. Nielsen, Axel E. II. Walker, William H., 1964–
    E59.W3W375 2009
    303.60973—dc22
                                                        2009025766

Publication of this book is made possible in part by the proceeds of a permanent
endowment created with the assistance of a Challenge Grant from the National
Endowment for the Humanities, a federal agency.

♻

Manufactured in the United States of America on acid-free, archival-quality paper
containing a minimum of 30% post-consumer waste and processed chlorine free.

14 13 12 11 10 09   6 5 4 3 2 1

CONTENTS

Introduction: The Archaeology of War in Practice   1
Axel E. Nielsen and William H. Walker

PART I

1   Variation in the Practice of Prehispanic Warfare
    on the North Coast of Peru   17
    Theresa Lange Topic and John R. Topic

2   Culture and Practice of War in Maya Society   56
    Takeshi Inomata and Daniela Triadan

3   War Is Shell: The Ideology and Embodiment
    of Mississippian Conflict   84
    Charles R. Cobb and Bretton Giles

4   Warfare and the Practice of Supernatural Agents   109
    William H. Walker

PART II

5   Warfare in Precolonial Central Amazonia:
    When Carneiro Meets Clastres   139
    Eduardo Góes Neves

6   Warfare and Political Complexity in an Egalitarian Society:
    An Ethnohistorical Example   165
    Polly Wiessner

7   Warfare, Space, and Identity in the South-Central Andes:
    Constraints and Choices   190
    Elizabeth Arkush

8   Ancestors at War: Meaningful Conflict and Social Process
    in the South Andes   218
    Axel E. Nielsen

9   Wars, Rumors of Wars, and the Production of Violence   244
Timothy R. Pauketat

Notes   263

References Cited   269

About the Contributors   317

Index   322

WARFARE IN CULTURAL CONTEXT

Introduction

The Archaeology of War in Practice

Axel E. Nielsen and William H. Walker

Current Perspectives on the Archaeology of War

Contemporary archaeological studies of war tend to rely on a priori pragmatic assumptions about the conduct of armed conflicts and their relationship to social processes. It is assumed that—within certain socio-environmental parameters—war is always essentially the same thing and follows a common logic, breaking out under similar conditions and having analogous effects on people involved. This logic leads to the search for universal models regarding the causes and consequences of war. Usually, these models are qualified only through the recognition that expressions of war vary according to societal types, as in the common distinction among tribal, chiefly, and state-level warfare (Carneiro 1970, 1981; Johnson and Earle 1987; Redmond 1994). In these studies, it is commonly argued that identifying the causes of war may help prevent or reduce violence in the present, as well as illuminating the role of past conflict in social evolution.

This perspective results in a two-step research program for the archaeological study of war (cf. Reyna 1994a). The first step is one of model building, a task that typically relies on ethnographic data and cross-cultural comparisons together with utilitarian, functional, and/or sociobiological arguments. These elements are combined into general propositions that seek to account for the occurrence of war in different times and places. In a second step, these generalizations are tested against prehistoric cases. This step is important, because archaeology gives access to the highest diversity of conditions and organizational forms, including hypothetically relevant ones that only existed before the expansion of Western powers and therefore lie beyond the reach of ethnographic and historical investigation.

Straightforward as it seems, this positivist program has proven difficult to apply, mainly because war seems to escape simple generalization. Warfare scholars have proposed a host of material and biological factors that could be responsible for intergroup violence, from population pressure and environmental stress to the search for economic or political gains in the form of land, prey animals, labor, mates, trade goods, tribute, power, or prestige (e.g., Chagnon 1988; Ferguson 1990; Haas 2001; LeBlanc 2003a; Meggitt 1977; Vayda 1961). Some authors think that certain social structures tend to generate conflict (Divale, Chameris, Gangloff 1976; Otterbein 1970; Reyna 1994b), while others see the roots of war in human (or male) dispositions born out of the evolutionary history of our species (Knauft 1991; Wilson and Daly 1985; Wrangham and Peterson 1996). These factors may be combined in complex, multivariate models (Otterbein 1994, chap. 13) whose predictive value, however, is hard to assess given the absence of an unambiguous causal hierarchy. The consequences of war seem to be equally diverse and unpredictable, including the possibility of completely opposite effects, e.g., population redistribution or aggregation, economic intensification or resource depletion, political integration or fragmentation, organizational complexity or evolutionary stagnation (Carneiro 1970; Cohen 1984; Earle 1997; Ferguson 1994; Haas 1990; Otterbein 1994). Although some authors see a theoretical convergence in the anthropology of war (Otterbein 1994:172), the fact is that of the many hypotheses that have been proposed about the causes or consequences of conflict, very few of them have ever been falsified or left the literature (Otterbein 1999:800).

The archaeological testing of these ideas has also been ambiguous. Beyond the notion that violence was present even in our most remote past and that it probably increased with the shift to sedentism and agriculture, there is little agreement concerning even basic issues—such as how frequent conflict was in prehistoric times—not to mention conflict's role in political change (Ferguson 1997; Guilaine and Zammit 2005; Haas 2001; Keeley 1996; Wright 1989). To some extent, this reflects differences regarding even more basic issues, like what constitutes appropriate archaeological evidence of war or how to go about comparing the relative frequency or intensity of conflict on the basis of material remains alone (Arkush and Stanish 2005; LeBlanc 1999; Martin and Frayer 1997; Topic and Topic 1987; Walker 2002; Wilcox and Haas 1994).

A common characteristic of most studies conducted according to this program is a lack of a serious consideration of agency and culture, a problem that in our view underlies many of the difficulties found by scholars seeking a universal theory of war and the application of its generalizations to prehistoric cases. Certainly, the existence of subjective motivations for fighting (revenge, fear of sorcery, attempts to satisfy blood-thirsty gods, etc.) are customarily recognized (e.g., as "efficient causes" [Otterbein 1994]), but they are usually treated as little more than misperceptions of material conditions, where the "true" causes of conflict are to be found (Ferguson 1990; Keeley 1996; LeBlanc 2003a). As Steven LeBlanc (2003:9) puts it, "people had to fight for real reasons," a statement that leads to the conclusion that the notions of prestige and revenge, and ultimately all values and dispositions that lie at the core of culture, however conceived, are unreal. On occasion, lip service is also paid to agency—for example, when the importance of decision making by political leaders, warriors, and other individuals is explicitly recognized. The logic of these "actors," however, typically follows the researchers' rationality and pragmatic assessments of objective structural conditions.

Religion and other cultural practices, beliefs, and institutions that are always woven into the conduct of war are also treated as epiphenomenal (Ferguson 1990:46; Le Blanc 1999:14), probably because they seem to offer little objective advantage to the fighting parties. They only seem to acquire relevance in relation to the question of whether "primitives" had "real war." According to some authors (e.g., Turney-High 1949), these cultural curiosities render "primitive" combat ineffective—unlike modern warfare, presumably more pragmatic—while others think that they do not alter the lethal nature and essential logic of war whenever it occurs (Keeley 1996). Some scholars acknowledge the possibility of a cultural approach to war, but the fact remains that this perspective is left out of most studies of the issue (e.g., Reyna 1994a, note 1). This is not surprising, since culture is typically conceived of in idealistic terms, as a "symbolic belief system" (Ferguson 2001:104) disembedded from its historical context and "practical entanglements," leaving room only for self-contained and nonexplanatory culturalist accounts such as: "people will fight for whatever culture tells them to" (Ferguson 2000:162).

A Practice Approach to War

One of the goals shared by the various strands of practice and agency theory is to overcome a number of conceptual dichotomies that have been pervasive in the Western intellectual tradition, such as those between subjectivism and objectivism, idealism and materialism, agency and structure (Bourdieu 1977; Giddens 1984; Sahlins 1981). Practice can be broadly characterized as culturally informed and historically contextualized action. This concept defines a standpoint that, without losing sight of material conditions and structural regularities, takes into consideration the role of culturally unique dispositions, proximate causation, and agency in the analysis of social process. By calling attention to what people actually do and how they think and feel about it, a strategic focus on practice can restore the weight that a positivist, utilitarian approach to the study of warfare has denied to these factors. Let us illustrate this point with a few examples.

Cross-cultural research by Carol Ember and Melvin Ember (1992, 1994) is often cited as strong proof of the importance of environmental factors in the explanation of war (Keeley 1996; Lekson 2002; Otterbein 1994:197). Based on Human Relations Area Files data, these authors analyzed the relationship between the frequency of armed conflicts and multiple variables across a number of ethnographic cases. Contrary to some straightforward, environmentally deterministic explanations, what they discovered was that the frequency of war primarily correlated with resource unpredictability due to natural disasters (such as those produced by pests or weather) rather than chronic or regularly recurring environmental stress. This finding supported the notion that the main cause of war was the perceived threat of resource scarcity. The second-highest correlation was found with "socialization for mistrust," a variable that the authors consider likely to be a consequence rather than a cause of war. Although these results were translated into "ecological" and "psychological" explanations of war (Ember and Ember 1992), the lack of correlation between conflict and actual resource scarcity led the authors to conclude unambiguously that *fear* was the real cause underlying both predictors of war: "fear of nature and fear of others" (Ember and Ember 1992:258; 1994:194).

From our perspective, what these results actually indicate is that the causes of warfare cannot be found exclusively in objective or subjective conditions, but instead lie at the "intersection" between them; they are to be found in the ways in which certain external, material conditions (e.g., environmental fluctuations, previous conflict) are subjectively construed within specific societies (e.g., fear, witchcraft beliefs) and by certain individuals (leaders, warriors, parents, and children). Needless to say, these perceptions of a threat—or the ways in which they are trans-mitted in the socialization process—cannot be understood without ref-erence to specific understandings of the forces of nature, other people, and violence; of existing forms of resource appropriation; of notions of landscape and place; and of shared memories of past interaction—to mention just a few important aspects involved in actual practice. Ignor-ing them would be just as arbitrary as pretending that these culturally construed dispositions are unrelated to any objective, environmental, or political reality, as "culturalist" explanations presumably do.

An emphasis on practice also avoids reifying analytical categories such as infrastructure-superstructure or economic-political-ideological, typical of so-called "materialist" accounts of war that currently dominate archaeological inquiry on the subject. According to R. Brian Ferguson (1990, 2001)—one of the leading advocates of a materialist approach—warfare is best explained by a nested hierarchy of constraining factors that gives causal priority to infrastructure (demography, technology, organization of labor, and environment) followed by structure (kinship, economy, politics), and that only at the end takes into consideration the influence of superstructural variables (ideology). "Wars occur when those who make the decision to fight estimate that it is in their material interests to do so" (1990:30), where material interests can be access to fixed or movable resources, other people's labor, decision makers' power within their own communities, or forestalling attacks by others. The author is aware that the distinction among different levels in his causal hierarchy is problematic, since he recognizes that the superstructure defines how war is conducted and how material interests themselves are construed, but curiously, this idea is dismissed right away:

> Superstructural patterns shape the way individuals perceive and act on conditions related to war. Calculations of material loss and gain neces-sarily must consider relevant properties of the existing social universe, and that includes the values and rules by which individuals are expected to live. . . . But independent of infrastructural and structural patterns

conducive to war, superstructural elements have a very limited effect. (Ferguson 1990:31)

Certainly, we do not intend to make the case for a causal preeminence of ideology over technology, or to argue that "ideas" by themselves can push people into organized violence without any commitment of material factors. Instead, we want to question the relevance of this and similar dichotomies, especially the dilemma of choosing between ideal and material explanations of conflict or any other practice. There is no such thing as a technology or interaction with the environment devoid of cosmological beliefs and cultural dispositions, let alone a religion devoid of materiality and bodily action, including use of resources and organization of labor. The speculation about the "independence of infra- and super-structural patterns," we contend, is a theoretical construct that only obscures how people actually experience the world and act upon it in any situation, violent or not.

A practice approach does not begin dividing systems or social "wholes" into artificial subsystems or levels in order to explain one of them by reference to the other, e.g., the superstructure by reference to the base or ideology as an outcome of social structure. Instead, it seeks to understand the patterns we call "system," "structure," or "culture" by reference to practice itself (Ortner 1984:148). We can think of practice simply as what people do. A strategic focus on this continuous flow of acts, materials, and ideas, on the actors (human or not) and collectivities it assembles (Latour 2005), on the unique practical logics it traces, and on the contingencies it experiences over time raises a different set of questions, expanding archaeological studies of warfare into areas that have been largely ignored by the positivist program.

The point of departure is warfare itself—how it is actually conducted and understood by specific people. This calls attention to a number of other practices that are intertwined with armed confrontations, forming unique "fields of action." We may then explore how these configurations came into being through a history of intersubjective encounters and negotiations, a contingent sequence of practices in which acts, materials (objects, spaces, bodies), ideas (knowledge, beliefs, values, attitudes), and dispositions shaped each other under changing socio-environmental conditions.

A practice approach also requires addressing the role of agency in the explanation of war. Although Ferguson presents his synthetic model of war as an attempt to take agency into consideration, his decision makers are just rational problem solvers or maximizers of infrastructural possibilities (see the "motivation premise" [1990:29]).[1] Certainly, people pursue interests, but this is just another way of saying that human action is intentional. To bring agency into play, in contrast, requires dealing with the complex ways in which "interests" and perceptions of conflict are formed in different cases. Moreover, since agency involves structuration, it is also important to consider the ways in which these inherited value frameworks, beliefs, and dispositions are recursively transformed through fighting or avoiding conflict, a point that is especially relevant in accounting for the historical dynamics of violence (e.g., feuding or cycles of war and peace).

Going back to the list of "material interests" pursued by war makers, one may find that some of them are self-evident (e.g., land, tribute, or labor), but others (e.g., valuables, people's positions in their own communities, or security) are difficult to assess outside specific contexts of practice. For example, how can we determine whether a person will improve his social standing by participating in war, by waging war only in certain ways, or by avoiding conflict all together, without reference to cultural values and dispositions and to the history that brought them into being? Could anyone make a contribution to understanding present-day conflicts or to world peace—as it is often declared in the "anthropology of war" literature—without taking seriously into consideration how collective projects (including economic or political ones) or perceptions (e.g., of threat) were developed, or how specific actors on all sides (presidents, soldiers, "the public") interpret—often religiously—their own interests?

If we accept the idea that purposeful action is a significant factor in the study of conflict, we need to pay attention to the ways in which motivations develop in practice and in specific contexts (e.g., Albert 1990; Harrison 1989). What is the role of violence in the construction of prestige or senses of self in particular societies? How do those who make decisions regarding war (political leaders, warriors, people who support them) construe their identity, their hopes, and their life projects? How do they experience their condition, their problems (scarcity, threat), or possibilities? How do they assess alternative courses of action, such as fighting, moving, surrendering, forging new alliances, peacemaking?

A practice research strategy can also take archaeological methods for the study of prehistoric war beyond the mechanical application of checklists of "diagnostic" traces. Warfare is commonly defined as a state of armed hostility between politically autonomous social units and distinguished from other forms of violence, such as feuding, human sacrifice, domestic violence, or criminal behavior. Such definitions may be necessary to single out activities, artifacts, and activity settings for analysis, but they often lead to the arbitrary isolation of intergroup hostility from other actions, institutions, and beliefs that are always—but contingently—related to it, forming a seamless web. These relationships should not be ignored because they are critical for understanding how armed conflicts were conceived and conducted in specific cases, and therefore for establishing the relevant archaeological evidence for studying war in the past. Practice theory, then, demands "thick," richly contextualized analyses aimed at understanding the role of conflict in historically constituted fields of action. Only this kind of approach may be able to explain why in some cases the "classic" traces of war are conspicuously absent (e.g., Robb 1997). It will also throw light on the role played by other conjoining practices that seemingly lack utilitarian value in combat (e.g., Kolb and Dixon 2002; Topic and Topic 1997), and it may demonstrate that some kinds of evidence commonly attributed to war are sometimes the result of other activities.

Focusing on practice is also a way of foregrounding the particular (place, time, context, situation, actor), a point that marks a significant contrast with the recent trends in the archaeological literature on war, as outlined previously. This is not to say that a practice approach to war needs to be purely descriptive and cannot generalize or engage in cross-cultural model building. Rather, it is a reminder that practice itself, with all its relevant specificities, is the arbiter of our theories. It is our impression that some of the theories regarding the "causes and consequences of war" that thrive in the archaeological literature have achieved their universal status at the expense of crucial details of practice (cf. Thorpe 2003). In this context, a tactical emphasis on specificity seems in order.

Archaeological Case Studies

This volume is the result of a symposium organized for the sixty-ninth annual meeting of the Society for American Archaeology, held in Montreal in 2004 and expanded later that year into an advanced seminar sponsored by the Amerind Foundation in Dragoon, Arizona. Participants in seminar sessions were asked to explore the possibilities (or drawbacks) of a practice approach to the archaeology of warfare, taking advantage of archaeological, historical, and ethnographic data from societies of different scales and time periods around the world. The chapters of this book offer a good illustration of the multiple avenues for archaeological research that open before us by looking at armed conflicts from the perspective of practice. To pursue the more contextual demands of practice-oriented research, all the papers embrace the importance of detailed study of ethnohistoric and ethnographic data in their respective culture areas. All seek to move beyond dichotomous thinking and to apply synthetic understandings of war, religion, politics, and technology. Despite this underlying unity, the papers congealed into two groups, allowing us to divide the volume into two parts.

In Part I (chapters 1–4), papers focus more directly on the relationship between practice and cultural codes, warrior identity, and religion. How the practices embody or link people's activity to the realms of mythology, supernatural power, and the animacy or agency of artifacts and architecture are important concerns. The chapters in Part II incorporate many of these themes as well but highlight the relationship between practice and political change.

Part I

In chapter 1, Theresa Topic and John Topic review the long history of prehispanic conflict on the north coast of Peru from 3500 BC to the time of the Inka expansion in the fifteenth century. Taking an "anthropological history" stance, they combine the analysis of archaeological evidence (architecture, weapons, iconography) with ethnohistoric and ethnographic data on Andean cosmology to approach the changing cultural motivations that led to those conflicts and the associated rules that dictated how war was waged in each case. They argue that combat and warrior sacrifices depicted in Moche iconography were highly ritualized confrontations among members of different communities belonging to the same ethnic group. In contrast, earlier Late Formative fortresses may have served as stages for pre-arranged ritual battles between local communities. Despite their different social connotations and archaeological expressions, war-related practices in both periods show analogies with ethnographic ritual battles or *tinku*. These actions, they propose, were central to the regeneration of life and political order within a long-lived Andean cosmology that emphasized the need for energy flows among different worlds, especially the reciprocity between humans and supernatural entities that regulated this process.

Codes of conduct in warfare of Preclassic, Classic, and Postclassic Maya civilization are the focus of the next paper, by Takeshi Inomata and Daniela Triadan (chapter 2). Their study highlights the importance of ritual practice for regulating and transforming conduct of war. They emphasize that practice theory is useful for understanding how, in specific historical contexts, cultural codes are maintained as well as transformed. They call for greater attention to the relationship between the conscious and unconscious elements of practice-shaping cultural codes. Similar to what the Topics do in chapter 1, Inomata and Triadan identify long periods of relatively stable forms of conflict that emphasized "ritual battles" in which warriors focused on gaining prestige through the capture of enemies for sacrifice in communal ceremonies. They also identify shorter-term cases in which such codes of warfare change to more destructive forms, leading to the rapid construction of defensive architecture, evidence of the fiery destructions of centers, and the collapse of regional polities. They argue that these data suggest that the landscape of Maya warfare was dynamic in all periods. They conclude that it is important to consider how self-interested decisions of leaders and participants in war articulated with the unconscious routines of daily life and communal ceremonies that enacted, re-enacted, and imagined Maya war.

In chapter 3, Charles Cobb and Bretton Giles provide another example of the relationship between warrior identity and the conduct of war. They explore the temporal and spatial variation in ideology associated with the pursuit of war in the Mississippian cultures of southeastern North America. Their approach emphasizes the importance of understanding the experience of warrior identity as it was embodied in

Mississippian practice. They focus on representations of mythological warriors in the imagery and objects of the Southeastern Ceremonial Complex (e.g., maces, shell gorgets, shell trumpets, copper plates), a collection of related religious practices that reached its apogee in the thirteenth and fourteenth centuries. They argue that representations of the human body, body parts, human-animal chimeras, and weapons in this complex embody beliefs about status, gender, power, and the cosmos that contributed to the production and reproduction of warfare in the later prehistoric Southeast.

William Walker (chapter 4), in the final paper of Part I, attempts to reorder thinking about the relationship between religion, war, and practice by analytically extending the notion of practice to nonhuman agents. He notes that in many human practices, including warfare, people attribute agency to artifacts and architecture that embody nonhuman social actors such as deities, ghosts, witches, and ancestors. He argues that archaeologists should consider the implications of such practices in their methods and interpretations of prehistoric warfare. Building on work by Alfred Gell (1998) and Michael B. Schiffer and Andrea Miller (1999), he suggests that we can measure the practices of spirits indirectly by measuring the practices of people. In a case study of warfare in the American Southwest, he illustrates how pueblo peoples' understandings of the practices of nonhuman agents can explain archaeological patterns of human remains, ceremonial architecture (namely kivas and pit houses), and defensive structures such as towers.

Part II

Eduardo Neves (chapter 5) summarizes the evidence of prehistoric war in the Amazon and Orinoco basins. He recognizes that the archaeology of Amazonia is still largely unknown but notes that some initial patterns of warfare are emerging in association with fortified, often ditch-encircled sites. In contrast to scholars who argue that the ethnographic record of the region is not applicable to prehistoric cases, he finds inspiration in contemporary Amazonian ethnography. Drawing on the work of Pierre Clastres (1974), he situates Amazonian practices of war in a context of centralizing ideologies and household-based economic production. He argues that the evidence of warfare he has identified corresponds to periods of settlement aggregation and offers the hypothesis that war was a catalyst in the cyclical rise and fall of regional polities.

The rich oral history of the Enga of Papua New Guinea highlands offers Polly Wiessner (chapter 6) the possibility of exploring cycles of war and their political significance in an egalitarian society. She analyzes how economy, political institutions, religion, and cultural values interacted in complex ways to shape the ways in which armed conflicts were conducted and their changing role in social reproduction over the past three hundred years. Her findings contradict Mervyn Meggit's (1977) thesis that the Mae Enga fought over land that was scarce due to population pressure. By demonstrating the close association between armed conflicts and the Tee exchange through peacemaking practices, Wiessner argues in support of the Enga's own explanation of war as a last resort to restore honor to their clans and the balance between groups so that exchange might flow. The marriage of warfare and trade in turn altered egalitarian structures, leading to a new configuration associated with the Great Ceremonial Wars, that allowed "big-men" to circumvent egalitarian structural constraints and harness warfare toward their economic and political advantage. She concludes that, in order to understand the political impact of warfare, we need to look at its relationship with other dimensions of culture.

The case study presented by Elizabeth Arkush (chapter 7) is situated in the period of endemic warfare before the expansion of the Inkas. Using geographic information system applications, she analyzes the distribution of fortified hilltops (*pukaras*) in the central Andean Titicaca basin, identifying several clusters of intervisible sites that she interprets as networks of allied communities. This interpretation is further supported by the association of some of the sites with different ceramic types. She also finds a correspondence between some cluster boundaries and the frontiers of ethnic groups recorded in sixteenth-century written sources, which formed the basis of colonial administrative units. The combination of these multiple lines of evidence leads her to infer that the ethnic and political mosaic found by the Europeans when they arrived in the region had its origins in the time of endemic conflict three centuries earlier. Although she perceives the actions of pre-Inka communities as highly constrained, practice theory allows her to approach this social landscape as the outcome of cumulative choices and negotiations concerning place and identity.

The armed confrontations discussed by Arkush in chapter 7 also characterized the late prehistory of the south Andes addressed by Axel Nielsen's chapter (chapter 8). Nielsen's goal is to look at the ways in which a focus on practice can inform our understanding of the relationships between warfare and political change. He takes advantage of the multiple uses and meanings of weapons and other war-related objects as a window into ancient Andean understandings of conflict. In this way, he identifies the close relationship between war and other semantic domains, including authority, fertility, transmutation, and ancestorship. Nielsen argues that ancestor worship—which had a long tradition in the south Andes—provided the cultural logic through which political integration and institutional complexity developed in the face of escalating insecurity. This contingency simultaneously affected a shift toward a corporate mode of political action, characterized by low levels of economic and political inequality, that discouraged conspicuous consumption and other expressions of individuality. Ancestors, as referents of collective identities and interests, rather than war leaders, were the ones to accumulate the prestige and power born out of conflict.

The closing chapter, by Tim Pauketat (chapter 9), discusses some common themes that run across the chapters and highlights some new questions and avenues of research that are raised by a practice approach to war. In Pauketat's view, it is particularly important to situate war and peacemaking in a broad context of culture production, calling for a social archaeology of violence that looks beyond the battlefield.

During the seminar that resulted in this volume, it became clear that although all participants agreed on the importance of a focus on how war is conducted and experienced by different people, they had different ideas regarding what a practice approach should look like and how it can contribute to an archaeology of war. While some participants embraced the approach as a full research program, others treated it only as a source of useful conceptual tools that can complement current approaches—for example, by putting more emphasis on cultural perceptions or individual motivations—without challenging their theoretical foundations. Still others were skeptical of the very notion that

practice can articulate a distinctive theoretical framework, using it only
as a heuristic device to formulate different kinds of questions about war,
material culture, or the past.

"Practice" has become a key word referencing multiple bodies of
post-structuralist social theory, so appeals to practice in archaeology
are associated with many different things and draw on various bodies of literature (cf. Dobres and Robb 2000; Dornan 2002; Pauketat
2001). Not surprisingly, this diversity is also reflected in this collection
of papers (Pauketat, this volume), regardless of the different degrees of
importance the authors are willing to give this concept in their studies
of war. Drawing mainly on Pierre Bourdieu, most contributors understand practice as the consideration of embodied "cultural logics," while
for a few, it implies a focus on history, context, and the contingencies
of particular trajectories as central to explanation. Some of the authors
in this volume follow a phenomenolgical approach, using the materiality of practice (bodies, places, objects) as a window into the realm of
experience and the process through which subjectivity is constituted.
For most, practice also implies a shift from structural to agency-based
explanations. Certainly these various theoretical approaches are not
necessarily contradictory or incompatible; indeed, most archaeologists
interested in practice currently integrate elements from several of these
lines of thought.

PART I

1

Variation in the Practice of Prehispanic Warfare on the North Coast of Peru

Theresa Lange Topic and John R. Topic

The symposium on practice theory and warfare at the Society for American Archaeology annual meeting in April 2004 and the subsequent Amerind Foundation conference took place in the midst of much-heightened interest in conflict. In the last few years, there has been a series of escalating hostilities on the international stage, and new technology has allowed all of us to be constant witnesses to these conflicts. Warfare has come to seem ubiquitous because it is so present in public consciousness. On the anthropological front, not long before the symposium and conference in question, Steven LeBlanc had published a volume on warfare in prehistory that attracted a very high level of public attention, and his stance that warfare is a universal human condition had been widely communicated; his argument that the causes of warfare are simple and straightforward (LeBlanc 2003:9–12) and that people fight for "real reasons"—notably, scarce resources—had found many supporters.

As the editors of this book explain (Nielsen and Walker, this volume), the focus on practice theory allowed prehistoric warfare to be considered from perspectives very different from the positivist model that has recently predominated in most discussions of the topic.

We brought to the symposium considerable background in the archaeological record of the north coast of Peru and an interest in reconsidering from a practice perspective the corpus of data and interpretations relating to warfare in that area. We surveyed coastal and highland fortifications from 1977 to 1980 and subsequently studied adjacent sierra cultures. Our view of prehistoric warfare in the central Andes has changed considerably over the years (e.g., Topic and Topic 1987, 1997b), and we now consider ourselves to be working within a framework of "anthropological history." We argue that conflict is best understood within a holistic and historically contextualized cultural framework, supported by detailed observations of the evidence for conflict at specific times and places. This approach, of course, will emphasize variability in warfare as it is practiced and experienced by different cultures in the Andes—at the expense of commonalities in practice that, we argue, should be demonstrated and not assumed.

The Organization of This Chapter

There is a long record for warfare on the north coast that has been studied by a number of researchers; the intent of this chapter is to consider this record from a practice perspective and to explore variability in the actual practice of prehispanic warfare in this area. Rather than isolate one culture or another to receive our attention, we will consider warfare as it was expressed in several different cultural contexts.

In this introductory section, we will comment briefly on our understanding of practice theory. We will then present a justification for our utilization of the concept of Andean tradition as a unifying factor in the diverse practices we will explore, followed by an overview of Andean cosmology and its links to at least some kinds of warfare that were practiced. A key concept in Andean tradition, *tinku*—which shaped some forms of Andean combat—will then be explored, with consideration of the ethnographic sources for tinku-related combat in recent times.

A review of the archaeological record will then ensue, with brief consideration of Preceramic, Initial Period, and Early Horizon evidence for conflict. An in-depth consideration of the practice of Moche warfare will then be presented, followed by a brief summary of post-Moche evidence for warfare.

The final section will explore the variation in practice that is evident in the archaeological record for the north coast of Peru, presenting conclusions about the utility of a practice approach to the study of warfare.

Practice Theory as an Interpretive Framework

Practice theory provides the archaeologist with the means of considering a past that is populated with human actors who are accorded agency and at least some capacity to assess their surroundings and to make decisions based on this ongoing evaluation. In its broadest sense, a practice approach studies what people do (Ortner 1984:149) and explains action as proceeding from actors' knowledge of and experience in the social world in which they are immersed.

Practice theory is a good fit with an anthropological history approach; many of the key tenets of practice theory can be traced to the historical sociology of Max Weber (1976). Weber considered the motivation of actors to be important and situated his discussion of actors' motivation within particular cultural and historical contexts. Practice theory also intersects in interesting ways with the philosophy of history developed by R. G. Collingwood (1994). Collingwood's interest in the process of interpreting the past was at least as great as his interest in what actually happened in the past. He pointed out that analysts cannot separate their interpretations of a past cultural context from the mindset they bring to the process of interpretation, and that an analyst's mindset is historically and culturally contingent. Collingwood was one of the intellectual forbears of trends leading to postmodernism, which within history and archaeology led to a cultural relativism that at times left researchers who were studying cultural traditions not their own to despair over ever being able to interpret the past. That was not Collingwood's intent, of course, but he did insist that history (as an intellectual exercise, rather than a synonym for the past) existed only in the mind of the historian.

As an interpretive framework, practice theory encourages a focus on the historically contingent motivations of the actors whose actions and decisions shaped the context being investigated. The actors are knowledgeable and experienced participants in the social, material, and cosmological worlds they inhabit and reproduce. Practice theory recognizes the agency of these actors and their capacity to understand the fields within which they operate, to manipulate these fields, and to effect change within them. The archaeologist is compelled to adopt a more anthropological reading of the past as the sites, spaces, and times under investigation come to be peopled with active agents who participate in historically specific and continually changing social fields. In this interpretational sphere, accepted conceptual categories and boundaries shift; the categories into which analysis has tended to be channeled by archaeologists ("subsistence," "trade," "warfare," "ideology," and "politics") are suddenly broken open and become overlapping and multilayered. Similarly, from a practice perspective, it will be difficult to isolate specific cohorts as the subject of study when a multifaceted activity like warfare is under investigation, since the complexity of the interrelationships between those directly and indirectly involved will be acknowledged.

Working from a practice approach, the researcher is challenged to empirically determine the motivation of actors in each case under analysis. The credibility of the universalist stance that "real reasons" with cross-cultural validity can be defined for important processes and events is challenged, and the analyst realizes that he or she must first attempt to establish what is "real" in the subjects' minds. Linking the contingent motivations of past actors to the archaeological record requires both careful observation of the empirical evidence and a testing of that evidence against multiple frames of reference.

Andean Tradition

This chapter will discuss prehistoric warfare on the northern coast of Peru (fig. 1.1) over a long time span. A foundational premise of the chapter is that reference can legitimately be made to an Andean tradition with great time depth, and that important elements of this tradition were shared by groups who are otherwise defined as distinct cultures or ethnic groups throughout the archaeological record and into the early Colonial era. While recognizing the overwhelming disruption of the European conquest and the emergence of the modern nation, we argue that significant strands of the Andean tradition persist in the present day and that these strands carry explanatory weight.

Tradition exists at varying levels. At the "cultural" level, tradition is constituted by the beliefs, logic, history, memory, and preferences shared and learned, constantly reinforced (or subverted) by action and performance. It is dynamic, continually reproduced, and altered by historically contingent experience. All individuals learn a tradition as part of their socialization, through observation, instruction, and daily experience of the life way of the tradition as it is embodied in those who are participants in it. With increasing maturity, individuals become increasingly active participants in the tradition, agents of its reproduction and alteration. Their identity as individuals and as groups of increasing scale is grounded in that tradition.

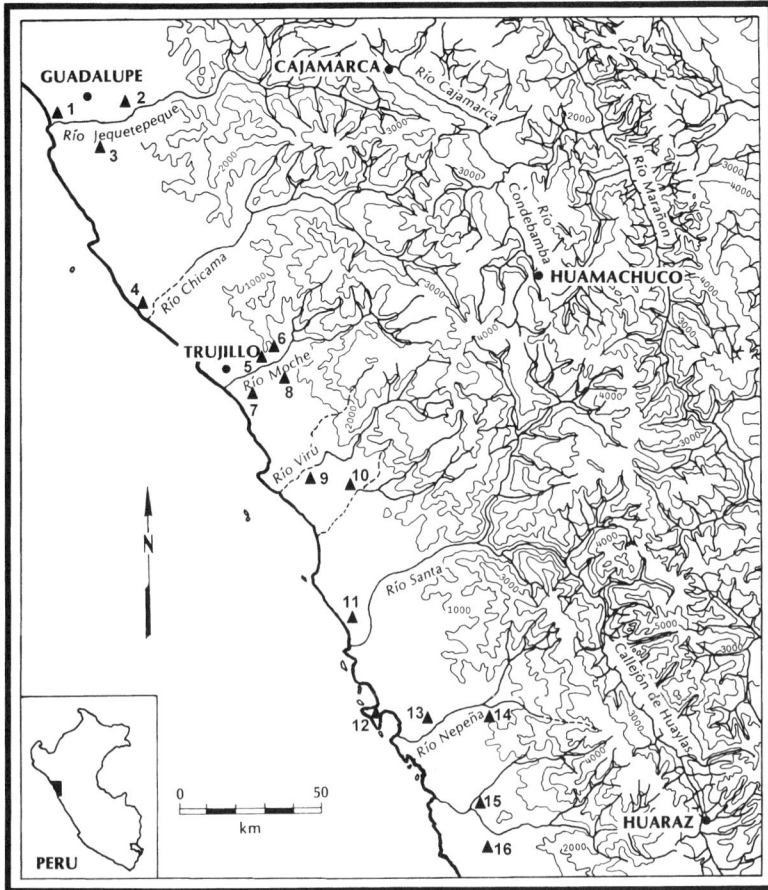

1.1. Map of the north coast of Peru showing sites mentioned in the text: (1) Pacatnamú, (2) Talambo, (3) San José de Moro, (4) Huaca de Cao Viejo, (5) Galindo, (6) Puente Serrano, (7) Huaca de la Luna, (8) Cerro Oreja, (9) Bitín, (10) Cerro de la Cruz, (11) Ostra, (12) Playa Catalán, (13) Pañamarca,(14) Quisque, (15) Cerro Sechín, and (16) Chankillo.

At a more inclusive or "macro" level, tradition is based on an array of perduring beliefs that will shape practice in similar ways in cultures widely separated in time and space. To state that peoples with common cultural ancestry will understand and interact with the world in similar ways is not particularly controversial; few would dispute the existence of a Buddhist tradition, a Judeo-Christian tradition, or a Confucian tradition, along with a characterization of these traditions as slow to change and deeply rooted. We, like many other Andeanists, recognize a macro tradition in the Andean region (Burger 1992:226; Silverman 2004:5) and consider it to be equivalent to other traditions recognized in the Old World. Many Andeanists have explored the specific elements of material culture, ecological adaptation, social organization, iconography, and cosmology that are widely shared within the tradition (cf. Bennett 1948; Moseley 2001:51–70). This tradition must not be considered immutable (Isbell 1997:16–27), and the fluidity and permeability of the tradition's geographical boundaries must be recognized (see Isbell and Silverman 2006). In general, the concept of tradition at this scale defies precise definition and analysis because of its highly variable expression and ambiguous boundaries. It is a fluid concept that can be utilized in both a deterministic and a permissive role.

While our focus will be on the north coast of Peru, we will range widely through space and time in the Andean area for explanatory and comparative material, drawing on archaeological, ethnohistorical, and ethnographic information. The empirical evidence for a culture's affiliation with Andean tradition is variable. An archaeologist must rely on material evidence, and this evidence is frustratingly uneven. Iconography, architecture, layout of sites, dispersal of features across the landscape, and burial patterns provide some of the best evidence. The ethnohistorian dealing with Andean peoples after the conquest has an extraordinary range of written documents to draw on, but interpretation is complicated by the ethnocentric lens through which writers (the vast majority of whom were working within the Spanish cultural and legal tradition) viewed indigenous practice. Ethnographic description and commentary provides much richer documentation of Andean peoples into the present day; these sources must be used with caution, given the changes of the last 475 years.

The themes running through Andean tradition that we consider to be of greatest importance for this paper are order and reciprocity. In the Andean tradition, order is highly desired. Humans play a very active role in creating and maintaining the conditions that foster order. The maintenance of appropriate relationships among all actors is a necessary precondition for order to flourish. Relationships must be maintained among human actors as individuals and as groups, as well as between humans and the array of nonhuman actors with whom they interact. Mutual obligations between any two parties must be carefully calibrated, frequently expressed, and periodically renewed. Order derives from the maintenance of proper relationships and results in economic and social reproduction.

The principle of reciprocity underlies all relationships, with a clear understanding that any service or gift or act by one party in a relationship will, in time, evoke a response by the other party (Mayer 2002:105). Not all reciprocal relationships are symmetrical, so the response may not be equivalent to the initial gift or act. The reciprocal relationship between those who govern and those who are governed was critical in Andean tradition; commoners gave service to their "lords" but expected to receive the generosity that was an obligation of lordship. There are, for example, numerous colonial petitions from north coast lords requesting authorization to maintain "taverns" where *chicha* (maize beer) was brewed and served to commoners in exchange for their labor. Many petitions were also filed with colonial authorities by north coast native leaders who sought permission to ride on horseback. Native leaders maintained that they needed to travel to their subjects and formally ask them to perform work. The proper behavior of a lord entailed first a formal request that work be done and then provision of food and beer while people performed the work (Netherly 1977). Spanish colonial authorities had great difficulty in understanding the reciprocity at the heart of the relationship between elites and commoners, since it contrasted so strongly with their own Spanish tradition of unquestioned hierarchical authority. An early indigenous writer, Guaman Poma de Ayala ([1615] 1980b), wrote a letter hundreds of pages long to the Spanish king, explaining that the Andean colony was being badly managed by Spanish authorities and recommending that greater attention be paid to Andean principles of organization. Guaman Poma assured the king that he would reap greater benefits from the colony if the Spanish adopted time-honored Andean patterns of reciprocity.

The ongoing definition and enactment of reciprocal relations is an integral part of the creation of order in Andean tradition, and this constitutes an important arena of practice.

Andean Cosmology

Cosmology is an elaborated version of what Westerners are more comfortable referring to as a "worldview"; however, it is a worldview that incorporates all the players and fields of endeavor, including nonhuman agents in those traditions that acknowledge beings or forces beyond the visible as active influencers of human social life. Cosmology provides knowledge of how the world works, including knowledge of all its social, material, natural, and supernatural elements. As such, it is the basis for understanding causality in the world (Marcus and Flannery 1999; J. Topic 1992). The cosmology that an individual has learned as part of his or her enculturation provides the conceptual framework within which he or she decides what constitutes a possible or an appropriate response in any given situation. Practice theory does not claim that an individual will always respond rationally to a situation, but the rationality of the response must be evaluated in terms of the actors' frame of reference (Cowgill 2000:54).

Viewed from any perspective, conflict is a complex field of action involving many actors and many stages of assessment, planning, preparation, enactment, and recovery. A cascading multitude of choices must be made, and these choices will occur within the explanatory framework of the culture—that is, within its cosmology. Like any other complex practice, warfare reflects the beliefs, attitudes, and experience of its practitioners. Examples abound from other traditions of the interrelationship between cosmology and the principles of warfare. In his introduction to a new translation of *Sun Tzu: The Art of Warfare*, Roger Ames notes that we cannot understand classical Chinese military principles without understanding classical Chinese thought, and that to understand the warfare of that place and time, "we must look to the dynamics of an underlying and pervasive conception of harmony (*ho*) that, for the classical Chinese world view, grounds human experience generally" (Ames 1993:43). Sixteenth-century Spanish combat was also conditioned by cosmological principles, in this case the need to offer salvation to the "infidels" they encountered. Before attacking indigenous forces, the Spanish were required to read the *requerimiento* to them, offering the possibility of salvation and eternal life to those who voluntarily accepted Spanish and Christian rule (Seed 1991:13, Sancho de Hoz [1534] 2004:48).We have argued above for an Andean tradition whose broad outlines endured for millennia, and cosmology is a key part of this tradition. Knowledge of Andean cosmology relies very heavily on ethnographies of Quechua- and Aymara-speaking communities, as well as on ethnohistoric information from the first two centuries following the Spanish conquest. While these sources often do not speak directly about the north coast, they do describe a cosmology that is widespread in the Andean area. Moreover, the antiquity of at least some aspects of the cosmology is supported by iconographic studies both on the coast and in the highlands (Burger 1992; Salazar Burger and Burger 1982; Cordy-Collins 1992; Lathrap 1973, 1982; J. Topic 2008; J. and T. Topic 2001).

Andean cosmology, like many other New World cosmologies, envisions a layered universe made up of several worlds (e.g., Nuñez del Prado 1974). The intermediate plane on which living people dwell is juxtaposed with both an upper world and a lower world. There is a flow of energy, or life force, between these worlds that allows for the renewal of the intermediate plane on which people live.

For example, animals are owned by the great mountain lords, the *apus*. The animals dwell within the mountains and the apus may either release animals to the terrestrial plane through springs or attract animals from the terrestrial plane into the earth, robbing people of their resources (Gose 1994; Flores Ochoa 1979). People hope to influence the apus by making offerings or *pagos* (payments) to them. The offerings are not simply gifts or bribes to all-powerful gods; rather, they serve to establish reciprocity between humans and the forces of reproduction that are embodied in the apus and other "natural" phenomena (such as the sun, moon, and springs)—they are part of an ongoing negotiation between actors in the social and supernatural worlds. The mountains are extremely powerful, but they are expected to behave in proper ways. If humans treat the apus with respect, the apus should respect humans. The offering given to an apu establishes a relationship between the giver and the receiver. The relationship is, of course, quite asymmetrical, but the actors consider it to be a reciprocal relationship nonetheless.

Reciprocity is as central a concept in Andean cosmology as it is in Andean social life (Mannheim 1991:19). Whether the relationship is symmetrical or asymmetrical, the giver and the receiver have complementary roles. The essential model for these roles is the reciprocal relationship between male and female. Hence the mountain apu, a phallic peak that is often snow capped, is contrasted with the flatter surface of the earth, which is *pachamama* (earth mother). Pachamama is the womb in which life is nurtured and seeds are germinated. The torrential rains from November to May, falling from the masculine heights of the celestial plane, are viewed as the semen that impregnates pachamama and leads to the renewal of the earth (Allen 2002; Platt 1986; Gose 1994). The renewal of the world is viewed as an exchange of life force between gendered elements, with masculine generally being equated with the upper and feminine generally equated with the lower. The point where regeneration takes place is at the surface of the terrestrial plane, where humans actively seek to ensure the smooth flow of life force.

The Quechua terms *yanantin* and tinku are very closely related to this cosmology of renewal. "Yanantin" essentially means "to serve together" and is applied to things that are the necessary complements of each other, such as a pair of gloves or men and women. "Tinku" is the point where complementary things join to create a larger whole; the term describes things like the center line of a person's face or the junction of two rivers. In cosmological terms, yanantin would refer to the masculine and feminine elements that are necessary for the renewal of the world, while tinku would describe the union of those gendered elements that enacts the renewal.

Tinku as Warfare

The term "tinku," however, also denotes warfare. In modern usage, the term is sometimes contrasted with the term *ch'ajwa* (Platt 1986, 1987; Hastorf 1993). Within this contrast, tinku is often glossed as ritual war, but as Tristan Platt points out (1987:164, 167) tinku is considered by participants to be an enactment of the balanced relationship between complementary forces that results in reproduction. Ch'ajwa, on the other hand, is often glossed as real war over real resources, such as land, but ch'ajwa disrupts the social fabric by transforming disputes about land and water into violent conflict. Tinku is a positive encounter, often between opposing moieties of a single community, reproducing social order; ch'ajwa, on the other hand, fractures the reciprocal bonds that bind complementary forces into a larger whole.

The distinction between ch'ajwa and tinku, however, does not exist in the early Colonial dictionaries (Bertonio [1612] 1984: *primera parte*, 273, *segunda parte*, 68, 350; González Holguín [1608] 1952:37–38, 293, 342–343; Santo Tomas [1560] 1951) and seems to be a relatively recent linguistic shift (see Topic and Topic 1997b).

Neither the Quechua nor the Aymara language in the Colonial period distinguished games from war—or ritual warfare from secular warfare—with the same precision that some anthropologists now ascribe to the terms tinku and ch'ajwa. If we are to better understand the actual practice of warfare by prehispanic Andean peoples, we must stop thinking in terms of the ritual/secular dichotomy, which seems to be a modern development. Instead, we should carefully consider the evidence available that speaks of the relationship between conflict and social and world order.

Tinku in the Ethnographic Literature

According to the ethnographic literature, participants continue to relate tinku to the reproduction of the natural world. The blood and energy poured out in tinku, they say, feed the earth and are eventually returned to the people in the products of the earth (Allen 2002:177, 183).

Leslie Brownrigg (1972; see also Hartmann 1972, 1978:210) provides a particularly vivid example of this from the Cañar area of Ecuador. There, each battle is called *el juego de pucará* [the fortress game] and can consist of pitched encounters involving multiple participants on each side. More properly, however, the juego de pucará consists of one or more pairs of combatants who face off against each other. The weapon used is a stone tied to a long cord, called a *huaraca*. The combatants stand face to face and take turns hitting each other with their huaracas, wearing large hats as shields to ward off the blows of their opponents' huaracas. This form of tinku has a certain similarity to Moche depictions of pairs of fighting men, in which each man is grasping the other's hair in one hand while the other hand is raised in a clenched fist (Bonavia 1985:49, 53). Brownrigg (1972:97) reports that "in the past, when a champion died, his conqueror cut his throat to collect the blood in order to sow it in his fields."

Another of tinku's connections with fertility is the relationship between men and women it implies. Catherine Allen (2002:160) notes that rivals in battle, like lovers, are considered to be yanantin, a matched pair or helpmates. Noting how the women whip the men fiercely during the Sargento (a dance related to the definition of "ch'ajwa" [*chahua*] in Bertonio's dictionary), Allen (1988:184) feels that sex must have been involved in the original tinku. Platt (1986:239–240) also relates tinku to sexual intercourse. In Platt's interpretation, combat involves a ritual copulation between males of opposing moieties.

In southern Peru and Bolivia, women who attend a tinku can be captured by the victors and, occasionally, raped (Gorbak, Lischetti, Muñoz 1962:250, 258; Platt 1986:240). Usually, the women who attend are unmarried, and in some cases, the captors may marry them. Inka soldiers, too, could keep women they captured in battle, and especially valiant soldiers were given *acllacona* (chosen women) as wives (Rowe 1946:280).

In addition to capturing women, the capture of booty is one of the goals of tinku. In Ecuador, fighters try to take their enemy's hat, jacket, or arms (Hartmann 1978). In Peru, the booty includes horses as well as arms and clothing (Gorbak, Lischetti, Muñoz 1962:255). In Peru's Azángaro province, prisoners are taken and held for a year before being released, and houses are sacked (Gorbak, Lischetti, Muñoz 1962:258). In some cases, booty can later be ransomed by its owner (Gorbak, Lischetti, Muñoz 1962:288). The taking of clothing and arms as trophies, and the taking of prisoners, is reminiscent of Moche battle scenes painted on ceramic jars (discussed below). Inka armies also took trophies and prisoners; some prisoners were sacrificed, while others might be marched through the streets in a triumphal procession (Rowe 1946:279).

Tinku is clearly related to both physical and social boundaries. Through the actions that constitute a tinku battle, the moiety system is reproduced and the moiety is resituated within the larger whole (Urton 1993; Allen 2002:177; Platt 1986, 1987). As noted above, the term "tinku" has the connotation of the joining of opposing halves to form a whole.

Some tinku battles take place on the boundaries of the combatants' territories. Thus, the Macha fight in the town square, through which runs the intermoietal boundary (Platt 1987:167). Allen's (1988:183) informants, who come from two different provinces, also say that they are defending their boundaries. In Azuay, Ecuador, participants shout "*¡Quitemos la raya!*" and fight over the dividing line or frontier between the villages (Hartmann 1978:210–211). In Otavalo, the goal is to capture the principal chapel (Hartmann 1978:204). Opposing moieties in Yucay each construct fortifications on their own territory, from which each side tests the strength of the other (Molinié-Fioravanti 1988:54).

All modern sources concur that participants consider one of the major functions of tinku to be the prediction of harvest results. The actors in tinku battles consider much bloodshed and many deaths to be positive outcomes that will result in a good harvest; the actors also believe that those on the winning side will have a better harvest than will those on the losing side. Participants consider a tinku battle to be both a sacrifice and an act of divination, with forces beyond the combatants directing the result. Participants do not mourn the dead and do not equate violent death in tinku with murder. The individual who wields the sling that causes the death is not responsible for the death itself. The individuals who participate in a tinku battle are, however, fulfilling a social responsibility to their group by playing an active role in the regeneration of the earth and thus contributing to the well-being of the community.

Tinku and Prehispanic Warfare on the North Coast

The ethnographic descriptions of tinku come after nearly five hundred years of colonial and national dominance and concerted efforts by both religious and secular authorities to abolish the battles. Yet the descriptions of the battles are widespread throughout the central Andes and rich in detail. They describe a practice that is closely tied to both the reproduction of society and the renewal of the world. It is a practice that is deeply rooted in the cosmology of the area. One does not expect prehispanic warfare to be exactly equivalent to any of the ethnographic descriptions of tinku. Nor, as we have pointed out, does one expect the practice of prehispanic warfare to be static through all time periods. But it is reasonable to expect that the descriptions of modern tinku retain some of the cosmological significance of prehispanic warfare, highlighting the reciprocal relationship that people have with the earth, the need to feed the earth, that the clash of bodies in battle is related to sexual reproduction, and that fighting puts segments of society in relationship to each other. It is this last element that may be the most difficult for the Western mind to comprehend, since we tend to think of warfare as the most egregious example of the breakdown of society. Tinku, however, brings the two complementary halves of society, the moieties into which all Andean societies are divided, into union. Moreover, it is a union that is reproductive and fertile.

Early Combat on the North Coast of Peru

There is tantalizing evidence for at least potential conflict as early as 3500 BC at the preceramic site of Ostra (see fig. 1.1), located on the ancient shoreline of an uplifted bay just north of the Santa Valley (J. Topic 1989). The site is a small shellfish-collecting station probably used intermittently as part of a strategy utilizing seasonally available resources in different parts of the valley. At Ostra, fifty-four piles of slingstones are arranged in two lines reaching inland from the beach line on either side of the site. The stones were carefully selected, slightly smaller than a human fist and relatively spherical. Each pile contained about one hundred stones. The piles were spaced about three meters apart, the minimum distance necessary between slingers. The stones are not found on the beach; they may have been brought from alluvial deposits a considerable distance away (the nearest terrestrial source we located was seven kilometers to the south) or collected by diving in the shallow bay from the offshore zone where cobbles eroded from the river zone are carried along by the currents.

There is no direct evidence of fighting here, but clearly a threat was perceived by the residents of the site and they prepared a response to the threat. The response was a simple one—to assure an ample supply of a projectile that the occupants of the site would have used regularly in everyday life, though perhaps less frequently in this beachfront location than on sojourns further inland.

A similar situation has been reported from the southern edge of the Santa Valley by Vincent Chamussy, who has found slingstones in association with two probable defensive walls at a preceramic site at Playa Catalán on a peninsula protruding into the ocean (Chamussy, personal communication 2005).

In the centuries succeeding Ostra, there is very little evidence for conflict on the north coast. The Cerro Sechín site has large stone carvings that have been dated to ca. 1200 BCE within the Initial Period (1800 to 900 BCE). The images seem to represent warriors and dismembered bodies (Tello 1956; Topic and Topic 1997b; Urton 1993; but see also Cordy-Collins 1983). Clearly, the builders of the site had experience of combat and its outcome, and that experience was remembered and memorialized in monumental fashion. Cupisnique iconography from the late Initial Period includes depictions carved on stone bowls of a spider carrying on its back a web bag filled with human heads (Salazar-Burger and Burger 1982), and this theme relates to "sacrificer" imagery that continues through the Early Horizon (ca. 900 to 200 BCE) into the Moche period (Cordy-Collins 1992); this imagery indicates the presence of a particular kind of violence, human sacrifice—a theme that will be repeated at intervals throughout Andean prehistory.

During the latter part of the Early Horizon on the north coast, a new type of evidence points to increasing significance of warfare in the lives of many groups. This evidence is primarily architectural and locational. The residents of some coastal valleys invested heavily in the construction of fortresses of considerable complexity and scale. The forts were often, though not always, located on mountaintops in the desert far from the irrigated valley bottoms and settlements where people lived and worked (fig. 1.2).

An interesting example of multiple fortifications from the Santa Valley deserves comment. During the Cayhuamarca phase (1000 to 350 BCE), a cluster of fortresses was built in one sector of the valley. Nearly every mountaintop has a fortress, most located an hour's climb or more from habitable areas on the irrigated valley bottoms. David Wilson (1988), who conducted extensive survey in the Santa Valley, considered these sites to be strategic fortifications, built to protect the valley from invasions from the south. The strategic utility of these fortifications, however, is unclear. The longest-distance weapon coastal Andean peoples had at this time was the sling, with an effective range of approximately fifty meters. The gaps between the forts would have allowed an invading force to bypass them and attack settlements directly. These were not effective strategic fortifications; they are isolated from settlements and it would have been very difficult and costly to garrison them permanently in anticipation of an unannounced attack. Fortification of the settlements themselves would have been a more practical response to a threat of invasion.

And yet, the valley residents *did* decide to allocate resources to the construction of these quite elaborate structures. They were placed as prominent features on the landscape, no doubt with the expectation

1.2. The location of Quisque on a desert ridge far from the irrigated valley bottom is typical of many of the late Early Horizon fortresses in the Santa, Nepeña, and Casma valleys. The fortress is built around a rock outcrop that is enclosed within a megalithic wall.

that they would be used. Many features are repeated from one fortress to another—the inclusion of a prominent rock outcrop within the fort walls (a feature associated with prisoner sacrifice in later Moche times at Huaca de la Luna) suggests that there was a template shared among residents of the valley for the construction of this type of building. We must seek an Andean motivation for the construction of the fortresses.

The fortifications are logical within an Andean framework if tinku battles were fought between moieties, communities, or sectors of the valley. These fortifications might have been the setting or objectives for battles, like the chapel in Otavalo or the flattened mountaintops and fortified constructions used in modern tinku in Ecuador and Peru. They may have been placed to mark boundaries between communities and to be visual statements of community identity, solidarity, and prestige. In this context, combat may have been prearranged, with the day and location negotiated by representatives of opposing sides. None of these fortresses have been excavated, and it is not clear whether they have been used; round stones (possible slingstones) are found in the vicinity of some, but it is not clear whether they relate to combat. In this case, we must deduce past practice from the architecture and from reference to patterns of combat elsewhere on the north coast.

Contemporary fortresses in the Nepeña and Casma valleys have drawn considerable attention. Quisque was built several kilometers up the Nepeña Valley on a steep hill overlooking the irrigated valley bottom, and it is a most impressive megalithic construction with massive stone blocks, roughly shaped, incorporated into the lower courses of its stone walls (see figs. 1.1 and 1.2). The fortification walls enclose an unusual outcropping of bedrock, which is surrounded by a well-built megalithic wall, recalling the Santa Valley fortresses and Huaca de la Luna in the Moche period. Chankillo in the Casma Valley is a much larger fortification, lying out in the desert, isolated from the irrigated valley bottom. Planners and builders provided the structure with many good "fortification" features, such as baffled entries and broad, flat wall tops that would allow defenders to challenge attackers. Multiple stairways on the wall interior provided defenders with access to the wall top; the innermost of three walls was provided with parapets, and Ivan Ghezzi (2006:76) suggests that the two outer walls may also have been parapetted. The builders also provided Chankillo with an unusual feature known as "barholds"—stone pins set into a niche or recess placed alongside a doorway (Topic and Topic 1997b: figs. 2.2a and 2.2b). Barholds are also known in later Wari and Inka architecture, in which they are used to lash a wooden framework against an entranceway in order to close the entrance. Each of the twenty-two doorways at Chankillo was provided with four barholds, two on each side, one near the top of the entrance and the other closer to the bottom. In every case, the barholds are located on the exterior of the doorway. With barholds located on the exterior, the gates could only be closed and opened from the outside (cf. Arkush and Stanish 2005:21 for a contrary view). From a Western standpoint, a proper fortress must have gates that can be securely closed from the inside and are difficult to breach from the outside. Structures that are closed from the outside are those that are left unoccupied for a period of time and must be secured (physically or symbolically) against human or nonhuman intruders who have no right of access.

We think it is quite likely that Chankillo was used as a staging point for prearranged combat and associated activities, like the Cayhuamarca phase fortresses. Ethnohistory provides some useful analogies. An indigenous chronicler, Juan Santa Cruz Pachacuti Yamqui Salcamayua ([1615] 1928), describes a battle that was staged at the Inka "fortress" of Sacsahuaman in Cusco. In that example, the emperor Tupa Inka had returned from an expedition to Ecuador and had brought captured Ecuadorian troops back to the capital. Topa Inka's father, Pachacuti, decreed that a battle should be staged between the Ecuadorian prisoners and his young grandson, Wayna Capac. Wayna Capac was put in charge of fifty thousand soldiers armed with gold and silver weapons; this force attacked the Ecuadorians, who had been placed inside the fortress. Wayna Capac captured the fortress and beheaded all the Ecuadorians. A wide range of military equipment was stored within Sacsahuaman (Hyslop 1990:54–55; Sancho de Hoz [1534] 2004:127–128). This situation was replicated in the provinces: the Augustinian priests report that weapons were stored at shrines associated with warfare in Huamachuco (San Pedro [1560] 1992). Gates might well have been open when battles were staged at Chankillo but closed during the times between battles, when the structure was not in use and participants would have stored equipment, paraphernalia, and supplies at the site.

Another intriguing detail is that many of these Early Horizon fortresses have large numbers of panpipe fragments on their surfaces. The fragments are a reminder that words for the musical instruments used in Inka battles appear in the dictionary of Diego González Holguín and that battle is an experience that engages all the senses, including hearing. Chants, battle calls, and sound from musical instruments would have combined to heighten the atmosphere for combat.

While all these late Early Horizon fortresses may have been used for some form of staged battles, there is interesting evidence for variation. Residents of the Santa, Nepeña, and Casma valleys invested heavily in the planning, careful construction, and maintenance of fortifications of modest monumental scale at this time. But further north in the Viru

Valley, the two contemporaneous hilltop redoubts reported by Gordon Willey (1953) are much more basic constructions. The builders selected hilltops near settlements and enclosed the top with rough stone walls. The lesser effort in construction may signal that this particular type of project was not as closely linked to community prestige and that the builders were preparing a response to a different kind of threat than their contemporaries further south; these less elaborate and more "practical" fortifications may have been a response to raids rather than to tinku-related combat.

Still, these Viru Valley fortifications have their ceremonial features. Within the enclosed area in each of the two redoubts are two platform mounds situated on the higher eminences, which Willey (1953:95) interpreted as shrines. These structures recall Chankillo's two round towers in the central enclosure. This pairing of mounds and towers may relate to moiety social structure, since tinku are often staged between two moieties. The Cayhuamarca fortresses in the Santa Valley do not seem to have these dual structures.

The number and distribution of fortresses in these valleys point to another dimension of variation. In the section of the Santa Valley just discussed, every local community may have maintained its own small fortress, assuming responsibility for its maintenance and supply (as well as utilization). The concentration of fortresses suggests that they were the scenes of battles involving relatively few combatants, perhaps participants from two adjacent communities. The large number of fortresses might also suggest a scheduling or rotation of battles through the year.

The military landscape of the Casma valley, on the other hand, is dominated by Chankillo; Chankillo is not the only fortress in the Casma Valley (Wilson 1995), but it is the predominant one, and it is much larger and more complex than the Santa Valley examples. This suggests that it was the scene of larger battles, perhaps involving participants from the entire valley or from adjacent valleys. The situation in the Nepeña and Viru valleys seems to be intermediate between these two extremes.

Finally, Ghezzi notes the presence of a temple within the walls of Chankillo and that structure's alignment with the December solstice; he characterizes the site as a "fortified temple" at which battles were fought (Ghezzi 2006:79). The scheduling of battles at Chankillo might have been dependent on calendrical observations made at the temple.

The Moche Case for Tinku Battles

There is some suggestive evidence for the linking of combat and human sacrifice in the Initial Period and Early Horizon, as well as for tinku battles in the Early Horizon. But this evidence pales in comparison to the very persuasive body of evidence for tinku battles in Moche culture that has been steadily accumulating on the north coast for the last fifteen years. Andean archaeologists with an interest in warfare have long been interested in the Moche because of the wealth of information provided by the very detailed iconography produced in some valleys between approximately 200 and 750 AD on a variety of media. The fine-line painted ceramic vessels from this period are especially abundant and informative, with complex scenes related to combat and sacrifice repeated frequently. The battle scenes include men paired in combat, armed usually with maces and shields and wearing elaborate costumes that identify them as warriors (fig. 1.3). The combat scenes merge into sacrifice scenes in which captured warriors, stripped of costumes and arms, are presented to elaborately garbed figures (fig. 1.4). The prisoners are sacrificed by having their throats slit; the blood is collected in a cup and presented by a female figure (who has been dubbed the "priestess") to an especially elaborate figure (the "warrior priest") who consumes the blood, while the "bird priest" and another central figure observe (Alva and Donnan 1994:131–133). Similar scenes have been discovered painted and modeled as murals on some of the most important monumental structures in valleys throughout the Moche culture area.

Throughout the twentieth century, these representations excited considerable commentary by Andean scholars. An early tendency to read the images as literal representations of history was succeeded in the 1970s and 1980s by the view that the ceramics and murals narrated episodes from Moche mythology. The key scenes in Moche iconography were considered to be clear evidence of the importance within Moche culture of warfare and the offering of human blood. Given the ubiquity of the scenes on ceramic vessels and the prominent positioning of murals on temple walls, most archaeologists concurred that knowledge about the themes would have been widely shared within the culture. However the scenes were considered to depict mythic reality, not everyday experience.

1.3. Moche combat scene with two warriors paired off. The warriors wear similar elaborate costumes. Other Moche battle scenes depict more warriors engaged in combat, but the men are usually shown fighting in pairs rather than in formations. Fortifications are never shown in battle scenes, which seem to be set in the desert that would have been located between adjacent valleys. (From Donnan 2004. Drawing by Donna McClelland.)

But this interpretation has been radically transformed in recent years as the iconographic images have been supplemented by a stream of exceedingly interesting archaeological finds. These new finds have led many archaeologists to note the strong parallels between the ethnographic descriptions of tinku and the Moche depictions of combat and its aftermath (Bourget 2005; Donnan 1978, 2001, 2004; Hocquenghem 1987; Topic and Topic 1997a; Verano 2001).

The archaeological finds that have led to this reinterpretation of the iconography include the uncovering of deposits of multiple sacrificed individuals at Huaca de la Luna in the Moche Valley, as well as burials of elite individuals whose grave goods have led to their identification as the "priestess" from the sacrifice scene (at San Jose de Moro in the Jequetepeque Valley) and the "warrior priest" (at Sipan in the Lambayeque Valley). These finds have persuaded many scholars that the practices encoded in Moche art were actually performed with some

I.4. Moche sacrifice and presentation scene. Prisoners are sacrificed in the lower register, while in the upper register, a goblet of blood is being presented to the warrior priest. (From Donnan 2004. Drawing by Donna McClelland.)

regularity by the Moche (Alva and Donnan 1994; Bourget 2001; Verano 2001). Combat for the purpose of taking prisoners, the sacrifice of captives, and the supervision of the sacrifice ceremony by a formal hierarchy of individuals with well-defined roles are now considered to have been historical reality for the Moche.

This reinterpretation of Moche combat is supported by other lines of evidence. During the period when battle and sacrifice themes dominate Moche art, and when there is good osteological evidence for sacrifice of the captured prisoners, there are no fortifications. At the very end of the Moche period (discussed below), fortifications became common; this change signals a significant change in the patterns of warfare practiced by the Moche. But during the 200 to 650 AD period, there is strong evidence that the Moche placed a high value on a form of combat closely linked to tinku principles and that this activity was highly publicized among the Moche.

What can practice theory contribute to this emerging reinterpretation of Moche warfare? Practice theory encourages us to work conceptually not only with "armies" or with mega-categories like "elites" and "commoners," but also at a level where we can accord importance to the actions of individuals and appreciate that there will be varied individual agendas for the actors. It is legitimate to ask who the individuals are who are fighting and being sacrificed.

Who is fighting? This is both a simple and a complex question. Clothing and costume in the Andes is highly correlated with ethnic identity (e.g., Cieza [1553] 1985; Guaman Poma [1615] 1980), and the costumes depicted on the warriors in Moche iconography are rich in detail. There is considerable variability in the tunics, headdresses, and ornaments of the warriors, but in the vast majority of cases, the style of the costumes is very similar. While one case has been noted of combat between opposing groups with very different costumes and arms (e.g., Lau 2004; Reichert 1989; Verano 2001:113), it is now generally conceded that the uniformity in costume indicates that most battles were between members of the same ethnic group. That is, Moche warriors were fighting other Moche warriors and not invaders from the highlands or from outside the north coast.

There is abundant (and not surprising) evidence that warriors were young adult males. The bodies from the Huaca de la Luna sacrificial deposits were young men, mostly in their early to mid-twenties, healthy and robust, some with well-healed injuries that have led John Verano to suggest they are "professionals," not "weekend warriors." Some writers (e.g., Donnan 2001; Topic and Topic 1997b; Verano 2001) have suggested that only the Moche elite were engaged in warfare, since the depictions on fine-line vessels show elaborately garbed and armed men. The fact that presentations of prisoner scenes show different ranks (for example, some naked prisoners are carrying other naked prisoners in litters) does suggest that there were status differences among the warriors involved. The blood of men of higher status might have carried greater value in sacrifice ceremonies and might have been preferred by the gods. On the other hand, the stripping of prisoners and depiction of them as uniform subjects speaks to a symbolic homogeneity once they are captured; they are now all material for sacrifice.

If we are to consider Moche combat as a form of practice, we will want to know what factors shaped the behavior of the various actors in this complex drama. Of a cohort of potential warriors, what decisions would make an individual more or less likely to become a combatant and to risk becoming a sacrificial victim? These individuals would have been steeped in the cultural values that were embedded in the daily life around them and enacted in the larger-scale social and religious pageants that served to reinforce social identity. But what degree of intentionality and choice did these agents display as they made decisions and laid plans? Was the status of warrior hotly contested? Was it restricted only to those of elite lineage, or was it a means of achieving higher status? Was it one that most fit young men of the appropriate age filled? Was the role desired or dreaded, anticipated or avoided?

We cannot answer all these questions, but we can note that the scenes of naked prisoners carrying other naked prisoners in litters show no guards. Prisoners are often also shown with ropes around their necks, but these ropes are not pulled taut, nor are prisoners shown struggling against their captors. These elements of the iconography suggest that the individuals participating in combat fought willingly. Social expectations may have been very strong, and the choice to become a warrior may not have been entirely free, but warriors do not seem reluctant. The ideal seems to have been for captives to meet their fate in a stalwart fashion, and there would probably have been considerable cohort-specific reinforcement of that ideal.

Christopher Donnan (2004:113ff) has recently presented intriguing evidence that some individual warriors were well-known champions. He has identified specific individuals depicted in three contexts: on portrait head vessels, as fully modeled figures armed as warriors, and as naked or semi-naked captives. The portrait head vessels, as the term implies, depict the modeled and painted heads of identifiable individuals. The vessels are hand painted but mold made, so there were many copies of the same portrait in circulation. The implication is that these individuals were indeed very important, highly recognizable "celebrities." We do not know if all the individuals depicted as portrait heads were warriors who participated in the battles leading to the sacrifice ceremony, but Donnan's analysis shows that some certainly were. It is not clear whether warriors tended to come from the elite class or whether enhanced status could be achieved through combat, but clearly some warriors achieved iconic status even though they suffered sacrifice. They were "champions" not (necessarily) because they were "winners" but because they championed their communities. The display of individual skill by champions seems to have been a key part of Moche tinku combat. Steve Bourget (2005:77) suggests that the weapons used by Moche warriors were designed *not* to be robust; they used small shields and maces with relatively fragile shafts that would perhaps have allowed their individual expertise to be displayed.

Unfortunately, most of the Moche pottery that displays combat and sacrifice has been looted and lacks any context. Hence, we can say the warriors had iconic status within their communities, but the lack of provenience information for the pottery makes it difficult to define the scale of the "communities" involved. There is evidence, iconographic and osteological, for the sacrifice ceremony at a number of different archaeological sites (e.g., Sipan, San José de Moro, Huaca de Cao Viejo [El Brujo], Huaca de la Luna, and Pañamarca), encompassing the north coastal valleys from Lambayeque to Nepeña. It seems clear that most warriors were "Moche" culturally or ethnically, but it is difficult at this point to be precise about the communities they represented. Opposing factions might represent moieties of a single community, different communities associated with a single temple pyramid complex, or competing temple pyramid complexes.

Some vessels imply that combat took place in the intervalley desert, suggesting that communities from different valleys were fighting.

Participants might represent two different temple pyramid complexes, with each side intent on capturing prisoners for sacrifice at its own temple.

Participants on both sides were still "Moche," however, sharing the same cultural identity. In fact, it was probably quite important that they were members of the same culture. The capture and sacrifice of prisoners is *the* central theme of the iconography of Moche phases III and IV; if this theme is related to the concepts of renewal that we described above, it is logical that the community would offer sacrifice. By participating in battle and sacrificing members of their own culture, communities establish reciprocal relationships with the regenerative powers of the earth. But in the Andean context, community membership is nested—that is, each moiety is paired with its complement and then that pair, in turn, is paired with a complementary pair, and so on (Netherly 1990; Platt 1986). Thus, political expansion does not involve the subjugation of the "other" but rather the expansion of the community by increasing the scale at which moiety divisions are still recognized as complementary units bound by reciprocal obligations. Warfare, which may be viewed on the one hand as increasing the fertility of the earth, nevertheless also puts communities in relationship to each other. The concept of tinku is, in fact, the idea of creating a whole from two parts.

The motivations of the participants in Moche warfare derive from the role of bloodshed in the cosmological structure of renewal. Participants performed two essential services for their communities: not only did they offer themselves for the renewal of the earth in a confrontation of complementary social units, but their combat also then led to the reproduction of the social fabric at a more inclusive level that combined the complementary parts into a larger whole. As Donnan (2004:113ff) has documented that warriors enjoyed much respect and prestige; there may have been other rewards that often go hand in hand with increased status, such as women, land, or offices. Warriors' kin groups would also have shared in their prestige.

Warfare may have been a quasi-public event, and young men may have been witnesses before they were participants. This would accord with current practice; modern sites for tinku battles are open to observers but are not particularly safe venues. In the Moche representations, the arraignment of prisoners also seems to have taken place in relatively public areas. The sacrifice of captives seems to have been a less public event, carried out in enclosed spaces before a restricted number of observers. At Huaca de la Luna, captives were sacrificed in an interior walled courtyard that incorporated a rock outcrop. No public is shown in the fine-line drawings on ceramic vessels, and the only women shown in the drawings are the priestess and other females who seem to be preparing the prisoners for sacrifice (Arsenault 1994).

The evidence from Moche constitutes an especially clear case of ritualized warfare—warfare with culturally specific rules of engagement with direct links to sacrifice. There can be no doubt that a cosmology of renewal was a key motivator for the participants and for other actors.

Later Coastal Warfare

During the Moche V/Late Moche period (ca. 650–750 CE), there are important shifts in the practice of warfare on the north coast of Peru. Fortifications become a notable part of the built environment once again. The location of the major Moche V site of Galindo in the Moche Valley (T. Topic 1991) is a break from past tradition, crowded against a steep hill and provided with extensive fortifications. Massive parapetted walls of adobe and stone were stocked with piles of slingstones spaced about every three meters (fig. 1.5). The walls were rebuilt three times after massive rains damaged segments of the fortifications. Puente Serrano is another heavily fortified settlement just upvalley of Galindo. Further north, fortified hillslope settlements appear in the Zaña and Jequetepeque valleys during Late Moche times (Castillo 2001:325–326; Dillehay 2001).

A new level and type of threat was being perceived by the populace of Moche V/Late Moche centers. It is not clear whether the classic tinku-related combat so prominently depicted in earlier Moche times continued. The fine-line painting on Moche V pots is very different from Moche III–IV examples, and new themes are popular. One combat-related theme that occurs on several vessels shows the very specialized weapons associated with tinku battles in revolt, capturing human warriors (Quilter 1990, Bourget 2005). The meaning of this scene is opaque, but it certainly suggests a change in perception of ritualized combat.

1.5. One of the major settlement fortification walls at the Moche V site of Galindo. This view shows the wall top with hundreds of slingstones stockpiled on top of the parapet behind the breastwork. The walls at Galindo were damaged by El Niño events at least three times and repaired after each event, indicating that the defense of the settlement was a long-term concern.

During Moche V and into the succeeding early Chimu time period, north coast cultures are redefining the nature of combat and the roles of those responsible for it. One interesting change is in the weaponry. The clubs and shields used by champions of the earlier Moche times may have continued to be used, but non-elite warriors may have used a less specialized type of mace—essentially a clod breaker used in agricultural work. However, the quantity of slingstones on parapetted walls indicates clearly that defenders expected to be able to hold off attackers for some time with this longer-range weapon. Combat became less personal, since the distance at which opponents engaged with each other was expanded considerably from the earlier hand-to-hand encounters.

It is not clear where the perceived threat to Late Moche settlements came from. The residents of the fortified settlements could have feared attacks from other Moche or from groups living further up in the narrow valley necks, below the sierra rainfall agricultural zone. Hence, it is not clear whether Late Moche people were raising offensive armies or simply defending their settlements; there are somewhat different implications for the two scenarios.

However, we do know that they were defending at least some settlements. As fortified settlements were once again built in some north coast valleys, the combat role must have shifted from that of a community champion to one encompassing a larger cross-section of adult males, with concomitant changes in cultural attitudes towards war and warriors. As a larger segment of the adult male population became involved in combat, there may have been a lessening in the prestige accorded combatants (though exceptional skill in combat might well have served as a means of individual advancement). If much of the fighting was done with slings, the level of skill and training required of combatants might have been less than that required of the earlier champions, who used specialized weapons. Still, many, most, or all able-bodied adult males might have undergone some training for potential defensive and offensive roles in combat. Men experienced in war would have emerged in leadership positions, assuming responsibility for training, logistics, and oversight of both defensive and offensive projects. The construction, maintenance, and provisioning of defensive walls required coordination and would have involved the labor of most of the community, including older men, women, and children who might not otherwise have played a major role in combat itself.

In the succeeding Late Intermediate Period (1000 to 1470 CE), there is much evidence for fortifications in the early Chimu period (to about 1200 CE), with both fortified settlements and fortifications located on isolated hilltops occurring (T. Topic 1990). One isolated hilltop fortress is located at the summit of Cerro Oreja in the Moche Valley, a high, steep hill at the point where the valley narrows. Fortification features include walls, salient angles overlooking the valley, and residential areas with sleeping platforms, suggesting that soldiers were barracked atop the hill.

At least two sites, Talambo in the Jequetepeque Valley and Cerro de la Cruz in the Chao Valley, demonstrate clear evidence for battles having been fought or threatened. Cerro de la Cruz was an early Late Intermediate Period fortified settlement with enclosing wall, parapet, slingstones, and dry moat (fig. 1.1). Our survey of this site in 1980 identified numerous piles of slingstones on the ground outside the dry moat

(T. Topic 1990). At Talambo, a contemporary walled site where a battle was actually fought, slingstones are dispersed over a broad band below the site walls, but at Cerro de la Cruz, the neat piles of slingstones on the parapets inside the walls and on the ground outside the dry moat had not been used. We interpreted the pattern as evidence that a Chimu army laid siege to the settlement but did not actually attack. The leaders at Cerro de la Cruz may have capitulated in the face of a threat that encompassed all members of their settlement.

The evidence from Cerro Oreja, Cerro de la Cruz, and Talambo indicates clearly that the Chimu were raising offensive armies and maintaining them in the field for at least limited campaigns. How were the soldiers recruited? We have no direct evidence for the Chimu, but ethnohistoric sources tell us that in the succeeding Late Horizon (1470–1532 CE), the Inka required that all male heads of household were subject to a labor tax that included military service (Murra 1982). In addition, the Inkas also mandated training exercises for all adult males, and the Chimu may also have done this. The sling was still the principal weapon for the initiation of hostilities (based on the prevalence of slingstones at fortified sites), but we don't know how commonly combatants closed to hand-to-hand combat. It is worth noting that hand-to-hand combat is much deadlier than slinging stones at a distance. In hand-to-hand combat, the odds of serious injury to a combatant are about 50/50: one or the other of a pair of combatants is probably going to suffer serious injury. Successful hand-to-hand combat would require more specialized training and perhaps greater skill than the use of the sling as a primary weapon.

While we have only a few indicators of Chimu strategy, it is likely that logistics and negotiation loomed large in that strategy. In an offensive campaign, fielding the largest possible number of soldiers and being able to maintain them in the field for a period of time was probably much more important than the level of skill of individual soldiers. Logistical support required the provisioning of food, and possibly weapons and clothing, to the soldiers. Some logistical support may have been provided by women accompanying the army (a pattern noted for the Inka by Pedro Pizarro [1571] 1978:201).

Negotiation was also surely a key strategic element in a successful campaign. In the cases of Cerro de la Cruz and Talambo cited above, one settlement decided to capitulate while the other decided to fight. The conditions under which the leaders and members of each settlement made their decision would have included assessments of the number of opposing soldiers facing them, as well as of the extent of logistical support available to maintain the opposing force in the field. But the decision on whether to resist or not would also be based on culture-specific expectations surrounding the practice of war, and no doubt a careful calculation of the benefits of allying with those offering a new relationship. We know that the Inka visited reprisals on groups who resisted annexation to the Inka empire but provided rich gifts and favorable alliances to those who voluntarily accepted Inka rule, showing special favor to the local rulers of groups who capitulated. Again, we don't know whether the Chimu followed these practices. We do know, however, that the Chimu, during the later part of the dynasty, controlled huge amounts of prestige goods but did not have the logistical capability of the Inka to maintain armies in distant localities for long periods of time (J. Topic 1990). Fortifications are not known for the later part of the Chimu period in the core area of the north coast (ca. 1200 to 1470 CE), and there is little other evidence to speak directly to any aspect of warfare during this period.

A state the size of the Chimu state might be expected to maintain a professional army, but evidence for it is ephemeral, and again, the Inka analogy suggests that warfare was seasonal, adapted to the agricultural cycle. Much of Chimu political expansion may have been achieved through negotiation, pledges of alliance and strong reciprocal relationships, and conferring gifts and status on elites willing to be encompassed within the Chimu political and cultural orbit. Implied threat was clearly part of this process.

As different as Chimu warfare was from the Moche tinku-like battles that we described earlier, there is evidence suggesting that the Chimu continued some earlier practices. There are painted textiles from the late Middle Horizon/Late Intermediate Period that show prisoners with ropes around their necks (Nancy Porter, personal communication, 2009). At Pacatnamú (in the Jequetepeque Valley), there may be evidence for the sacrifice of prisoners following capture in battle. Remains of fourteen young males were found with severe traumas, some combat related but others presumably inflicted at the time of death. The bodies were left exposed for an extended period of time before being covered over (Verano 1986).

The Chimu were conquered by the Inka in about 1470 CE. The Inka were a highland people who expanded rapidly out of their capital, Cuzco, in the southern sierra of what is now Peru, in the middle of the fifteenth century. Although they incorporated soldiers from many conquered peoples into their army, early colonial testimony suggests that coastal peoples were excluded. A rebellion of north coastal people is cited as one reason that the Inka forbade them to bear arms (Rostworowski de Diez Canseco 1988:125–126).

Conclusions: Exploring Variation in Prehispanic Warfare

This chapter has undertaken a review of prehispanic warfare on the north coast of Peru, attempting to understand how combat was practiced in a number of specific, historically contingent contexts with good archaeological documentation. When we attempt explanation from a practice standpoint, causality is situated closer to the actors than it is when we use evolutionary or processual approaches, and observed variability is accepted as a predictable outcome of variation in practice. A dialectical tension is created between particularist and universalist explanatory poles. A practice approach does not preclude the definition of universal responses to specific situations, and in fact it will lead to greater confidence in those commonalities that can be recognized.

Because practice theory focuses on how actors behave in particular situations, it forces us to ask more detailed questions about warfare in the past. We must consider all the agents involved in warfare, not just the ones actually participating in combat. Those who make the weapons and costumes, those who train and counsel combatants, those with responsibility to ensure that necessary supplies and support staff are available, and those who make the decisions about whether an encounter will proceed or not are all part of the action. In addition, a wide circle of actors less immediately connected with combat will behave agentially as they decide on the nature of the support and encouragement they will give to sons, brothers, partners, lineage mates, neighbors, companions, and associates. These individuals constitute the community in which the warrior is embedded. We have explored the motivations and the reasoning that might lead to individual and group decisions for combat.

In the first part of this chapter, we discussed Andean tradition and Andean cosmology with the intent of illustrating the conceptual parameters within which prehispanic actors probably operated. We do not claim that there is a millennia-long mental template, unchanged, for Andean peoples. But we find the evidence extremely strong for a deeply rooted tradition valuing order and proper reciprocal relationships. Humans are necessarily linked to the forces that sustain life, and warfare is a means of producing and reproducing reciprocal relationships that expedite the recycling of life-force through the various levels of the universe (Topic and Topic 2001). In the cases considered here, there is almost constant, though variable, evidence for this cosmology from at least the Initial Period up into the present day.

We then considered the archaeological evidence for warfare, noting variability in the record on the north coast but arguing that variability must be understood in the context of Andean concepts of proper relationship. There are many indicators that the actions of participants in prehispanic warfare and the communities that supported them were motivated by an acknowledged and widely shared belief in the value of combat as a means of maintaining relationships with humans, as well as with nonhuman beings and forces.

In some cases, a key motivator for those contemplating combat seems to have been the benefit to the community and to the earth of the shedding of human blood. Modern tinku battles retain vestiges of this tradition. There is much variability in modern tinku, but death and/or bloodshed is an immediate outcome of the battle, and actors concur that this is an important and desired result. We have argued that the Moche practiced a form of warfare that is analogous in many ways to modern tinku battles, but while bloodshed is shown on the field of combat (especially nasal bleeding), the ideal scenario was for death to follow at a later stage as part of an elaborate ritual in the major temple complexes.

Tinku battles were probably fought by the groups who built fortresses in the Early Horizon, and the presence of shrines and rock outcrops within fortresses suggests the possibility of some similarity to

Moche practice. But without any iconographic or osteological evidence, it is not clear whether death on the battlefield or capture of prisoners and subsequent sacrifice would have been considered best practice by participants. From this period, the architectural evidence stands on its own and is not remarkably revealing. It does, though, indicate considerable variability in some aspects of the practice of warfare within a relatively limited area of the Andes.

Tinku-related concepts directed some Inka combat, according to ethnohistoric sources. The Inkas certainly, at times, killed in battle on a large scale with no apparent attempt to link deaths to nonpolitical benefits. But the account of Santa Cruz Pachacuti Yamqui cited earlier suggests that another pattern was also in place, and that a more ritualized capture and sacrifice of prisoners was also practiced regularly. This pattern is distinct from both ancient tinku combat and modern tinku combat, and it marks the meeting of Andean cosmology with imperial political structure and motives.

Andean cosmology, which places a high value on warfare as a means of building relationships and offering sacrifices to the gods, must be taken into account when considering Andean warfare, and its potential for explaining the actions and motivations of participants must be kept firmly in mind.

Variation in architectural evidence on the north coast is a clear indicator of differences in practice relating to warfare. The presence or absence of fortifications, and the degree of elaboration of defensive architecture, tells us much about combat but must also be understood in the context of prehistoric Andean peoples, who from a very early date built nonfortified, massive monumental structures, especially mounds and mound complexes (Donnan 1985; Moore 1996). On the north coast of Peru, as in other parts of the central Andes, the construction of monumental structures was an important social process that built relationships within local and valley-wide communities and was an important means of transmitting messages of strength and solidarity to other groups. The elaborate fortifications in the Cayhuamarca phase of the Santa Valley and in the Casma and Nepeña valleys are part of this tradition of public construction as social statement. The fortifications document the importance that the cycle of activities related to combat held for these groups, but they also (like earlier platform mounds) were a material embodiment of community pride, expressed in scale and quality of construction.

These late Early Horizon fortifications isolated from population centers contrast both with the absence of fortifications in Moche phases I–IV and with the heavily fortified Late Moche settlements. It is clear from the iconography that warfare was a major activity in the social sphere of Moche peoples, but the absence of fortifications until Late Moche times results from the community's assignation of responsibility for warfare to groups of champions. It is interesting to note that combat based on tinku principles may result in both the presence of forts (e.g., in the Early Horizon) and the absence of forts (in Moche phases I–IV). Just as in modern tinku, specific places are sometimes the object of attack and defense but at other times not. The importance of tinku combat comes not from where it takes place, nor from its specific relationship to material remains, but through how it tests and affirms social relationships.

We have discussed cases related to tinku combat in which fortresses are isolated from settlements, as well as other cases in which a link to tinku is not obvious and the settlements themselves are fortified. The Ostra site, with its slingstone piles, is a very early case of preparation for site defense. Most fortified settlements on the north coast, however, date to the Middle Horizon and the early part of the Late Intermediate Period (ca. AD 650–1200). Some of these are quite large settlements, such as the Moche V site of Galindo described above, with its fortification walls provisioned with slingstones. As we have noted, slingstones are common at fortified settlements, both early and late; however, they are rarely shown in Moche battle scenes, where hand-to-hand combat with special shields and maces is preferred and the emphasis is on the individual skill of warriors.

Fortified settlements are evidence of conflict very different from that depicted in Moche iconography or implied by the isolated Early Horizon fortresses. A much broader range of the population would have been affected, and all residents of a fortified settlement would have been participants in the conflict in one way or another. Adult males would have been primary defenders, but the women and children in a settlement under attack probably had active roles to play. There is no evidence in the archaeological record of Andean women playing combat roles, but this is a suggestive rather than a conclusive pattern. It is likely that women and children at the Ostra site participated in the selection, transportation, and arrangement of slingstones, and they may have been active in the building and provisioning of defenses at other sites.

Strategies for response to attackers may have varied with the makeup of the defending body; if a fortified settlement under threat of attack had the full spectrum of the population within the walls, negotiation or capitulation might have been preferred over a vigorous defense that could lead to a massacre.

The presence of fortified settlements demonstrates that the decision makers (whether at the local, regional, or state level) had perceived a threat, assessed its nature and significance, and decided that the best response to the threat was to rally resources and labor to build barriers and to stockpile weapons and other provisions. The impact of this decision on residents would have been profound. Some might have been relocated to other settlements, and travel to or use of spaces that had been part of the regular round of activity would probably have been disrupted. The sense of crisis might have been short term or long term, but all residents would have been affected.

In the face of evidence for fortified settlements, archaeologists usually attribute causation to environmental, economic, or political factors, often in combination. When we first reported the slingstone evidence from Ostra (J. Topic 1989), we suggested that the preparations for conflict at the site might have been related to demographic pressure. Coastal populations at that time were not large, but the importance of shellfish collecting meant that suitable locations for villages were limited; they needed to have beach access, and beaches that were near sources of fresh water were especially valued. At that time, an explanation of the rudimentary defenses as a result of competition for scarce resources seemed very logical. But practice theory reminds us that the definition of resource scarcity and the structuring of the response to it are historically contingent; while raiding or attempts to dislocate others from resources might form part of a response to a perceived shortage, we should not assume that it is an inevitable response or the preferred response of every group. It is as important to understand the social relationships among groups across a wide geographical area as it is to understand the resource base.

We have pointed out that the appearance of fortified settlements in Late Moche times and the concomitant reduction in emphasis on the warrior as champion mark a significant shift in the Moche social and political landscape. Others commenting on this time period note clear evidence of environmental stresses (Moseley 1983). There is, for example, evidence for a number of El Niño events during the occupation at Galindo that washed out the site's fortification wall three times. Michael Moseley (1983) also argues that tectonic uplift of the coast set off sand invasions that overwhelmed the irrigation system on at least the south side of the Moche Valley. Changes of this sort would have affected all residents of this valley and probably of adjacent valleys, stimulating ongoing attempts to respond effectively to the deteriorating physical landscape. Is warfare an inevitable response to environmental degradation/change in a landscape peopled by settled agriculturalists who have invested heavily in irrigation systems and other infrastructure? Obviously, this is a complex question with many accessory factors, such as population size, alternative subsistence sources, the nature of alliances with neighboring groups, and the degree of political centralization. Warfare may not be inevitable, but the threat of warfare certainly did appear in Late Moche times in several valleys on the north coast, and it may have been an important part of the constellation of responses to environmental stress.

Certainly much Chimu warfare was related to political expansion, namely the examples given above; Theresa Topic (1990) has documented sequential expansion of the Chimu state to the north and south and linked that expansion to specific fortified sites (fig. 1.1). But there is also suggestive evidence that some form of tinku-related combat continued to be practiced alongside the expansionist warfare.

Settlement fortifications in late prehistory on the north coast were a response to threatened and actual warfare directed by the ruling elites of a state-level polity for the purposes of territorial expansion and political centralization. In contrast, the tinku combats that were prevalent during the Early Horizon and the earlier Moche period would not necessarily have led to increased political centralization. Any group that can wage war is an autonomous political unit; tinku puts communities in relationship to one another, but that relationship is not a relationship of political domination. Rather, it is a relationship of social and cultural affinity. It is not surprising, then, that warfare was one of the central themes defining Moche culture. While Moche warfare may have produced and reproduced a unity that was more cultural than political in nature, we have also pointed out that repeated success in battle could lead to the ranking of Moche political units and perhaps some shifting of allegiance from one temple complex to another (Topic and Topic 1997a). Chimu and Inka warfare placed communities in ranked relationships with each other, but they also resulted in a centralized, bureaucratic administration (J. Topic 2003).

From a practice standpoint, we will expect variability in warfare, as actors shaped responses to their particular circumstances—responses rooted in actors' history, tradition, and culturally informed assessment of best outcomes for themselves as individuals and as communities. An important methodological point underlies the expectation of variability. The material record of war is, of course, incomplete and imperfect. Accidents of preservation determine what sites, features, and artifactual assemblages are available to the archaeologist as raw material for interpretation. In the face of imperfect data with many gaps, it is all too easy to adopt simple models that purport to explain all cases. LeBlanc's assertion (2002:8–9) that conflict characterized nearly all of human history, and was invariably based on competition for resources, is an archetypical universalist theory, discounting the possibility of significant variability in the causes of and motivations for conflict; alternative scenarios are dismissed as a result of archaeologists naïvely accepting outdated Rousseau-ian models or misinterpreting the meaning of gaps in the archaeological record.

We maintain, however, that the material record must be given a higher level of credibility and must be accepted as reliable evidence for the actual practice of warfare. It is incomplete evidence, and sometimes frustratingly contradictory. But the evidence results from purposeful action—from what people actually did—and must be used with respect and sensitivity.

Consider, for example, the iconographic depiction of combat between champions in a context in which fortifications do not occur. Compare this scenario to one in which well-constructed fortifications have been built in isolated locations and iconographic depictions of champions are absent. These two scenarios contrast in fundamental ways and *must* lead us to infer significant differences in the practice of warfare in the two societies.

Rather than writing off variation as a fluke of imperfect preservation, a practice approach allows us to engage with variability as the authentic record of particular human experience. Improved understanding of the causes, impacts, and roles of conflict in human society through time is dependent upon a willingness to accept and learn from the variability that is so patently visible in patterns of human conflict.

2

Culture and Practice of War in Maya Society

Takeshi Inomata and Daniela Triadan

In warfare, codes of conduct prescribed by cultural logic and enforced by notions of warrior honor, often collide with goal-seeking pragmatism, driven by political and economic interests and a bare instinct for survival. Practice theory provides an important framework to analyze relationships between these contradictions and conflicting factors. Cultural codes and logic, held consciously and unconsciously in people's minds, shape their bodily practice, which in turn reproduces and transforms their notions and values. The practices that we examine in this chapter are not limited to physical battles in wars, which may not take place frequently during the lifetime of one individual. Equally important, or potentially even more important in this regard, are the routine practices of training, discourse, and rituals of war, in which battles are imagined, re-experienced, and re-enacted.

In Maya society, warfare occupied a prominent place in the mind and practice of many people. War was a common theme of art, inscriptions, and rituals. Hostilities often broke out between rival groups. The study of Maya warfare has traditionally focused on its causes and social effects. Some scholars have emphasized ecological and economic causes (Webster 1977), while others have pointed to cultural and political factors (Cowgill 1976; Demarest 1997). Although inquiries along these lines continue to be important, we also need to add the examination of recursive processes between cultural notions and practices of war. Warfare, whether real or imagined, played an important role in shaping values, meanings, and identities in the lives of the Maya, and such cultural notions, in turn, affected how war was fought or avoided. In this chapter, we focus on the Classic period (AD 250–950), in which archaeological, epigraphic, and iconographic evidence on the culture and practice of war is most abundant, but we also pay attention to the preceding and following periods (fig. 2.1).

2.1. Map of the Maya area with the locations of the sites mentioned in the text.

Culture, Practice, and War

The organizers of the Amerind seminar urged participants to examine warfare in cultural contexts and from the perspective of practice theory. These approaches require a critical evaluation of their theoretical bases.

The concept of culture, which has long been central to anthropology, has come under serious challenge in recent years (Dirks, Eley, and Ortner 1994). In the debate, three issues stand out. First, the traditional notion of culture privileges its homogeneity, boundedness, and coherence at the expense of internal contradiction and fluidity. Some anthropologists advocate paying closer attention to individual emotions and experiences, as well as to events and actions (Abu-Lughod 1991; Appadurai 1996:12; Barth 1994:358; Clifford 1986:19; Friedman 1994:207).

This leads to the second point. The focus on homogeneity tends to give culture a false appearance of timelessness. This presupposition of stability as the natural condition of culture does not leave any room for explanations of cultural change beyond external factors or extraordinary events (Aunger 1999:S94). Internal contradiction and fluidity imply that cultural continuity requires conscious efforts by the members of society (Brumann 1999:S11) and that culture always comprises a potential for change.

Third, the emphasis on coherence also translates into a view of culture as a neutral domain detached from politics. However, the contestation of diverse meanings in culture is inherently political, and politics inevitably involve the negotiation of meaning, which shapes one's perceptions of power (Wade 1999:450). This view is associated with the renewed interest in Antonio Gramsci and his concept of hegemony. As Raymond Williams (1977) noted, hegemony is a process of dominance and subordination, in which political, social, and cultural forces are interlocked in an inseparable manner.

These three issues do not necessarily mean that we should completely abandon the concept of culture. Nobody can deny that people in direct or indirect interactions share certain ways of thinking and doing things. Still, we cannot take this aspect of culture for granted. We need to ask how, by whom, and under what conditions culture might be shared and continued (Dirks, Eley, and Ortner 1995:3). In this inquiry, we need to recognize different layers of meaning, ranging from conventional meanings of collective symbols to diverse emotions harbored by individuals, including loyalty, hatred, and disinterest. Equally important are the varying degrees of consciousness in which such meanings are constructed and perceived. On one end of the continuum, culture involves unconscious daily routines, whose political implications people rarely recognize or evaluate. On the other end, there exist certain domains of culture to which people attach consciously recognized senses, including pride, identity, value, and objection, or for which they at least use emotionally or ideologically loaded language. These domains of culture are objectified cultures, which may be owned, displayed, and rejected (Wade 1999:452). It should be noted that objectified cultures do not necessarily have the same meaning for all individuals. Yet they may present common frames of reference with which people can construct narratives and upon which they can act. Both daily routines and objectified cultures provide fields of political negotiation, but people may engage in them at different levels of recognition of their political implications.

Practice theory, particularly the version elaborated by Pierre Bourdieu (1977), has had a critical influence on this trend in anthropology. Bourdieu addressed the political nature of social relations, often hidden under a stable, harmonized appearance. His theory overcomes a static view of society and culture by envisioning a political process in which social structures are constantly produced and reproduced through practice. A key contribution of Bourdieu, and practice theory in general, was to highlight the social significance of bodily practice as opposed to traditional anthropological approaches that privileged abstract ideas and language. However, there are certain limitations to his arguments.

First, as important as Bourdieu's perspective emphasizing bodily practice is, it appears to lead at times to an overemphasis on unconscious processes. This tendency may also derive from Bourdieu's focus on routine daily practice that does not involve much conscious self-evaluation. This problem is reflected in his concept of *habitus*, which, according to him, produces practice. He characterized habitus as a certain disposition or a principle of regulated improvisations without giving much credit to people's ability to evaluate the situation surrounding them and the potential consequences of their actions. Although Bourdieu recognized the possibility of strategic calculation, which may be carried on "quasi-consciously" (Bourdieu 1977:76), he gave it only halfhearted treatments. This theoretical formulation was rooted partly in his criticism of traditional Marxist theory, which tended to fall into a dichotomy of agents aiming consciously toward objectives and those mistakenly guided by false representations. Though this criticism is valid, denying people's ability to act consciously is not a solution.

Karl Marx's original notion of praxis continues to be important in this regard. We need to explore the relations between politically conscious acts and certain domains of practice that may be carried out less consciously, without falling into a simplistic dichotomy.

Second, despite Bourdieu's criticism of structuralism, the concept of structure continued to be central to his theory. As a result, his theory focused strongly on stability, and the issue of social change remained marginal. Bourdieu did argue that habitus, which only operates in relation to a social field, may produce significantly different practices according to the conditions of the field, and that habitus can be transformed in different circumstances. Still, he attributed changes mostly to external factors and gave little consideration to the potential and mechanism of internally generated changes (Jenkins 1992:82, 90). In this regard, another important practice theorist, Marshall Sahlins, more explicitly addresses social transformations through practice, as can be seen in his analysis of the encounter between the Hawaiians and Captain Cook (Sahlins 1985). His influence on archaeologists, however, has been limited in comparison to that of Bourdieu (see Ortner 2001). This may be partly because his rich historical contextualization, often in the settings of culture contact and globalization, discourages uncritical applications to different archaeological contexts.

Third, with his strong emphasis on political relations, Bourdieu often characterized social processes as if they were games or capitalist economic endeavors as reflected in his concepts of symbolic and cultural capital. People's belief in and commitment to social and moral values did not receive adequate attention. We need to recognize the importance of cultural and aesthetic values in people's minds without reducing such belief and commitment to false-consciousness (Bloch 1986:177). Following such values can even become a political objective for certain people.

This discussion of cultural concepts and practice theory calls for comments on the concept of agency, which was an important subject of debate in the Amerind seminar. In our view, the notion of agency concerns individuals' power to act or not to act in their social surroundings. It is misguided to consider a group, an institution, or a supernatural being to be an agent, precisely because this perspective ignores contradictions and contestations within culture, hidden under a seemingly coherent process. We need to recognize covert dissent and disinterest among multiple individuals, as well as different modes and levels of engagement and commitment, which constantly present possibilities for ruptures and transformations of collective actions and notions. We should note that our view of agency neither assumes a free-willed, pragmatic individual nor requires a methodological focus on individuals. Agency is always embedded in specific historical and cultural contexts (Dobres and Robb 2000:4), and the notion of agency is relevant to the study of macro-level social processes.

In the archaeological study of warfare, the contributions and problems of culture concepts and practice theory become particularly clear. War, in contrast to daily routines, is an extraordinary event, in which much is at stake. While success may lead to territorial expansions, material gains, political power, and sexual access, participants risk the loss of lives and the devastation of the community. Thus, warfare typically involves highly conscious, calculated acts through intense evaluations of past events and future outcomes. Extreme emotions caused by war, such as rage, fear, and desperation, may also override socially sanctioned behavior. In addition, we need to recognize that warfare is a field in which remarkable individuals, such as Alexander the Great, Genghis Khan, and Hernán Cortés, may affect the course of history in far-reaching, long-lasting ways. This understanding, however, should not lead to the great-man theory of history. Conducts of battle are shaped substantially by cultural concepts of war embedded in historical contexts. In this regard, goal-seeking actions in war, even those of great military leaders, are still rooted in historically contextualized dispositions.

Cultural factors associated with war include religious beliefs concerning battle and death; socially shaped roles in war related to gender, age, class, and other groups; and notions of honor and stigma tied to these roles. Also important are cultural codes that regulate and restrict conducts in war, as the explicit statements of the Geneva Convention and the more ambiguous notions of fairness and humanitarian ways do in today's world. Although such codes are often maintained and enforced consciously through public discourse and recurrent practice, their transformations can take place dramatically. Changes in conducts of war may be brought about by innovative individuals, new technologies, contacts with foreigners with different practices, or those who simply do not follow the rules. Deviations from shared codes are indeed common in history, implying that dissent to such collective impositions is constantly present. From the perspective of winning, those who do not follow the rules may have a distinct advantage in the outcome of battles. Breaking the rules, however, may come at considerable political and social risks. Those who win a battle ignoring the rules may have difficulty gaining followers, and in modern contexts, they may face international tribunals for war crimes.

Conducts of war, however, can never be understood only in terms of cunning calculations of weighing the advantages and disadvantages of following rules. Cultural codes of war are often tied to a strong sense of warrior honor and pride, which constitute an explicitly objectified domain of culture in many societies. Practices of war, thus, are shaped not only by conscious strategizing, but also through the performance of cultural notions of warriors. The creation, reproduction, and transformation of such codes, as well as other cultural notions of war, take place not only through participation in actual battles, but also through the discourse of war and the practices of training, playing, and rituals, in which war is imagined, remembered, and enacted. After all, even warriors usually spend substantially more time outside the battlefields than in actual combats, and many individuals, such as women, may never participate in, or see, real battles. Diverse experiences of multiple individuals merge and collide in the production and negotiation of cultural notions associated with war. In this process, material objects, buildings, and landscapes, such as fortifications, war monuments, and ancient battlegrounds, also shape people's experiences and imaginations as to what war means in their lives (Saunders 2003). In this sense, in the practice of war, the symbolic and the material are inseparably tied together.

Conducts of Battle in Maya Society

Some aspects of war practice among the Maya can be gleaned from texts and images that refer to conflicts and their outcomes. These data, however, come primarily from the Classic period, and we have little information of this kind from the preceding and following periods. We should also note that textual and iconographic data are not objective documentations of historical events but reflections of how the Maya perceived and remembered them. Another set of data includes material remains of war techniques and technologies, particularly fortifications and weaponry, which provide critical information on changes in war conducts before and after the Classic period.

Classic Maya Warriors

The epigraphic and iconographic record suggests that warfare in Classic Maya society may be characterized as ritualized battle. As indicated by numerous stelae depicting bound and submissive war captives, battles focused on personal engagements and the capture of prominent enemies for subsequent sacrifice (fig. 2.2) (Freidel 1986; Schele and Miller 1986). The emphasis on capturing opponents is not an effective strategy from the pragmatic perspective of defeating enemies, as the Aztecs learned the hard way when they faced the Spaniards. This practice was certainly rooted in the Mesoamerica-wide religious beliefs in the dedication of human blood and lives to supernatural beings, as many scholars have pointed out (Boone 1984).

This type of conduct on the battleground was also tied to the power and prestige of warriors. In the case of the Aztecs, it is clearly documented that warriors who made prominent military contributions could ascend the social hierarchy (Berdan and Anawalt 1992). Although social mobility among the Classic Maya was probably more limited, achievements in battle doubtless enhanced the prestige and status of individuals and groups. In colonial-period Yucatan, for example, the war captain called *nacom* was an elected position (Landa 1938:128). Military achievements of Classic Maya warriors appear to have been measured mainly in terms of the capture of enemies, as in the case of the Aztecs. Titles of Classic Maya warriors often proclaimed them to be the captors of prominent enemies or indicated the number of captives they had taken. It is suggestive to note that Plains Indians regarded taking coup, or touching the enemy, as evidence of warrior prowess and bravery higher than that associated with killing him. The Classic Maya may also have considered taking captives rather than slaying enemies as a demonstration of the skills and courage of warriors.

The cores of Classic Maya armies were elite males, often the kings themselves. Stelae commonly depict rulers in warrior attire, and accompanying texts often narrate their victories in war and the number of

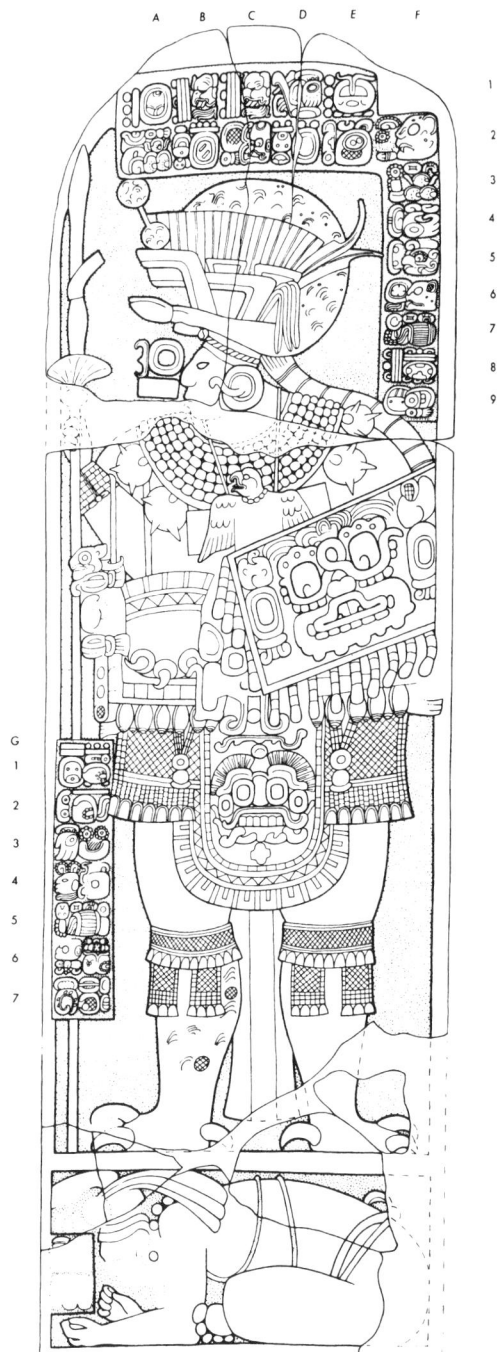

2.2. Aguateca Stela 2 depicting Aguateca Ruler 3 in warrior attire standing over the captured Seibal king, Yich'aak B'alam. (Reproduced from Graham 1967, fig. 5.)

captives they took. These captives occasionally included the rulers of enemy cities, indicating that these monuments were not totally ficti- tious glorifications but a reflection of the real risk that the rulers took through their physical participation in battles. Other elites are also depicted as warriors in murals and ceramic paintings. These battle scenes with multiple warriors seem to emphasize the individuality of warrior costumes and fighting actions rather than coordinated mili- tary formations. Their participation in battles and their achievements through personal engagements with enemies probably constituted a sig- nificant part of elite identities.

It is not clear to what extent nonelites participated in wars because stone monuments deal only with elite acts. Indirect yet indicative data are the dates of wars compiled by Mark Child (n.d.). He points out that many of them fall within the dry season, which may imply that battles involved nonelites who were freed from agricultural tasks during this time of the year. If so, Classic Maya wars were not just about duel-like combat among elite warriors, but they involved the organization and coordination of a large number of soldiers in strategic planning.

Warriors are in general depicted as males, though the gender of some warrior figurines is not immediately clear. A rare image on Naranjo Stela 24 shows a powerful woman standing on a war captive, but she wears female attire and holds objects associated mainly with a female iden- tity (Graham 1975:2, 63; Joyce 2000:63). Depictions of half-naked cap- tives are consistently males, but a unique monument appears to show a female captive with a torn dress (Houston, personal communication 2003). Together, these images indicate that battles were predominantly male activities but the division between gendered roles may have been crossed on rare occasions.

Texts and images also tell us about the outcomes of wars. In most cases, wars did not lead to territorial conquests, and the Classic Maya lowlands were never politically unified. However, we should not under- estimate the dynamic nature of Classic Maya warfare and the magni- tude of its political consequences. There is increasing evidence sug- gesting that the outcomes of battles considerably affected the political and economic fortunes of winners and losers, although in most cases defeated dynasties persisted at their centers. The winners of decisive bat- tles often enjoyed prosperity and appear to have commanded a certain level of political control over the defeated communities, in some cases driving their rulers into exile (Chase and Chase 2002; Houston 1993; Martin and Grube 2000).

Fortifications

Defensive features appear to have been more prominent, though far from prevalent, during the Late Preclassic period (300 BC to AD 250) than in Classic times. The central part of Becan in the Rio Bec region, for example, was enclosed by a moat and wall system, which probably dates to the Late Preclassic period (Webster 1976:2, 87, 103–113). Moats at Edzna and Cerros may also have served defensive purposes (Matheney 1983; Scarborough 1983), and the fortification of the small settlement of Muralla de León probably dates to this period (Rice and Rice 1981). The central precinct of the largest settlement of that time, El Mirador, also appears to have been defended.

Thus, significant transformations in conducts of war appear to have taken place during the Preclassic period. The first evidence of sedentary villages with ceramics in the Maya lowlands appeared around 1000 to 800 BC, substantially later than in neighboring regions. During the following millennium, lowland Maya society underwent a rapid process of population growth, social stratification, and political centralization, culminating in the large Late Preclassic center of El Mirador with enormous pyramids, which dwarfed any subsequent Maya buildings. Increasingly intensified and organized warfare probably played an important role in this social transformation. El Mirador and other large centers, however, declined at the end of the Late Preclassic period. The construction of fortifications at various centers may have been related to this social upheaval.

During the Classic period, there were only a handful of fortified centers. The fortification of Becan probably continued to function during the Early Classic period (AD 250–600) (Webster 1976). In the same period, the residents of Tikal in central Peten built an extensive system of ditches and walls (Puleston 1983; Webster et al. 2004), though their functions are still debated. The rival center of Calakmul may also have had some defensive walls around its central precinct. Most other settlements do not exhibit any defensive features, although some centers occupied high locations with some defensive advantage. Given the frequency of war recorded in the inscriptions, the easy access of many Classic Maya centers is striking and stands in clear contrast to examples from South America (see Arkush, this volume; Nielsen, this volume), the Near East, historic Europe, and East Asia. A relevant question concerns the locations of battles. Though we have little direct evidence, it is reasonable to assume that most combats took place in areas outside of major settlements. The rare depiction of a battle in the Bonampak murals seems to show an open field rather than a densely settled area (Miller 1986), and there are few indications of the wholesale destruction of settlements or commoner populations.

Towards the end of the Late Classic period, a more substantial change in the conduct of war appears to have occurred as reflected in the constructions of new defensive systems. Some of the best evidence comes from the Petexbatun region in southwestern Peten. After a probable defeat in a war, the dynasty that ruled over the region abandoned its primary capital, Dos Pilas, and moved permanently to its secondary capital, Aguateca, located in a more defensible location. During the reign of the last ruler, Tahn Te' K'inich, the Aguatecans built a series of defensive walls (fig. 2.3), and nonelites who had remained at Dos Pilas also constructed stone walls with blocks dismantled from temples and palaces (Demarest et al. 1997). The desperate situation at Aguateca is evident in the placement of walls blocking the former causeway and in the cessation of the construction of a major temple (Inomata et al. 2004). The walls at Aguateca were in some cases placed in patios of residential groups and abutted against existing buildings. Some standing structures left just outside the walls provided locations where attacking enemies could have hidden. In other words, the builders of the Aguateca walls did not have much previous experience in fortification designs and in siege warfare.

Aguateca was eventually attacked and defeated around AD 810. The elite residential areas in the epicenter were burned, and the commoners were probably forced to leave. It appears that the enemy intended to terminate Aguateca as a seat of political and economic power (Inomata 1997, 2003). This event marks a striking departure from the previous form of war in the Maya area. Still, some ritual aspects of war were retained. Following the battle, the enemies conducted rituals and deposited a large amount of broken objects as they destroyed the royal palace and some temples of Aguateca (Inomata et al. 2001).

In other areas of the Maya lowlands, we do not have clear evidence of fortifications at the end of the Late Classic period, but there are frequent references to wars in the monuments. For instance, the main

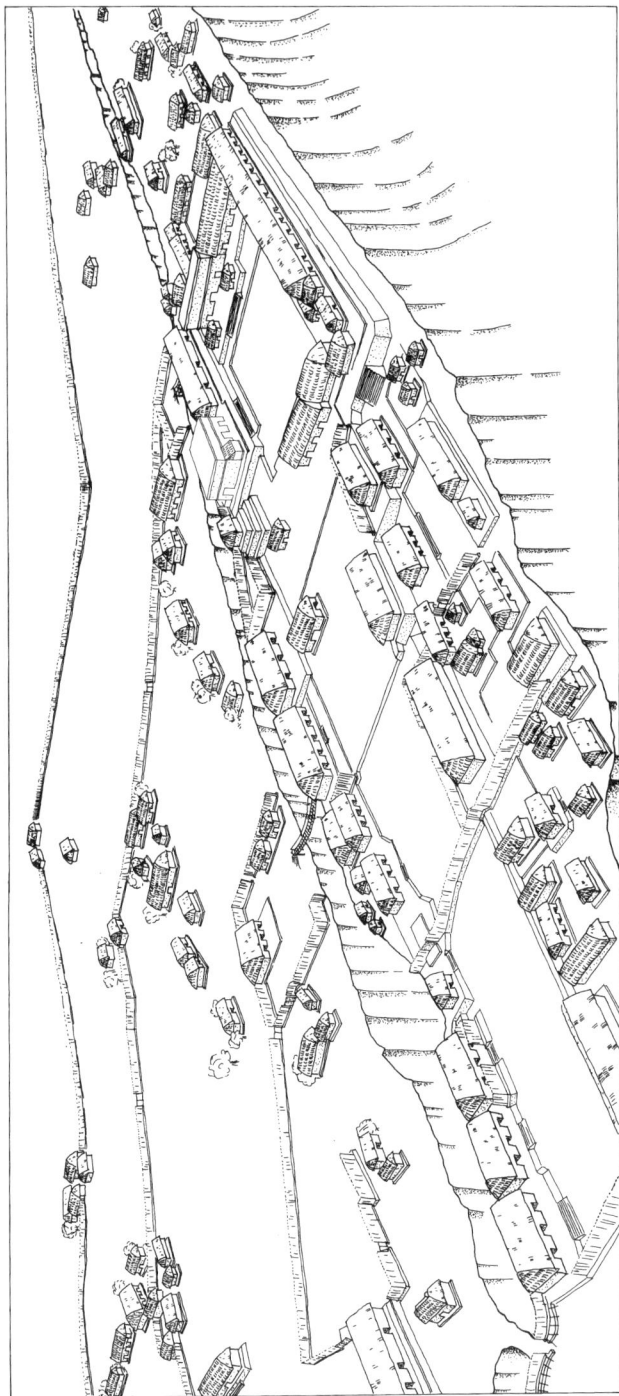

2.3. Reconstruction drawing of the central part of Aguateca with defensive walls.

themes of the Bonampak murals were war and sacrifice that took place in AD 792 (Miller 1986). They were the last monuments created at this center and were never completed. The last monument of Yaxchilan, Lintel 10—dating to AD 808—records the capture of the Piedras Negras ruler, which apparently triggered the fall of the Piedras Negras dynasty (Houston et al. 1999). Many other Maya centers in the southern lowlands also declined during this period in the phenomenon generally called the Classic Maya collapse. The intensification of warfare appears to be a factor in this major social change (Demarest 1997).

During the Terminal Classic period (AD 830–950), the focus of Maya civilization shifted to the northern part of the Yucatan peninsula. Fortifications became more common, as can be seen at the centers of Uxmal, Cuca, Chacchob, Dzonot Ake, Chunchucmil, and Ek Balam (Bey et al. 1997; Dahlin 2000; Graham 1975:4, 83; Webster 1979). Some of these centers may have fought destructive battles geared toward the annihilation of enemies (Dahlin 2000), comparable to the case of Aguateca. Yet, a substantial number of northern centers were not fortified. The emphasis on fortifications continued into the Postclassic period (AD 950–c.1530), as seen at Mayapan and Tulum in the northern Yucatan, Zac Peten in central Peten, and the hilltop centers of Iximche' and Q'umarkaj in the highlands. More destructive campaigns had become common practices, though they still retained strong ritual components.

Weaponry

We have little evidence on Preclassic Maya weaponry, but during the Classic period, the most common weapons—at least for high-status warriors—were probably handheld spears used in close combat, as depicted in art and reflected in the prevalence of chert spear points in the archaeological record. A potentially important change in weaponry, which may reflect different conducts of war, was the introduction of spearthrowers or *atlatls* and darts during the Early Classic period (see Schele and Freidel 1990:145–149). Spearthrowers extended the range and force of projectiles (Hassig 1988:75–79; Howard 1974) and allowed a warrior to carry more projectiles because of the lighter weight and smaller size of the darts in contrast to the traditional long-shafted spears. The appearance of spearthrowers was closely related to the influence of Teotihuacan on some Maya centers during the fourth and fifth centuries. Teotihuacan, located in central Mexico, was at that time the largest city in Mesoamerica, and its warriors probably fought more intense conquest wars than their contemporaries in the Maya region. Their weapons of choice, according to mural paintings, were spearthrowers and darts. An adoption of this new technology in the Maya lowlands could have promoted a shift in emphasis from close person-to-person combat to more anonymous long-range battles. The spearthrower, however, did not replace handheld spears as a predominant weapon in the Maya area.

The seemingly limited acceptance of spearthrowers in the Maya region is difficult to interpret, which is further complicated by the uncertain nature of the Teotihuacan influence. Some scholars propose a Teotihuacan takeover of the Tikal dynasty (Martin and Grube 2000:29–33; Stuart 2000). If this is the case, the strength of Teotihuacan warriors, along with the effectiveness of their weapons and tactics, must have impressed the Maya. Other researchers, however, argue for more diplomatic relations (Braswell 2003:25–26; Demarest and Foias 1993). Depictions of spearthrowers in Classic Maya art often appear to emphasize their significance as foreign symbols rather than as effective weapons. They tend to occur with other Teotihuacan imagery, such as shell helmets and Tlaloc (the Mexican rain god) masks (Taube 2000). These objects and their depictions may have served as the symbols of power for the Maya rulers who claimed exclusive connections or associations with the distant metropolis.

Another possibility is that in the Maya lowlands, which had many forested areas—as opposed to the fairly open landscape of central Mexico—the effect of these projectiles may have been limited. We should, however, note that substantial deforestation had already been taking place in the Maya lowlands. If Maya warriors indeed valued courageous acts, they may have preferred battles in open fields. Thus, the apparent reluctance of the Classic Maya to adopt the new form of battle remains problematic, but it may have derived at least partly from their strong commitment and belief in their traditional way of war that glorified hand-to-hand combat and the capture of enemies.

During the Terminal Classic period, spearthrowers and darts became more prevalent, at least in iconographic depictions. This trend is particularly prominent at the center of Chichen Itza. This may reflect another wave of central Mexican influence, and the more open, drier environment of the northern Yucatan peninsula may have favored this type of weapon. Highly indicative are the murals of the Upper Temple of the Jaguars at Chichen Itza depicting attacks on settlements with warriors carrying shields, spearthrowers, and darts (Coggins and Shane 1984:162–165, figs. 17–20). In one scene, the attackers use high scaffolds and a scaling ladder, suggesting that the settlement had fortifications or was located in a defensive place (fig. 2.4). The paintings also show the massacring of residents inside the settlement. Because depictions of battle scenes are extremely rare, these murals alone cannot be treated as evidence for a change in the conduct of war. Nevertheless, given the increase in fortified centers, it is not unreasonable to think that this form of battle involving attack on settlements became more common during this period.

The bow and arrow also appear to have been introduced to the Maya lowlands around the Terminal Classic period, although at the southern site of Copan, this technology may have been present since the Late Classic period or even earlier (Aoyama 1999:153–157; Inomata 1995:562–563; Rice 1986:340). The quantity of possible arrow points found in archaeological contexts, however, is relatively small. The bow and arrow, which played important roles in more intense wars in various societies of the Old World, never became a preferred weapon among the Maya.

Conservatism and Transformation

These data indicate that warfare in Classic Maya society had a strong ritualized aspect, in which conducts may have been significantly regulated by cultural codes. War was an important concern for the Classic Maya, particularly for elite males, and they did indeed fight rival groups frequently. Their focus on personal engagements, the capture of enemies, and subsequent sacrifice was probably sanctioned by religious beliefs and tied closely to the power, prestige, and honor of warriors. These cultural notions of war may have discouraged the wholesale destruction of enemy settlements, as reflected in the general lack of defensive features. Yet, the presence of fortifications at a small number of centers suggests that such cultural codes occasionally broke down. The tension between the attitude honoring the commonly held cultural codes and the one seeking pragmatic ways of defeating enemies also

2.4. Mural on the south wall of the Upper Temple of the Jaguars at Chichen Itza (Reproduced by permission from Coggins and Shane 1984, fig. 20.)

appears to have been expressed in the sporadic introductions of new weaponries and the persistent reluctance to utilize them in the most effective manner.

Despite this conservatism, we can observe certain chronological trends in the conduct of war, particularly when we examine the periods preceding and following the Classic period. The Late Preclassic and Terminal Classic periods may have witnessed the intensification of warfare, as reflected in the increases in fortified sites. The pattern, however, is far from clear cut. During the Late Preclassic period, in particular, there existed numerous centers without any evidence of defensive features. Arthur Demarest (1978) has argued that cultural codes regulating conducts on the battlefield were not honored in wars between different cultural groups, promoting the construction of fortifications along ethnic boundaries. The few Late Preclassic fortified sites, however, do not exhibit a clear spatial distribution consistent with this hypothesis. We probably need to pay greater attention to unpredictability and idiosyncrasy in the conduct of war.

It is conceivable that, during the remarkable social transformation of the Late Preclassic period, heterogeneity and fluidity were particularly salient aspects of the culture of war. Various activities—constructing enormous pyramids at El Mirador and other centers, living in these large settlements with these large monuments, participating in and viewing the mass spectacles of rituals taking place there, and receiving guidance and orders from rulers and priests—were all relatively new, unprecedented experiences and experiments for the Late Preclassic Maya. Though we still know little about Preclassic Maya warfare, we can imagine that this rapidly changing society most likely involved competition between groups seeking power, wealth, and prestige with the help of armed forces. In addition, there probably existed contestations and contradictions between the opinions and attitudes of people seeking new forms of war on the one hand and those honoring more traditional, less intense forms of fighting on the other.

The transition from the Classic to the Postclassic period exhibits a somewhat more consistent trend with a stronger emphasis on fortifications and direct attacks on settlements. An increasing number of scholars argue that the social change during this period, as well as that at the end of the Late Preclassic period, was triggered by severe droughts (e.g., Gill

2000). It is beyond the scope of this chapter to evaluate this evidence. We simply note that, to understand the social outcomes of such climatic changes, we still need to examine actions of agents who evaluated and acted on such external factors and were conditioned by their historical contexts. Changes in conducts of war during the Terminal Classic period were rooted in the history of practices during the preceding Late Classic period, which had constantly presented a potential for change.

In comparative terms, Maya wars—even those during the Late Preclassic and Postclassic periods—appear to have been less intense and their techniques less sophisticated than devastating battles waged in the Near East, China, Japan, and historic Europe. In this regard, Maya warfare exhibited a precarious balance between cultural codes of conduct, with a tendency toward conservatism on the one hand and political and economic motivations—as well as personal agendas that may have encouraged deviations and innovations—on the other. This balance did break down occasionally.

Practice of War at Home during the Classic Period

To understand the stability and change in Maya warfare, we need to examine aspects of the culture of war that were expressed, maintained, and probably internalized to a certain degree, not only through repeated participations in combat but also through discourses and enactments of war away from the battlefields. We have particularly rich archaeological, epigraphic, and iconographic data from the Classic period to address these issues.

Images of Warriors

Among the practices that took place outside the battlefields were image-making and image-viewing of war. These images, materialized in stelae, ceramic paintings, and figurines, are parts of the process and reflection of how different groups and individuals engaged in the culture of war. The presence of stone monuments in plazas, presumably viewed by many elites and nonelites alike, was a constant reminder of the warrior ideal. Through these sculptures that strongly emphasized the capture of prominent enemies and their sacrifice rather than the destruction of enemy cities or territorial conquest, the Classic Maya

expressed and imagined not only the idealized notion of successful warriors but also specific ways of conducting war, which became a part of
the kingly image and a base for their political support.

War was also an important theme of figurines. For example, at the
site of Aguateca, 32 percent of male figurines found in elite residences
represent warriors, and male figurines outnumber female ones two to
one (fig. 2.5) (Triadan 2007). It is probable that, for the Classic Maya
elite, war was a common motif of daily conversation and play by children. Elite men may also have spent significant time training as warriors. In other words, activities associated with war were largely naturalized parts of daily life, through which most Maya elites created,
performed, and were exposed to their cultural notions of war and ways
of being warriors. Although some of the figurines may represent specific
historical figures, many of them appear to show more generic images of
anonymous warriors. Unlike the images on stelae that narrate historical
events and acts of specific individuals, these figurines, along with associated daily acts, reflect more generalized notions of war and warriors.

Women were also depicted in various war-related contexts, in addition to the above-mentioned rare images of females as protagonists or
captives. In the captive presentation scene of the murals of Bonampak, two royal women in courtly attire are present among numerous
male warriors. In the scene, which depicts an elaborate dance, more
courtly women conduct bloodletting rituals by piercing their tongues
(Miller 1986). In other images, royal women assist their husbands in
putting on their warrior costumes (Schele and Miller 1986:211). Thus,
elite women participated in the creation and reproduction of the culture of war through their ritual performance and possibly through their
daily discourse of war. These images imply that women's contributions
were expected and recognized. It is conceivable that they also supported
male warriors through the elaboration of battle costumes and the preparation of food.

Theaters of Battle and Death

Although most iconographic representations concerned elite activities, nonelites were most likely involved in the creation and reproduction of the culture of war. This community-wide process was particularly evident in mass spectacles of the sacrifice of war captives and other

2.5. Figurine representing a warrior excavated from Structure M8–8, an elite residential structure at Aguateca.

theatrical representations of war. Although Classic-period depictions of such rites on stelae, ceramic paintings, and murals do not include non-elites, historical documents indicate that communal festivals during the colonial period often involved numerous commoners (Barrera Vásquez 1965; Inomata 2005; Landa 1938:125–126). Various lines of evidence suggest that during the Classic period, large numbers of nonelites often participated in rituals of war in a manner comparable to mass ceremonies conducted in the colonial times. Through the spatial analysis of plazas, Takeshi Inomata (2006) has argued elsewhere that the large public plazas of each Classic Maya center served as stages for many of these rituals and were in most cases designed to accommodate the center's entire population. In addition, warrior kings depicted on stelae placed in these plazas wear elaborate attire, including large headdresses of bird feathers, emphasizing their visibility for a large number of spectators.

On some occasions, rulers and their war captains, along with their captives, presented themselves on a wide stairway facing a plaza, as seen in the Bonampak murals (Miller 1986:115; Schele and Miller 1986:218), on stelae, and in ceramic paintings (e.g., Kerr no. 767).[1] The use of stairways facing plazas as theatrical stages further enhanced the visibility of such acts for a large audience. Other images show sacrificial victims tied to wooden scaffolds, probably placed in plazas (Coe 1975; Trik and Kampen 1983: fig. 38; Taube 1988; Kerr nos. 206, 2795, 2781). Charles Suhler and David Freidel (2000) suggest that the postholes found in the Great Plaza of Tikal in front of Structure 5D-32–2nd were traces of such a scaffold (see Coe 1990:518–521, fig. 168b). These temporary structures also provided effective theatrical stages in front of a large audience. Moreover, sculptures of chacmools at the Terminal Classic center of Chichen Itza, representing reclining captives (Miller 1985), are found in highly visible locations of temples and platforms in plazas (Ringle and Bey 2001:278).

Some ceramic paintings of war rituals seem to represent more closed settings inside or in front of palace structures (e.g., Kerr nos. 680, 5850, 6650, 7516). The East Patio of the Palace of Palenque, for example, contains sculptures of captives and may have been a place for presentations of defeated enemies. In these cases, the rituals were more exclusive, with a smaller number of participants than those in plazas. Still, we should note that palace complexes of many Maya centers maintained a certain degree of visibility from the outside. The Palace Group of Aguateca, for instance, was located at the end of a wide causeway, and spectators standing on the causeway could observe ceremonies taking place in this compound (Inomata 2001). Thus, although presentations of captives and their public executions took place in various spatial settings, many of them were theatrical events sponsored by the rulers and elites and were probably attended by a significant number of nonelites.

Not only did these sacrificial rituals represent enactments of religious beliefs shared by many members of the community, but they also served to demonstrate the results of war and the achievements of warriors to the rest of the community. As many battles likely took place away from settlements, many community members probably did not witness the combat. The sacrificial rites following combat were in a sense re-enactments of battles and constituted a phase of prolonged war. The climax of armed conflicts—that is, the humiliation and execution of important enemies—was reserved until the final stage of war so that the entire community could participate in this memorable moment. Victorious warriors re-enacted and re-experienced the excitement of battle, and those who stayed in the settlements could witness the triumph of their group. Thus, these ceremonies were occasions for the affirmation of victory and the creation of shared experiences and memories.

Certain rituals took the more explicit form of a physical fight. In the case of the Aztecs, there exist historical documents that describe gladiatorial battles in which a captive, tethered to a large stone and given a sword edged with feathers instead of obsidian blades, had to fight several Mexican warriors who attacked with real weapons (Nuttall 1903:30). Stephen Houston (2002) suggests that comparable gladiatorial battles existed in Classic Maya society, although evidence for them is scarcer. An alabaster vase shows two men fighting with pointed weapons (Kerr no. 7749) and a ceramic vase painting depicts two groups of fighters holding circular knives (Kerr no. 700). A significant number of figurines may also represent combatants in such battles.

A particularly common form of ritual battle for the Maya was the ballgame (Colas and Voss 2000; Gutierrez 1990). We should note that the use of athletic competitions as a metaphor for war and as a form of controlled violence is prevalent in many societies of the world. Today, the connection between sport and war becomes clear in the popularity of professional wrestling on TV, the surge of nationalism in the Olympic Games, and the riots following many soccer matches. In Maya society, the sport metaphor is used in the Popol Vuh, the creation myth of the colonial-period K'iche', in which the Hero Twins fought the gods of the underworld in ballgames (Tedlock 1985). During the Classic period, the hieroglyphic stairway of Naranjo mentions a military conflict with Caracol, followed by a ballgame, which indicates a close connection between the two events (Graham 1975:2, 107; Schele and Miller 1986:250). The ballgame was also associated with sacrifice, as seen in the Popol Vuh, in which the Hero Twins' father and uncle lost a game to the gods of the underworld and were sacrificed (Tedlock 1985). Sculptures in the ball-court of Chichen Itza show the decapitation of a player following the competition. Some games involved victorious rulers and captives of war. Hieroglyphic Stairway 2 of Structure 33 at Yaxchilan depicts the ruler Bird Jaguar hitting a captive who is bound like a ball and falls down the stairs. Other rulers also possessed the title of ballplayer, indicating that such demonstrations of battles in front of the community were an important act in creating the images of warrior kings.

Theaters of battle and death, along with real combat and daily practices, contributed to the processes in which the objectified cultures of warrior honor and codes were produced and reproduced. Through theatrical performance, such objectified cultures were presented to, and to a certain degree shared by, various members of the society, including women and nonelites. In addition, the enactments of battles and the sacrifice of enemies brought the community together and highlighted its collective identity by producing shared experiences and by showing others as inferior.

The Politics of Human Sacrifice

We also need to examine negotiations and contestations in the culture of war, which took place among elites and warriors, as well as between different groups. One aspect of such negotiations and contestations relates to the power and prestige of warriors. In examining the practice of human sacrifice with this focus, data from medieval Japan is suggestive. Japanese warriors commonly decapitated enemies, even though they did not have any religious beliefs associated with trophy heads and were aware of the impracticality of carrying around heavy heads on the battlefields. The practice of decapitation in war had its own logic in the political and military system of Japan. Land, as the economic basis of the samurai class, was distributed according to contributions in battle. Particularly important was to kill adversaries of high rank. The clearest way to secure the recognition of such achievements was to cut off the heads of victims and present them to superiors.

In a similar manner, bringing captives home in Maya society meant the public announcement of war achievements by the captors, though this political implication was probably not as consciously recognized by the Maya as by the Japanese. If warriors left the bodies of enemies in the battlefields, it would have been difficult to determine whether they really had killed important adversaries or whether they were lying. Public presentations of captives gave the captors proud moments in which their work was acknowledged and rewards were promised. We should emphasize that our point here is not to make a cross-cultural generalization. Nor do we suggest the applicability of the Japanese model to Classic Maya society. Cultural notions surrounding Japanese and Maya practices are vastly different. Instead, we attempt to illustrate through these examples the political significance of physical acts and witnessing, which is emphasized by the performance and theater theories (Inomata and Coben 2006).

Theatrical events associated with war also provided occasions on which rulers and elites claimed their authority and power over the rest of society. The imposition of death on individuals and its presentation in front of spectators were explicit manifestations of elite power sustained by political institutions. In these spectacles of pain, suffering, and death, the superiority of the executors was expressed (Futrell 1997:47). Thus, the theatrical roles of organizers, warriors, victims, and spectators defined and constructed the social order between the ruler, nobles, and commoners. Moreover, practices of war away from the battlefields had significant political implications for the continuing tension between victors and losers. In Maya society, in which ancestors were sources for the legitimization of the dynasty and for the integration of the community, the capture and sacrifice of rulers and other prominent individuals meant the negation of proper mortuary treatments, causing as much damage as the loss of their lives.

There is evidence suggesting that the victors sometimes conducted rituals in the settlements of the defeated. Archaeologists have identified "termination rituals" at sites, such as Yaxuna and Aguateca, in which victorious enemies ritually destroyed or "sacrificed" buildings and other monuments (Freidel and Schele 1989; Inomata et al. 2001). The symbolic killing of buildings and other materials representing the value and tradition of their community forced the unfortunate losing groups to witness and experience their defeat. At the same time, these practices imply a certain level of culturally mediated mutual understandings between combating groups about procedures after the battles, which probably contributed to the avoidance of the wholesale destruction of settlements.

The effects of these practices on the political relations among elites, nonelites, men, women, victors, and losers derive partly from the magnitude of emotional reactions. Strong emotions caused by violent death are probably not limited to modern Western societies. The prevalence of the theme of death in arts, stories, and ceremonies in any society indicates a profound preoccupation with this fundamental phase of human existence. These emotions may also translate into an intrinsic interest that draws people to spectacles of terror, as modern Westerners who watch movies and TV programs of similar kinds can attest. Human sacrifice, however, may cause such diverse reactions as horror, hatred, desperation, excitement, satisfaction, erotic feelings, and joy (Kyle 1998:7), depending on cultural contexts and individual dispositions. Thus, the social process related to sacrifice touches on cultural particularities and human commonalities. It is probable that different and conflicting emotions and attitudes coexisted within a group and sometimes within one individual. In Aztec society, ambivalence toward sacrifice is evident in the legend of Quetzalcoatl (Demarest 1984:232), in which Quetzalcoatl, a revered king who opposed human sacrifice, was defeated by Tezcatlipoca, who demanded bloody rites. Similar conflicts probably existed among the Maya.

Such ambivalence and diversity in emotional reactions most likely led to the complexity of the political processes related to the culture of war. The institutionalized imposition of death through ritual battle could have provoked not only admiration but also antagonism toward the political authorities. Although the elite may have tried to enhance their prestige through these rites, their political outcomes could sometimes have been unpredictable.

Conclusion

Conducts of war are shaped and affected by diverse factors and processes, including innovative individuals who adopt new techniques and technologies or who do not follow existing conventions; consciously and

unconsciously held cultural notions of war that may be shared, negotiated, and contested; and external conditions, such as contacts with
foreigners and environmental changes. The dynamics of war is particularly difficult to study because, on the one hand, strong values attached
to cultural notions of war may be resilient and, on the other hand, the
high stakes of war constantly present a potential for drastic transformation. The stability and change in Maya warfare, like that in any other
region of the world, resulted from such complex interplays. Practices of
war in Maya society constantly comprised contradictions, in which a
tendency toward the adoption of new techniques and technologies col-
lided with the conservatism of maintaining cultural codes of war. There
is a chronological trend in Maya warfare during the Late Preclassic and
Terminal Classic periods with a tendency toward an intensification of
war. The Classic period may be characterized by stronger adherence to
existing cultural codes, though it did not preclude certain deviations. In
comparative terms, Maya warfare appears to have been generally conservative with rather unsophisticated techniques and technologies.

The study of Maya warfare has traditionally focused on identifying
specific causes of change in warfare, but we also need to examine the
nature of recursive interactions between cultural notions and practices
of war, which can potentially lead to stability or transformation. Such
studies will also provide a necessary basis for a better understanding of
the effects of possible climatic changes and other external factors. Cultural notions of war were established, challenged, and acted on not only
through conduct during combat, but also through diverse practices that
took place away from the battlefields. In Classic Maya society, various forms of interaction, including daily discourses at the household
level and community-wide mass spectacles, provided opportunities for
people to enact, re-experience, and imagine warfare. Through the rep-
etition of rituals and daily practices, certain notions—such as specific
ways of being a warrior—were established, reinforced, and performed
as objectified cultures by certain groups. Some of these cultural values were probably internalized and naturalized in the minds of many
individuals, substantially shaping their conducts of battle. In addition,
there were political motivations of various groups to continue such
practices, including rulers who needed to legitimize their authority and
warriors who desired social promotion through military achievements.

These ritualized and routinized practices away from the battlefields may have contributed to the maintenance of cultural codes of war to a certain degree. At the same time, practices outside the battlefields constituted arenas for negotiations and contestations among diverse groups, reflecting the diversity in their interests and emotional reactions. The process of contestation constantly presented a potential for change in conducts.

Archaeological studies of warfare need to place more emphasis on the culture and practice of war. As warfare highlights contradictions and conflicts between unconscious routines and conscious calculations, as well as those between diverse groups of people, it presents a challenging, yet productive field of study for refining archaeological theories of culture and practice.

Acknowledgments

We thank Axel Nielsen and William Walker for inviting us to the seminar and for their thoughtful comments on an earlier draft of this chapter. We also thank John Ware and the staff at the Amerind, who were wonderful hosts, as well as the seminar participants for stimulating discussions and suggestions.

3

War Is Shell

The Ideology and Embodiment
of Mississippian Conflict

Charles R. Cobb and Bretton Giles

The recent popularity of such works as Stephen Ambrose's (1997) *Citizen Soldiers* is due in no small part to the ability of modern historians to portray the experience of warfare from the perspective of those who did most of the fighting and dying. Prior to the nineteenth century, the prevalence of illiteracy among common soldiers left a very slanted perspective on life in the trenches, one provided by the likes of Julius Caesar's sketches of Gaul or Thucydides's account of the Peloponnesian War (Keegan 1976:32, 64–69). Hence, our ability to somehow connect with those who experienced premodern warfare and its consequences can be elusive. Certainly, as archaeologists, we can avail ourselves of the numerous ethnographic accounts of aggression and violent encounters in non-Western societies, but to merely impose these characterizations on the past is to subscribe to a somewhat sterile, neo-evolutionary modus operandi that is at odds with the passions evoked by armed conflict.

The Mississippian Southeast in North America (ca. AD 1000–1500) presents an inviting challenge to us in this context (fig. 3.1). During the first millennium AD, Native American societies in the region displayed a broad pattern of increasing sedentism and dependence on indigenous cultigens. Around AD 1000, there was a marked shift to the incorporation of tropical domesticates (especially maize) into the diet, substantial towns numbering in the hundreds to thousands of individuals appeared across the landscape, the ancient tradition of earthwork construction underwent a significant expansion, and social stratification became pronounced in many localities. Archaeological and ethnohistoric evidence leaves little question that widespread conflict was part and parcel of this larger sweep of change. Southeastern archaeologists have documented a wide range of material signatures of physical trauma and defensive architectural features. In addition, there is a broad can-

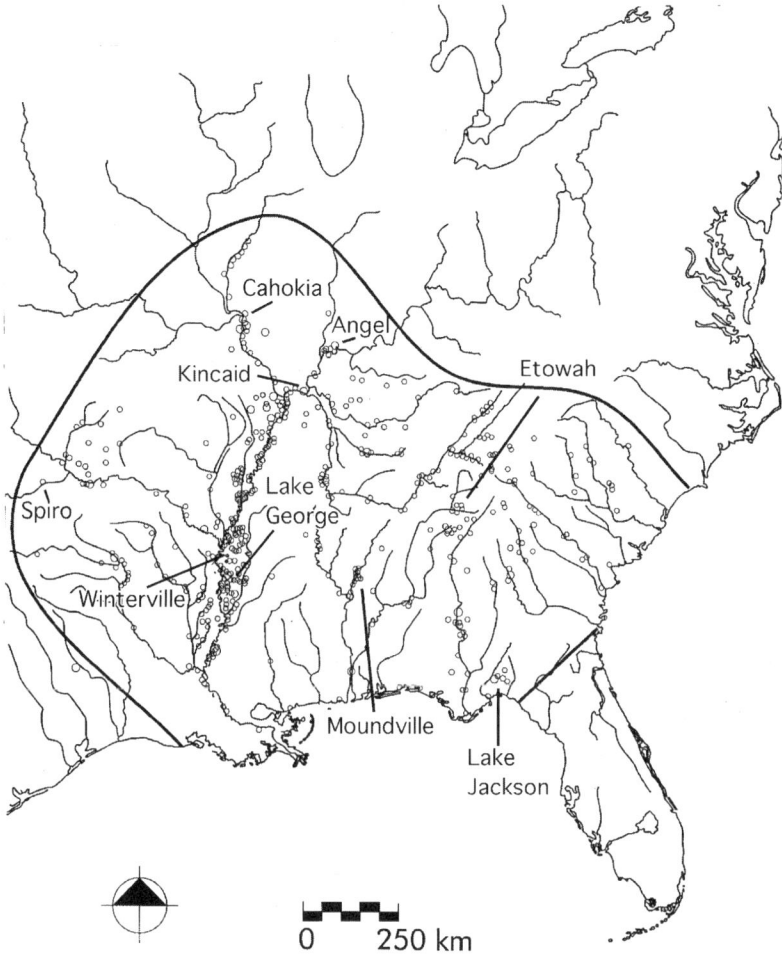

3.1. Major sites in the Mississippian Southeast, highlighting some of the largest mound centers. (Adapted from Lewis and Stout 1998, fig. 2.4; used by permission from the University of Alabama Press.)

vas of myth and symbol rendered in shell, copper, and other materials that linked human ambition, privilege, aggression, and the cosmos with ideological perceptions of the male. It is with this pattern that we concern ourselves. There is a vigorous literature (much of it drawn from ethnohistory) describing the underlying reasons and methods of Mississippian warfare. Yet there is relatively little that addresses how the archaeological record itself may inform us about the meshing of the ideological beliefs and secular pursuits of war, and how these may have varied through space and time.

Here we focus on expressions of myth and war found in a suite of Mississippian art and prestige objects, often associated with a phenomenon known as the Southeastern Ceremonial Complex. Representations of the human body, body parts, human-animal chimeras, and certain weapons like the mace convey how Mississippian warfare embodied beliefs about status, gender, power, and the cosmos. This portrayal of idealized violence through the body was an essential component in the social reproduction of warfare through two sets of dialectical relations: (1) a tension between the state of the perfect warrior and the fear of bodily defilement, and (2) the juxtaposition between the mortal male and the warrior who transcends boundaries between mortality and divinity during times of conflict.

Archaeology, Embodiment, and Representation

While the body has been extensively debated, discussed, and theorized about in the humanities and social sciences (e.g., Butler 1993; Csordas 2000; Lock 1993; Turner 1996), embodied approaches have only relatively recently entered into archaeological debates, beginning in the early 1990s (e.g., Hamilakis 1999; Hamilakis et al. 2002; Joyce 1998, 2005; Kus 1992; Lucas 1996; Marcus 1993; Meskell 1996; Montseratt 1998; Rautman 2000; Tarlow 2000; Treherne 1995; Yates 1993). Building upon an embodied perspective, archaeological studies have attempted to move away from objectification of the passive body and toward understanding how the situated body socializes and evokes subjectivities. In addition to its usefulness for addressing major social changes, the embodiment perspective is particularly important for apprehending those practices and occurrences that are intimately linked with the course of life (Turner 2003), such as birth, death, and as we will propose, the founding of warrior identities.

One of the advantages of embodiment approaches is that they attempt to dissolve the arbitrary divide between agency and structure (Meskell 2000), since constructions of the self can only be understood by reference to the social body and vice versa. Helle Vandkilde (2003:127) emphasizes this point with regard to organized conflict:

"Warfare is a flow of communally based social action aimed at violent confrontation with the other. Being a warrior is consequently a social identity founded in warfare." Representations of the body are particularly important in promulgating cultural ideologies surrounding identities such as that of the male warrior. The dissemination and circulation of idealized or iconic bodies in material forms (e.g., depictions on pottery, statuary, and modern visual images in a variety of media) are effective as "public representations" that legitimate and promote particular masculine/feminine bodies and performances (Bordo 1997; Joyce 1998:148; Moore 1999; Sperber 1992:60–64). Such representations contribute to the pedagogic action necessary for the reproduction of habitus (Cicourel 1993:100). To name an archaeological example, Michael Shanks (1995) proposed that from the eighth to the seventh century BC, Grecian worldviews underwent significant ideological shifts, in part related to a rising militarism. These changes in subjectivity were both signified and reproduced by new images of the body, such as the prevalence of men at arms and iconography that suggested the liminal space between mortality and divinity occupied by the warrior.

Certainly depictions of warriors or warfare are not direct reflections of lived experiences. Instead, public representations of idealized bodies and bodily performances imply particular social "positions" that can be taken up by men and/or women in certain situations and spaces (Butler 1993; Connell 1995). Such social positions are neither completely arbitrary nor uncontested. The meanings attached to normative bodily performances as they are manifested in art and similar media are embroiled in attempts to naturalize relations of power that may be embedded in gender, racial, and class distinctions (Bordo 1993:339; Joyce 2005:145–147; Lesure 2005:246).

Archaeology provides a particularly useful vantage point on the entanglement of the production and reproduction of bellicosity with the constitution of masculinity. Depictions of individuals (or victims) in a posture of armed conflict have much to tell us about the somatization of oppression, ritual, and belief systems and how these changed through time. As we will argue, Mississippian renderings of the body and body parts were critical for constructing the ideal warrior and the privileging of particular ways of being-in-the-world, forms of embodiment that in turn played a pivotal role in establishing and normalizing hierarchies.

The Practice of Mississippian Warfare

European accounts amply describe the military capabilities of sixteenth-century southeastern chiefdoms (e.g., Clayton et al. 1993). Further, there is archaeological evidence from the Great Plains eastward of a long-term pattern of chronic raiding punctuated by intermittent large-scale conflicts well before the Entrada—with the scale of violence taking an upswing with the rapid spread of maize agriculture and sedentary societies around AD 1000 (Anderson 1994; Bamforth 1994; DePratter 1991; Dye 1995; Steinen 1992).

Small-group ambushes likely typified Mississippian violent encounters, but on occasion, groups were apparently capable of larger, coordinated assaults. The archaeological record shows a sharp upsurge in the construction of fortifications and in skeletal trauma from impact injuries after AD 1000, and Mississippian societies seemingly entered into a new phase of even more heightened aggression sometime around AD 1300 as manifested by a strong trend towards nucleation into fortified villages and towns (Anderson 1994:136–137; Brose 1989:29; Milner 1999; Morse and Morse 1983:283). The importance of defensive works was attested to by sixteenth-century Spanish and French explorers in the Southeast, whose descriptions of fortified towns as places of refuge accord well with the archaeological evidence. Mississippian palisades were impressive wooden-post constructions, extending distances of up to two to three kilometers around some towns, and they were often enhanced with features such as ditches. Often, stockades had projecting round or square bastions at intervals of twenty to forty meters (DePratter 1991:43), a distance that was comfortably encompassed by overlapping bowshot cover fire (Lafferty 1973).

Bow-and-arrow technology appears to have been introduced to the Southeast in the middle of the first millennium AD (Blitz 1988; Nassaney and Pyle 1999). It is questionable as to how dramatic this new technology may have been for the rest of society even if there are clear advantages to the bow and arrow over the *atlatl* in terms of accuracy, range, and rapid-fire delivery. In the Southeast, these advantages may have been more important originally for hunting game, given the lack of evidence for a spike in effective violence with the appearance of the bow and arrow. Whatever the reasons underlying the later spread of Mississippian warfare, the bow and arrow appear to have been readily adapted to the predation of humans. In Mississippian cemeteries, stone arrowheads are frequently found embedded in human bone or in locations around a body that are suggestive of having been lodged in soft tissue (e.g., Bridges et al. 2000; Milner et al. 1991). As Lawrence Keeley (1996:49–54) points out, however, if one's objective is to kill or severely injure, projectile or fire weapons are far outclassed by shock weapons such as maces, axes, and bladed weapons.

Popular shock weapons among Native Americans included the celt, mace, and a variety of bifacial blades (Van Horne 1993). As with the bow and arrow, there is enough evidence of trauma in skeletal remains to indicate that these weapons were employed with considerable success during the Mississippian period—typically manifested as skull depressions or fractures, or else as parry fractures on the forearms (e.g., Bridges et al. 2000; Milner 1995; Milner et al. 1991; Smith 2003; Tallman 2004). Southeastern ethnohistoric accounts indicate that the ability to personally subdue a person was much more likely to confer prestige than a long-distance kill (Dye 2002; Van Horne 1993). The personalized nature of a close confrontation was viewed as an important gauge of status and power. It also increased the odds of taking a scalp or head as a trophy of one's success, practices reflected in some burials by cut marks on the cervical vertebrae and the absence of skulls, or cut marks on the skull indicative of scalping (e.g., Bridges 1996; Bridges et al. 2000; Milner et al. 1991; Smith 2003).

Southeastern military groups apparently used fire and shock weapons both in small-group, surprise encounters and in well-orchestrated maneuvers that likely reflected a longer tradition of larger-scale warfare. Even considering the exaggerations and biases inherent in explorers' narratives, there is sufficient correspondence in the sixteenth-century de Soto accounts to make the case that many southeastern Indians worked as choreographed units in battle, following a hierarchy of leadership and communication that transcended common conceptions of "tribal" warfare—even though there is no evidence for standing armies or an institutionalized military (DePratter 1991:43–47; Dye 2002). Karl Steinen (1992:134) refers to the larger-scale Mississippian attacks as "raids in force" that were limited (relative to those of warring states) but well organized. Several of the Spanish military debacles on the de Soto route (1539–1542) were at the hands of Native American groups fighting in what were described as squadrons that worked together to make feint attacks, flanking maneuvers, and other developed forms of engagement.

Given the vagaries of preservation in the archaeological record, it is a challenge to assess the scale and intensity of warfare in the Southeast— or any other part of the world, for that matter. For regions like Fiji, ethnographic descriptions suggest an atmosphere of almost constant bellicosity among numerous competitive chiefdoms (Carneiro 1990), and Napoleon Chagnon (1968) made the issue of sustained tribal war a centerpiece of his studies of the Yanomamo (however, cf. Ferguson 1997). It would not be difficult to extend this image of steady inter-group aggression to the Mississippian world and its neighbors (Dickson 1981). The Norris Farms site in the Illinois Valley (ca. AD 1300), for example, appears to have been situated on an unfriendly Mississippian and Oneota frontier characterized by chronic raiding (Milner et al. 1991). Despite the substantial palisade surrounding the community, the cemetery (with a total of n=264 individuals) has yielded a significant number of individuals who were killed or wounded with fire and shock weapons (n=43), apparently over the course of several generations. The violent deaths stand out in particular for the recurrent pattern of scalpings (n=14) and decapitations (n=11) associated with them. The Koger's Island site in Alabama had a formal cemetery arranged in rows. Of 108 individuals analyzed, 5 percent had scalp marks, 5 percent had cranial fractures, 46 percent had upper body fractures, and 22 percent had lower body fractures (Bridges 1996; Bridges et al. 2000).

Although not every injury at these two sites can necessarily be attributed to warfare, the frequent evidence of bodily trauma and the nature of bodies' injuries indicate that numerous individuals were likely the victims of raids or battles. An extreme example of the scale and impact of war is possibly seen in southeast Missouri, a very fertile bottomland with a high concentration of Mississippian sites. Fortifications are commonplace in the region, and in one locality, a number of hamlets and villages appear to have been razed and deserted around AD 1300 (Price 1978). James Price (1978) believed that the burning was likely conducted by the residents, but it is difficult to believe that an external threat did not play some role in their abrupt departure.

All of this evidence does not necessarily mean that communities literally lived under a daily fear of attack. Instead, the threat of major assaults appears to have been intermittent. The palisade at the Kincaid mound center in the lower Ohio Valley fell into disrepair at least once before a new one was erected (Muller 1978). At Cahokia, the palisade surrounding the central precinct was built relatively late in the occupation of the site (and rebuilt several times thereafter), and it appears to have been constructed rapidly and arbitrarily, even cutting through extant neighborhoods, perhaps in response to a looming crisis (Pauketat 1999:54; Trubitt 2003). Yet even episodic warfare takes its toll, particularly if interspersed within a more chronic pattern of raiding and ambush. How such threats affected everyday life remains an important topic of inquiry.

Causes of, and Motivations for, Mississippian Warfare

The reasons underlying Mississippian warfare have been scrutinized from the same perspectives applied to warfare elsewhere in the world. Scholars have attempted to distinguish ultimate from proximate causality, debated univariate versus multivariate stressors, and questioned the relative importance of social versus ecological factors. Early attempts at broad, explanatory arguments in the 1970s differed in the specific causes that were earmarked but were similar in the sense that warfare was seen as adaptive in some sense (at least for the winners). Lewis Larson (1972) saw most Mississippian warfare as being due to competition over arable land, and he proposed that warfare was an adaptive "predatory expansion" either to gain new agricultural lands or to reduce populations that might vie for those lands. Jon Gibson (1974) countered that this model perhaps was applicable in some locales, but in others (notably the lower Mississippi Valley), conflict was also important for regulating social relations within hierarchical societies. Relying on French descriptions of Natchez social and political organization in the early 1700s—probably the last eyewitness accounts of a surviving Mississippian society—Gibson observed that the Natchez practice of generational demotion in kinship could be offset by personal sacrifice and valor in battle.

Southeastern archaeologists have to a large degree relied on historical upstreaming to expand on Gibson's arguments, working from eighteenth- and nineteenth-century accounts of social and military institutions to further explore the social motivations for war. In many societies throughout the Southeast, warfare was indeed a key means for elevating status and gaining power for young males who excelled in battle, as well as for extending the scope of chiefdoms (Anderson 1994:134; Dye 2002). Exactly how powerful an inducement personal and political gain was in actually fostering hostilities is difficult to assess, but revenge, feuds, and other actions apparently supplied ample opportunity for a person seeking status through valor. The balance between war and peace was such an important structuring principle for many southeastern groups that certain societies (such as those of the Choctaw and Chickasaw) were framed by a complementary duality of red (war) and white (peace) organizations (Dye 1995; Lankford 1992).

Although descriptions of motives for warfare among southeastern groups post-date the European arrival by at least two centuries, there is sufficient evidence from sixteenth-century histories to give some credence to their general applicability to earlier times. These histories also provide some basis for the notion that Mississippian warfare had an expansionist component (Hudson 1976:239–240), which may account for the high degree of military organization described by the Spaniards. The de Soto and Luna chroniclers describe punitive actions against troublesome chiefs (Steinen 1992), while control over people and their production is a leitmotif of Spanish accounts (Anderson 1994:133).

Most of the ideas enumerated here are to a large degree uncontroversial. Researchers agree that warfare was commonplace during the Mississippian period, and that a number of reasons—social, ecological, and political—can be invoked to account for its prevalence. Bruce Dickson's (1981) thesis that Mississippian warfare resulted from a complex interaction of resource and social stress is still attractive to many. But despite the general agreement over the broad outlines of the physical conduct and impact of Mississippian warfare, there is still much we do not know about the everyday conduct of conflict and its relation to the exercise of power, which should spur us to re-examine Mississippian conflict in this light (Schroeder 2003:12; Trubitt 2003). One intriguing case is made by David Dye (1995), who suggests that many of the prestige objects circulating through Mississippian rituals that were tied to feasting, smoking, and exchange were also essential to the reproduction of practices and beliefs surrounding war. As we argue, this rhetoric was played out with particular emphasis through iconography and the embodiment of male warriors and weapons.

Embodying Mississippian Warfare

The most provocative images of warfare and the Mississippian body appear in the Southeastern Ceremonial Complex (SECC). The SECC is a mix of motifs that include human figures, animals like birds and snakes, symbols with known meanings such as the cross-and-circle (which likely represents the sun and four quadrants of the world), and other elements that are more ambiguous, such as the hand-in-eye (Brown 1976; Galloway 1989; Knight 1986; Waring and Holder 1968). This imagery is frequently rendered in raw materials that have their own symbolic import. Shell and copper are prominent in this regard, due to the spiritually charged regions from which they originate (water and the earth), and their natural colors, white and red, which have a long history of ceremonial connotations in the Southeast. SECC motifs can also be found in other media, however, such as ceramics and stone.

It is very unlikely that the SECC represents a coherent belief or religious system practiced across the Southeast because of the chronological and regional variability evident in the iconography (Muller 1989). There is a clear upsurge in the display and exchange of the symbols during the Mississippian period, with a notable peak occurring around AD 1200 to 1350, yet many motifs have roots as early as the Archaic period. Nevertheless, there is sufficient overlap of an array of distinctive and repeating elements after AD 1000 to suggest that Mississippian communities across the Southeast to some degree shared ritual and cosmological narratives, as well as ideas about how these narratives should be rendered symbolically. Further commonality is evident in the depictions of warlike iconography, including representations of warriors, weapons, and victims of dismemberment. These occur with sufficient frequency to be recognizable as a central theme in SECC art (Brown 1976). In fact, Vernon James Knight (1986) argues that there is a widespread warfare-cosmogony "sacra" transmitted in symbols and motifs that goes well beyond the limited bounds of the "classic" SECC, and that this was a pivotal organizing principle in Mississippian ritual-social-political systems.

Rather than taking a scattershot approach to the myriad depictions of Mississippian symbols of war, we limit ourselves to the mace—and those symbols and figures associated with it—as a point of departure. The mace was an apparent shock weapon, characterized by an elongated handle and a crescent head. Stylistically, there appear to have been several varieties of the mace, although they all share these same general attributes (Brown 1996:474–476). We know from surviving examples that maces were produced from both stone and wood. James Brown (1976) has argued that maces and certain other shock weapons should be viewed as prestige as well as utilitarian items, an idea that has support in studies of Mississippian war and its accoutrements (e.g., Dye 1995; Van Horne 1993). This interpretation is further buttressed by evidence for the exchange of maces over considerable distances. At Spiro, Oklahoma, for example, several caches of maces included specimens made from cherts native to southern Illinois and western Tennessee, some five to six hundred kilometers to the east (Brown 1996:475). Still, although stone maces may have been largely used for ritual and display, the ethnographic accounts do describe Native Americans carrying wooden clubs into battle (Van Horne 1993:171–172).

The mace or war club appears to have been a preferred weapon by AD 1200 (Dye 2002:128). Mace iconography is of particular interest because it seems to be a strong marker of the zenith of SECC art in the twelfth and thirteenth centuries AD (Muller 1989:15). Focusing on the mace may thus allow us to consider the embodiment of warfare within one horizon during the Mississippian period, in a sense holding chronology constant. In the close of our discussion, we will use our results to consider diachronic and spatial dimensions of the embodiment of war.

There are a few SECC pieces displaying maces that have achieved canonical status in southeastern archaeology because they draw together so many elements of Native American cosmogony and warfare—and they do so with considerable skill and aesthetic achievement. Here, we briefly review two of these objects. The first is one of the so-called Rogan plates, attributed to a mound burial at Etowah, a large mound center and town in northwest Georgia (fig. 3.2). The second is a shell gorget from Castalian Springs, a mound center in Tennessee (fig. 3.3). Both of these objects derive from burial mound contexts and were interred with individuals traditionally presumed to hold some sort of elite status (see Brain and Phillips 1996:133–134, 253–254).

3.2. Copper plate from Etowah, Georgia (height: 50 centimeters). (Reproduced from Thomas 1894, plate xvii; used by permission of the Bureau of American Ethnology Bulletins.)

The copper plate is a display ornament worn on the front of the head. It is a mélange of iconography that includes symbols of debatable meaning—such as the bi-lobed arrow and the bellows-shaped apron—as well as more obvious features like the mace held in the right hand of

3.3. Shell gorget from Castalian Springs, Tennessee (diameter: 9.7 centimeters). (Courtesy of Jon Muller, Southern Illinois University, Carbondale.) Note that the position of the gorget holes suggest that the body was displayed horizontally rather than as shown.

the central figure and the decapitated head held in the left. The individual's pose is suggestive of dancing and celebration, a fitting denouement for the successful warrior returning from combat. The protagonist's profile has facial painting or tattoos that are also often the mark of a warrior. The bird motifs are often explicitly associated with a bird of prey, interpreted as a symbol of strength (e.g., Strong 1989).

The Castalian springs gorget has many structural similarities with the copper plate. A central figure extends a mace in one direction with a severed head balancing the other half of the artistic field. Once again, a bi-lobed arrow caps the central figure's head, while a bellows-shaped apron adorns his waist. The forked eye of the presumed warrior is another common SECC motif that is believed to represent the eye-surround of a bird, perhaps a falcon. Like the copper plate warrior, the person in the gorget appears to be engaged in a motion of running or stylized dancing. Overall, the figures on both the plate and the shell show a strong sense of oppositional symmetry, and the pose of the figures seems to evoke the division of the background field into the quadrants that constituted the cosmological system.

Both of these visually arresting artworks are representations of what we know to be characteristics of Mississippian warfare: the centrality of the male, the importance of shock weapons, the penchant for decapitation. Yet it is unlikely that these objects simply depict the celebration of actual events. Much of SECC artwork—bellicose and otherwise—appears to allude to the cosmos, with undertones that are otherworldly in nature (Hall 1989; Knight et al. 2001). Knight et al. (2001) in fact downplay the "quotidian realities" of the art, maintaining that SECC battle imagery symbolizes a charter myth (or myths) narrating the triumph of ancestral heroes. Possession of such imagery was likely an important validation of chiefly power, linking possessors and their lineages to charter myths (see also Dye 2004).

The similarities between the plate and gorget do seem to argue that reference is being made in both to a widely shared account rooted in cosmology. As for the specific individual represented in shell and copper, he most likely represents a spirit being or mythical hero (Thunder or Morning Star) who occurs in a variety of Native American stories (Brown 2004; Knight et al. 2001; Van Horne 1993). Thunder had the ability to transform into a falcon (a representative of the Upper World in the cosmos), and this would support the notion that the bird motifs are not costumes or ornamentation, but in fact are extensions of an individual who is simultaneously human and raptor (Knight et al. 2001). The Upper World symbolism of the bird is continued with the mace, which often is depicted with feathers attached (as seen in the Etowah copper plate). The upper part, or striking end, of the mace is commonly semi-circular in shape, perhaps reminiscent of the sky vault (Van Horne 1993:132). Further, maces are often stylistically halved (as seen in the copper plate) or quartered with lines or paint (Sievert 1993:191; Van

Horne 1993:132), suggesting the dualities of social organization or the
quadrilateral sectioning of the cosmos. The celestial connotations of the
mace, combined with its upheld position in many representations, may
also suggest the axis mundi—a connection between This World and the
Upper World.

Other elements of these two examples of Mississippian art may have
cosmological referents as well. There is a widespread pattern among
peoples of the Americas, including well-known southeastern groups
such as the Natchez (Gillespie 1991), of using the head and other body
parts to index important stars and planets. Sacrifice is also a recurring
theme in cosmic renewal. Thus, decapitated heads may be a symbolic
replacement for celestial phenomena, as well as a reflection on hapless
warriors.

Given the preceding artistic and ethnohistoric evidence, it would
seem that representations of warfare involving the body alluded to
knowledge and narratives familiar to viewers of the imagery in its original contexts. Yet it is important not to compress belief systems. We must
approach with caution the idea that depictions of warriors in southeastern art were direct and transparent references to well-known stories or
cosmological beliefs. This structuralist argument glosses over the cultural and social variability inherent in the creators and spectators of that
art, as well as the complexities of the Native American narratives. While
representations may reference elements of particular myths and narratives, they are not static dioramas. Rather, they are subject to discourse
and ideological workings and re-workings in both their manufacture
and their artistic consumption (Pollock and Bernbeck 2000).

Granted, many of these variable understandings of the representations of warfare elude the archaeologist. However, we can take that next
step toward understanding the process of embodiment by moving a
notch down from the cosmological or mythological referents of the warrior representations, and by considering how the imagery evoked practices and beliefs revolving around the actual lives of individuals who
engaged in conflict. We suggest that Mississippian representational art
focused on the mace, warriors, and body parts conveys two major paradoxes revolving around identity and participation in armed conflict: a
dialectic of disruption involving wholeness:dismemberment, and a dialectic of ontology centered on human:hybrid. The apparent oppositions

in each of these relationships can be interpreted as tethered unities that gave rise to Mississippian understandings of warfare as a constitutive part of peoples' embodied existence in the everyday world.

Importantly, these relations emphasize that male participation in conflict was not necessarily a completely naturalized and unquestioned drive somehow rendered inevitable by cultural values and practices. It has been observed that—despite a few notable exceptional sites—the overall osteological evidence for conflict in eastern North America is relatively spare, given the prevalence of fortifications and bellicose iconography during the late prehistoric era (Lambert 2002; Smith 2003). Although sampling bias and taphonomic factors seriously impede our ability to resolve the question of the prevalence of intergroup conflict, it must be kept in mind that the decision to embark on war was frequently a contested issue within historic southeastern societies: the Natchez were known to offer peace calumets (a form of pipe) to avert hostilities (Swanton 1946:699) and among the Creek, the decision to go to war typically was divisive (Swanton 1928:428–429). Depending upon the status of a community's extralocal relations, the respective chief for the red or white group was in at least nominal command. The process of negotiating peace and maintaining civil inter- and intracommunity interactions was an additional path to social status for some individuals and lineages, although this route was often contested by warrior groups (e.g., Lankford 1992). Mississippian representational art highlights some of the conflicting perspectives that may have been held concerning the violent encounter.

The Dialectic of Disruption:
The Perfect Warrior and Dismemberment

According to the adage, there are worse things than death, and the anxiety tied to battle in the Southeast appears to have been not so much about confronting death as the manner in which one died. A widespread if not universal fear among communities of warriors is defilement and dismemberment of the body. In a pre-Cartesian understanding of the world that makes little distinction between body and spirit, the violent death of a foe does not guarantee his extermination. In fact, his body itself must be ravaged to extend the corporeal disorder to the soul (Cassin 1981; Gillespie 1991; Meskell and Joyce 2003:149; Redfield

1994; Shanks 1995). This appears to be a fundamental reason why mutilation so often follows death in battle, and why disfigurement becomes
the metaphorical antithesis of the complete warrior. In Classical Greece,
this anxiety was made prominent in the *Iliad*, as heroes such as Hector,
Ajax, and Achilles constantly threatened to feed their enemies to the
dogs. James Redfield (1994) refers to this kind of defilement as an "anti-
funeral," in which the destruction of the enemy's body is the perfection
of victory for the warrior. Likewise, Lynn Meskell and Rosemary Joyce
(2003) see mutilation in Dynastic Egypt, particularly the severing of
penises and decapitation, as a form of disrupted perfection.

In ethnohistoric accounts, southeastern Indians expressed a particular dread for the defilement of their bodies, particularly warriors who
faced the elevated possibility of being killed away from home with
their physical body at the mercy of the enemy (Hudson 1976:327–328).
Describing Choctaw treatment of captives, Bernard Romans ([1775]
1999:132) observed, "The body [was] cut into many parts, and all the
hairy pieces of the skin converted into scalps." A dramatic realization
of the fear of post-death disfigurement was found in a cemetery at the
Lubbub Creek site in Alabama. One burial contained several bodies:
at the bottom was a supine male with a copper plate depicting a raptorial bird, another male with a projectile point in the rib cage was laid
on top of the first male, and the two males were accompanied by the
arms, legs, and feet from one to two additional individuals. A plausible
interpretation of this juxtaposition of bodies is that the internment of
a male of some rank was commemorated with a human sacrifice, followed by body parts that may have been war trophies (Bridges et al.
2000:39).

Given the anxiety surrounding dismemberment, why take the
chance in battle? One answer lies in the fact that military prowess was
a significant source of power and status for a male hoping to elevate his
social position. Until achieving such distinction, men could expect to
be relegated to a mundane life about the village (Hudson 1976:240).
For Creek society, "Young men remain in a kind of disgrace . . . and
perform all the menial services of the public square until they shall have
performed some warlike exploit that may procure them a war-name . . .
This stimulates them to push abroad and at all hazards obtain a scalp,
or as they term it, bring in hair" (Swanton 1928:426). Thus, violent

encounters with the enemy were something to be sought rather than avoided. At a very personal level, the motivations for participating in raids and battles could be compelling.

Mississippian images of warfare are replete with dismemberment. We have already seen in the well-known Etowah copper plate and Castellian Springs gorget that the decapitated head on one side of the warrior balanced the mace on the other. Other objects with these elements occur relatively commonly. As distinctive examples, a carinated jar from northeast Arkansas displays a scene of alternating broken and intact maces with human heads (Galloway 1989:34); "floating" maces and human (and avian) body parts occur on a number of engraved shell cups from Spiro (e.g., Phillips and Brown 1978: plates 55, 56, 57); and a well-known shell gorget from southeast Missouri shows a figure with a mace in his belt and holding a severed head (Phillips and Brown 1978:177). These images are those that are restricted to the mace as a weapon. If other weapons were considered, the sample would broaden to include objects (particularly gorgets) displaying individuals holding knives or axes in one hand and the now-familiar head in the other. As opposed to boasts in the *Iliad* that were rarely carried out, we have already seen from examples of imagery in Mississippian cemeteries that dismemberment was not an idle threat. Yet despite the fears of disruption faced by an individual, the anxiety surrounding failure in battle could be offset by the ability to bridge another dialectical relationship, that between the opposing categories we recognize as mortal and divine, human and animal.

The Dialectic of Ontology: Humans, Chimeras, and Valor

Although a given conflict (the so-called "Just War") may be considered worthy by a wide segment of society, it is not necessarily easy to convince an individual to pull the proverbial trigger on another. Even for societies that may be engaged in a near-perpetual state of war, individuals may have to be subjected to some form of physical, social, and/or ritual regimen to effect the transformation into a warrior who is capable of killing. This process typically involves elevating a person temporarily to a liminal state, in which he can have capabilities and powers that are distinct from those of the "normal" condition of everyday life.

While modern war demands the drilling of numbers of individuals into a cohesive unit provided by the boot camp experience, in many premodern societies the focus was on the individual (Ehrenreich 1997: 10–11). Societies worldwide devote special rituals to moving a person from one status to another, to transforming a peaceful person into a being who is capable of killing. This transformation often consists of both rites of passage and rites of intensification. Rites of passage may provide the initial conditions to create the warrior stepping out of the innocence of adolescence. Rites of intensification are communal and recurrent and focus on tangible threats; they are rituals engaged in by those who have already become warriors, meant to segregate the individual temporarily from the quotidian world so that he may enter the world of passion and the supernatural engendered by conflict with a foe.

In many southeastern societies, rites of passage related to success in war conferred official status on the fledgling warrior and heralded promotion for those established warriors who continued to distinguish themselves in battle. Typically, war names or titles were assigned to boys-become-warriors; additional names and titles could be acquired after successive, successful raids (Hudson 1976:325; Swanton 1946:695). Rites of intensification could be prolonged affairs as men prepared mind, body, and soul for an upcoming conflict. Natchez males on the brink of war convened in the war chief's cabin for a period of three days, fasting, drinking special potions, and abstaining from sex in order to purify the body (Hudson 1976:243–244). Warriors regaled one another with death songs, dances, and stories from past successes at war. Similarly, as a prelude to war, Creek men secluded themselves in sweat lodges for several days, taking only certain foods and drinks (Swanton 1928:429). These ritual practices also served as a form of purification through spatial and temporal separation (Dye 2004:197–198).

As a result of such preparations, warriors on the eve of battle were often viewed as occupying a passage between mortality and divinity (Ehrenreich 1997; Redfield 1994; Shanks 1995), or in southeastern cosmology, This World and the Other World (i.e., either the Lower World or the Upper World). Although the guarded nature of southeastern rites hinder eyewitness accounts, it appears that the ceremonies in part were directed toward breaking down the barriers of ontology, so that individuals became capable of assuming the strengths and qualities of the world about them. Following their emergence from the sweat lodge war-preparation rituals, Creek men prepared a bundle that included a charm or "war physic" made from snake and mountain lion bones—both powerful animals in cosmology as well as in everyday life (Swanton 1928:429). Humans in Mississippian war iconography are commonly depicted as chimeras (particularly with avian traits), melded with animals that had characteristics or histories that empowered the warrior for success in battle. Similarly, in the American Southwest, animal-human hybrids are pervasive in war-related art, with felines particularly common, and war priests may have embodied the powers of the mountain lion specifically (Schaafsma 2000:137–139).

Southeastern Indians were apparently fascinated by what Charles Hudson (1976:139–148) refers to as anomalies: "beings which fell into two or more of their categories." These included bears, which were four-footed but could walk on their hind legs like humans and were similarly omnivorous; and bats and flying squirrels, which were four-footed but had wings. Such animals were often attributed with special meaning and powers. Although humans were recognized as a distinct category, the lines between human and animal were often blurred. But this was not a randomly porous edge, where all animals and all humans were potentially equally interrelated. Kinship and special circumstances such as war directed hybridity in certain directions. Individuals belonging to animal clans, for instance, were assumed to have properties associated with those animals.

It is in this light that we can view the falcon-warrior imagery prevalent in SECC art that appears to be closely tied to adroitness in war. There are differences of opinion over how to interpret these twofold entities, whether they represent individuals adorned in bird costumes or portrayals of supernatural hybrids (Knight et al. 2001; Phillips and Brown 1978; Strong 1989; Van Horne 1993; Waring and Holder 1968). In our minds, there is no clear resolution to this debate because the depictions could speak to different dimensions of Mississippian society (Phillips and Brown [1978], for example, stressed the multivocal nature of Spiro shell art). While the hybrid figures may be representational in their depiction of deities such as Thunder or else in their reference to charter myths, they also embody the knowledge and capabilities requisite to entering the frontier between This World and the Other

World. These chimerical images incorporating raptorial birds were the ideal emotional and ritual template of any man who was poised to go into battle.

Hybridity also spanned the worlds of animals and objects. The rendering of social and cosmological divisions onto the mace described earlier (e.g., the sky vault) is likely a symbolic referent. But maces are commonly depicted with avian characteristics, such as feather tassels, or the head of the mace is painted to resemble a falcon's tail (Phillips and Brown 1978: plate 121; Van Horne 1993:146–148). One cave-art rendering appears to show a mace transforming into a bird (Simek et al. 2001:144–145). In essence, the mace appears to have become a symbolic prosthesis and metonym, an extension of the warrior-become-raptor.

The Rise and Fall of the Ideal Warrior

Despite the timeless, sacramental quality of the hero, public perception of the rewards due the warrior may vary considerably with the passage of time, shifting social priorities, and the ideology of war. As a result, warriors encounter different opportunities and constraints accruing from their success, and it is important not to homogenize a warrior class at any one point in time, or through time within the same society (Vandkilde 2003). Given sufficient chronological resolution, archaeologists also can detect changes in the ideological status of the warrior in prehistory. The Neolithic of northern Europe provides one such instance, in which there was an apparent emergence of an individualizing ethos centered on male warriors that was reproduced through a bodily aesthetics of warfare and violence (Treherne 1995). "Consumables" in burials— weapons, drinking vessels, riding and chariot paraphernalia, body ornamentation—all seem to reflect a rise in a warrior ideology focused on powerful individuals that was at odds with an earlier communal focus. A broadly similar ascendance of a distinct warrior ideology is posited for the American Southwest ca. AD 1300 (Schaafsma 2000).

A comparable pattern took place during the Mississippian period. There was an apparent climax in war iconography rendered in exotic materials that is defined by the culmination of the SECC phenomenon of the thirteenth and fourteenth centuries AD. A number of individuals have proposed that this zenith corresponds with a relatively brief and widespread episode when individuals were able to leverage cosmology and ritual to usurp the fetters of kinship and community and emerge as very powerful rulers of polities based in large mound centers like Spiro, Etowah, and Moundville (King 2003; Knight 1986:683). The reasons underlying this historical blip and why it should be so widespread are still poorly understood, but it did appear to rely strongly on the power vested in the control of a complex of portable, symbolically charged objects.

In our previous discussions of the mace—one focal point in this sacra complex—we relied primarily on artistic versions of this ceremonial and functional shock weapon. It should be pointed out, however, that the mace appears in three forms—first, as a representation in media such as copper and shell; second, as an actual, functional mace; third, as an ornamental or decorative mace (usually in miniature). Moreover, the mace appears in some depictions with elements such as arrows and snakes.

These different forms and contexts of the mace hint at more variability in embodiment, weapons, and a warrior sensibility than we have addressed, since we have focused on its representational contexts. We will not follow up on this theme here, except to observe briefly that small, decorative maces likely played an important symbolic role in the articulation of certain individuals with the power and ritual surrounding war during the thirteenth and fourteenth century apogee. At Etowah, several individuals were interred with headdresses composed of small copper versions of maces, arrowheads, bird heads, and bird tails arrayed in a fan-like display (Brain and Phillips 1996:162–163). A similar headdress was found in a burial at the Lake Jackson mound center in Florida, with maces apparently identical to those on one of the Etowah examples—suggesting that they were made by the same artisan or at least with the same template (Brain and Phillips 1996:181–182). These ornaments recall the feather headdresses common in Native American societies that were worn for different occasions. James Adair ([1775] 1930:397–399) provides a lengthy account of Creek warriors receiving feather "crowns" as they were awarded a new title following successful raids. The copper headdresses seem to re-emphasize the metaphorical ties of mace:bird:warrior and chiefly power. At a practical level, crowning the head is a convenient way to widely broadcast a social message about identity or status in a Wobstian sense. More abstractly, the mace appears to once again embody the paradoxes of a warrior's powers and fears, on the one hand as a physical and metaphorical extension of chiefly power, and on the other as a decoration of the body part whose loss in the field of battle was a source of acute anxiety.

War iconography and the media in which it occurred took on a more egalitarian cast after the 1300s AD. Democratization (i.e., the transfer of iconography to nonexotic media or contexts) appears at Moundville when these motifs began to appear on ceramics (Knight 1986:682–683), whereas at Etowah, objects that formerly occurred in elite burials (e.g., gorgets) are found in nonelite interments (King 2003:136). Unfortunately, the dating of specific displays of Mississippian artwork outside the SECC is imprecise. Yet there are some indicators that allow one to speculate that the tight interrelationship between the mace, humans, birds, and body parts was transferred to more accessible media in some instances, or that the relationship among the elements began to unravel. In northeast Arkansas, there are several examples of mace rim effigies on ceramic bowls that are late Mississippian (after the mid-1300s) and that lack other decorative motifs (Hathcock 1976, 1983). The mace occurs in a number of rock- and cave-art displays and is commonly associated with avian and/or human elements (e.g., Diaz-Granados and Duncan 2000; Faulkner 1997; Muller 1986:62–63; Simek et al. 2001). Nevertheless, there is considerable variety in the correlation between these key elements, and they lack consistent groupings similar to those seen in the plates and gorgets. On the other hand, cave and rock art is notoriously difficult to date, so definitive statements about changes in representations through time are difficult to support. Interestingly, one charcoal drawing of a warrior from Missouri has been recently dated to a calibrated interval (within two standard deviations) of AD 900 to 1280 with several intercepts in that range (Diaz-Granados et al. 2001). In this example, the coherence of the falcon-warrior and mace elements is more akin to the copper and shell art (fig. 3.4), and its early date may suggest that the emergence of the posited ideal warrior iconography was related to the growth of the huge mound center of Cahokia (Diaz-Granados et al. 2001:490)—although a single radiocarbon date is tenuous evidence, at best, for ordering events.

Although these temporal trends are not unequivocal, they are broadly suggestive of a breakdown late in the Mississippian period of

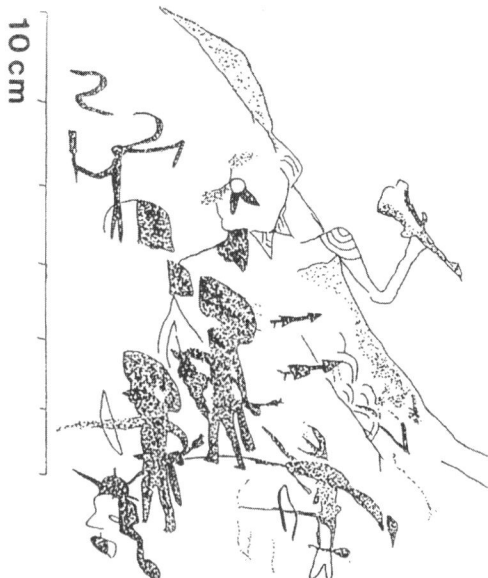

3.4. Pictograph from Picture Cave #1, Missouri (Reproduced from Diaz-Granados and Duncan 2000, fig. 5.48; used by permission of the University of Alabama Press.)

the links between the body, weaponry, and warfare, in conjunction with the decline of standard mortuary evidence for high rank. This decoupling may reflect the attenuation of the ability of individuals to achieve high positions of authority and power through the traditional manipulation of military prowess and the cosmos.

What else was occurring in this time frame that might account for such a transformation? We do have the aforementioned rise in warfare that seems to have occurred in the late Mississippian period over much of the Southeast, as seen in the increase in fortifications and nucleated villages. This may be paradoxical in that it would seem to suggest that an elitist warrior ideology was in eclipse as warfare was on the rise. However, it has also been postulated that the ritual underpinnings of chiefly power may have become less important through time, as seen in the decline of both moundbuilding and long-distance exchange in symbolic prestige goods in the latter part of the Mississippian era (Anderson 1999). Conversely, the secular authority of leaders appears to have been expanding to the point at which, at the time of the Spanish entrada, the chief of Coosa could reside in a relatively modest mound center while overseeing a polity that extended for hundreds of kilometers. While these interrelated variables do not explain the changes from one set of conditions to another, they do emphasize that the ideology reproduced by the idealized warrior of representational art no longer had broad currency by the late 1300s AD.

Conclusion

Motives for Mississippian warfare were possibly deeply personal, be they for achieving manhood, individual gain, punishing a rival, or exacting revenge. Yet by definition, warfare—at small or large scales—is also a communal enterprise. Warriors were recruited and organized within a milieu that drew deeply on ritual and historical tradition. It is within this personalized and socialized context that the embodiment of Mississippian warfare can be seen as an experience that was simultaneously embedded in the day-to-day experience, elevated in metaphor and allegory, and played out in material culture. At an abstract level that may have been widely comprehensible across the Southeast, heads and body parts substituted for celestial phenomena, while warrior figures may have referenced charter myths. At a more immediate level, objects like copper plates may have been strong reminders of the status of those who wore them. And, to the aspiring warrior, dramas rendered in copper and shell evoked both the promise and horror of warfare.

Moreover, the embodiment of Mississippian masculinity and warfare was not static. The ideology of the mythological warrior, offset by the specter of the anti-funeral, appears to have peaked in the AD 1200s and 1300s, then lost its power sometime after the 1300s. This does not mean that warfare, warrior cults, or even powerful polities diminished to any considerable degree—the ethnohistoric record presents too strong of a counter case to that argument. Instead, for reasons not clearly understood, the importance of foregrounding the body and conflict in representational art diminished even as traditional warfare appears to have continued unabated until the arrival of the Europeans in the sixteenth century.

4

Warfare and the Practice of Supernatural Agents

William H. Walker

The Practice of Nonhuman Persons?

Does study of the animism of artifacts and architecture have a place in an archaeology of warfare? Is it even possible to consider the role of nonhuman agencies in the study of prehistoric societies?

In 1990, Jonathan Haas (1990:171) described the possibilities for an archaeological analysis of warfare, noting that "archaeology lacks the rich descriptive detail of individual behavior and specific events provided by ethnographers" and that "we are largely unable to look at intention, individual motivations, emic explanations of events and patterns of behavior, kin relationships, and most other aspects of the ideological superstructure of long-dead societies." In the intervening years, archaeological method and theory has sought to overcome these obstacles, particularly with respect to ritual and religion—"ideological superstructure." In this chapter, I argue that we can and should consider animacy in our study of prehistoric war.

Practice and agency approaches to people and material culture (e.g., Bourdieu 1977; Gell 1998; Latour 2005; Sahlins 1976, 1999; Strathern 1988; Schiffer and Miller 1999) offer archaeology of warfare and religion an alternative means for conceptualizing relationships between beliefs, actions, and material objects. Analytical dichotomies (e.g., material/ideal, natural/supernatural, human/artifact, infrastructure/superstructure) are conceptual tools that in many cases have outlived their utility. It is important to look at the archaeological record differently.

Archaeologists in Europe and the Americas, for example, are increasingly describing and interpreting purposefully structured deposits resulting from specific events that were clearly intentional and tied to individual and group motivations (e.g., Bradley 2000; Chase and Chase 1998; Hill 1995; Marcus and Flannery 2004; Mock 1998; Nielsen

1995; Pauketat et al. 2002; Pollard 2001; Lucero 2006; Richards and Thomas 1984). These deposits are often ceremonial in origin and possess measurable traces of the consequences of animism for archaeological patterning.

In the American Southwest, deposit-oriented approaches to religion have been relatively robust. Southwest archaeologists now recognize ritualized closure activities involving the burning and burial of structures as well as sites (e.g., Creel and Anyon 2003; LaMotta and Schiffer 1999; Lightfoot et. al 1993; Montgomery 1993; Walker 2002; Wilshusen 1986; Wilshusen and Ortman 1999). Although some of these data may complicate inferences of warfare destruction (see Walker 1998, 2002), they suggest an important lesson for the archaeology of war: archaeologists interested in the study of violence have access to potentially rich evidence of the interplay of religion and war in prehistoric contexts.

To tap that potential, I suggest in the first section of this chapter that we expand our understanding of society to include human and nonhuman social actors. This redefinition would reflect the influence of animistic practices worldwide and bring the intersection of religion and warfare into sharper focus. In this theoretical section, I marry the practice orientation of Marshall Sahlins (1981) with more recent explorations of artifact agency in the social sciences and humanities, including anthropology (e.g., Gell 1998; Küchler 1988; Strathern 1988), sociology (e.g., Akrich 1992; Ashmore et al. 1994; Latour 1994), and archaeology (e.g., Boast 1997; Olsen 2003; Meskell 2004; Pollard 2001; Walker and Schiffer 2006; Yarrow 2003). I argue that much of this concern with the agency of material culture is a logical extension of practice-oriented perspectives and fits comfortably into what Graham Harvey (2006) has called the new animism.

In the second section, I describe an indirect method for measuring the practice of nonhuman agents. A central insight of Alfred Gell's (1998) study of the agency of art objects, also shared by Michael Schiffer and Andrea Miller's (1999) artifact-centered theory of communication, is that inferences about the practices of nonhuman social actors can be based on measures of the action and reactions of the humans they engage. I find this indirect method particularly useful for integrating oral traditions into archaeological interpretations. Starting with Sahlin's (1985) work and building on that of others interested in the intersection of history, social memory, and material culture (e.g., Connerton 1989; Küchler 1988, 1999; Nabokov 2002), I argue that aspects of narrative traditions are reproduced in ongoing practices involving interactions with nonhuman agents.

In the last section, I illustrate this approach with a case study from the pueblo Southwest. The American Southwest has been an important laboratory for the study of prehistoric warfare and religion (e.g., Adams 1991; Bullock 1998; Crown 1994; LeBlanc 1999; Lekson 2002; Haas and Creamer 1993; Schaafsma 2000). Building on that exploratory spirit, in this case study I rewrite the history of war and religion in the late prehistoric pueblo Southwest from the perspective of its nonhuman agencies.

Practice and the New Animism

Practice theory has enjoyed popularity in recent social theory because it offers an alternative to dichotomous approaches to society that unevenly elevate either its material or ideational aspects. By embedding or embodying both aspects within practice, it becomes possible to measure both differently using the materiality of artifacts, architecture, or even the human body itself. Perhaps an unintended consequence of the growth of such practice orientations has been a renewed interest in the role animate or agential material objects play in human practice.

In his early foray into practice theory, Sahlins noted that "the great challenge to an historical anthropology is not merely to know how events are ordered by culture, but how, in that process, culture is reordered." He asked a central question for those interested in the study of human history: "How does the reproduction of a structure [behavioral structure] become its transformation?" (Sahlins 1981:8).

The short answer to Sahlin's question is through practice. Agency and practice approaches to society (Bourdieu 1978; Giddens 1979) and culture (Sahlins 1981) emphasize that the actions of individuals and groups create historical order as they act it out. As Sahlins notes (1981:7), "Culture may set conditions to the historical process, but it is dissolved and reformulated in material practice." As a unit of analysis, "practice" provides a solution to difficulties generated by functionalist or structural approaches to culture (Ortner 1984). These approaches are often synchronic and depend on a series of dichotomous assumptions about human activity involving distinctions between mind and body, belief and action, material and ideal, natural and supernatural. In such social theory, one half of a dichotomy (mind, material, belief) always receives more attention than the other (e.g., the deep structures of the mind or the material base of social adaptation).

Although contemporary theoretical orientations prioritize different questions, it is clear that most of them (e.g., neosocialevolutionary, neo-marxist, contextual) share an anthropocentric paradigm whose unre-solved contradictions are reflected in these dichotomies. Anthropocen-tric social theories assume that only people can be social actors. Given that cross-cultural ethnographic data regularly record activities in which human participants assume they are interacting with nonhuman agen-cies, why is it so difficult to build new theory to model these societies more creatively?

With a few exceptions (e.g., Turner 1992), the reason seems clear. Social scientists know that nonhuman agents are not real and therefore do not take their practice seriously as they might other seemingly more material processes such as weather, animal and plant resources, exchange, or political organization. Instead, they try to account for such practices within other explanatory frameworks (psychological, linguistic, critical, functional). In effect, we put nonhuman agents or actors where they seem to belong in a scientifically defined reality—in the beliefs, opin-ions, or ideology of human agents. We literally embody them in people in a way that denies their nonhuman existence. Although social theories are designed to explain social processes, they are different from reality and it seems unnecessarily cautious to force them to follow it literally. After all, one corner of that reality—society—seems to change through time. Yesterday's classical evolutionary stages, culture traits, functional institutions, and behaviors are revamped by today's theoretical insights. The lesson is that theories are real only in the sense that they are tools with strengths and weaknesses relative to the questions they are asked to address.

Anthropocentrism forces analysts to construct inferences about actions, including their causal milieu, as beginning with either people or nature. This fundamental dichotomy then multiplies itself in other divisions. Nonhuman agencies such as gods, spirits, witches, ancestors, ghosts, mana, and tabu are not in reality part of either the human or the natural world and therefore must exist (to the degree they do) inside people's imaginations. They are transformed from actors to beliefs expressed in human actions, particularly rituals. Although artifacts and architecture could be treated as people's measurable bodies, houses, or tools, they are instead transformed into symbols of inner states or "ideological superstructure" to be contrasted with material things. When a theoretical perspective renders such large amounts of human activity immaterial and immeasurable, it seems reasonable to ask for more useful alternatives.

It has become clear to many ethnographers that people enter into social relationships with artifacts and act as if these objects have causal powers (see Gell 1998; Strathern 1999; Weiner 1992). "Spirits" and other nonhuman actors can possess, share, or reside in people as well as things. Alternatively, people's personalities and animating power can also circulate in artifacts, as they do in their corporal bodies (Strathern 1999). That artifacts can have agency like people is crucial for Gell's (1998) groundbreaking reconceptualization of the anthropology of art.

My longer archaeological answer to Sahlin's question, therefore, is that understanding the complex process of social reproduction and change entails beginning with a new understanding of society as an emergent property of practice that incorporates an active role for artifacts and architecture. When we peel away the veil of practical reason (Sahlins 1976, 1999) that preserves an anthropocentric paradigm and renders certain social actors immaterial and immeasurable, it becomes clear that nonhuman agents, often in the form of artifacts, are as important to war as warriors. Indeed, the point is that they are often warriors, albeit in multiple forms—some material and some not (see Nielsen, this volume). Edward Tylor (1889 II:108) originally defined spirits as personified causes. Robin Horton seized on this symmetry between human social relationships and those between people and nonhuman agencies to develop a clever definition of religion:

> The relationships between human beings and religious objects can be further defined as governed by certain ideas of patterning and obligation such as characterize relationships among human beings. In short, religion can be looked upon as an extension of the field of people's relationships beyond the confines of purely human society. (Horton 1993:31–32)

Such a definition makes apparent the logical conjoining of studies of religion and war. In a world where both human and nonhuman agencies are at work, we should expect that in war and other activities, these actors will be present. The Avatip of New Guinea, for example, are constantly at war, albeit often through sorcery. Their observable battles are only a small part of such conflict. As Simon Harrison notes:

> When men went on a raid, the spirits of the male initiatory cult, the men's agnatic ancestral ghosts, the tutelary spirits of their land, lakes and rivers, and all the other supernatural forces they controlled, were believed to go into battle with the men, protecting them from harm and striking fear into their enemies. A successful raid was not thought of as demonstrating superior skill, strength or tactics, but superior ritual power. (Harrison 1989:585)

A central component of religious practice includes the oral as well as embodied actions of social memories (Connerton 1989; Nabokov 2002:218–232). Oral traditions describe and proscribe practices in the present that create the future while drawing on the past. When considered in light of a new understanding of society as an emergent property of interactions involving human and nonhuman actors, the relevance of oral traditions and other seemingly immaterial forms changes. Although recent debate has focused on the question of historical accuracy in oral traditions (e.g., Echo-Hawk 2000; Mason 2000; Whiteley 2000), it seems equally important to ask how their organization, along with other aspects of human action, shapes memories and the past into relationships between people and artifacts in the present and the future. Two new approaches to artifact causality or agency offer possible solutions for integrating such practices into the archaeology of religion and war.

Measuring Supernatural Practice Indirectly

Behavioral archaeologists define behaviors or practices as interactions between people and objects (Walker et al. 1995). This relational definition of behavior analytically as well as physically overcomes the mind/body dichotomy that is inherent in more biologically based definitions of behavior that focus on the movements of organisms. It also overcomes the natural/supernatural dichotomy that artificially segregates ritual and religion from the physical world. I find this version of behavioral archaeology (Walker 1998, 2002; Walker et al. 1996) useful for implementing practice theory in archaeology.

This redefinition of behavior shares with contemporary practice approaches to material culture the assumption that the interaction between people and things mutually determines the identity of both and has causal consequences for their life histories (e.g., Meskell 2004). We live in a world of objects, and one obvious way to avoid the dichotomies between people and objects is to create units of analysis that involve a person and something outside of him or her (artifacts) and inside of him or her (beliefs). When this is done, what might be considered part of a person's environment is transformed into a unit of practice. If, through the magic of animism, a large portion of human practices include interactions with animated artifacts and architecture (Walker 1999), then taking those nonhuman agents' measure requires an equally creative solution.

Schiffer and Miller's recent study (1999) of the role of artifacts in communication and Gell's (1998) study *Art and Agency* possess an inspiring similarity. Each in their own way lays the methodological foundation for indirect or inference-driven studies of artifact causality that facilitate understanding of artifact agency.

The Interaction of People, Objects, and Nonhuman Agents

Schiffer and Miller (1999:65) propose an indirect method that highlights the role of artifacts in studying communication processes. Rather than building a model of inference focused on communication between people as message senders, they worked the problem from the other direction, addressing message receivers and how receivers interpreted and acted upon information gleaned from transmitters (often artifacts). They argue that through emphasizing receivers, it becomes obvious that material objects are an undervalued and critical component of all human communication processes. They name traditional models of communication "two-body systems" because those models focus on person-to-person scenarios emphasizing only senders and receivers. In contrast, they name their new model a "three-body system" because it necessarily requires the presence of a third body or transmitter. The third body—indeed, all three bodies (senders, emitters, and receivers)—can be either a person or an artifact. Schiffer and Miller note that this model was inspired by their own experiences in modeling the process of archaeological inference.

They argue that the interpretative process in archaeology is a special case of a more general information-transmission process involving people and artifacts. The communicating interactors in the case of archaeological interpretation are temporally separated, and typically—with some exceptions (as with time capsules or voyager spacecraft)—only the archaeologists planned on the communication. Through recognizing that practice—or in Schiffer and Miller's terms, behavior—can be conceptualized as a material form of information transmission, it becomes easier to envision theoretical abstractions like the interaction or communication between people and nonhuman agents in the form of animated artifacts and architecture. For example, in the three-body system, a shaman's (receiver's) treatment of his patient derives from inferences he gathered in an examination of the patient's (emitter's) body that emitted the tell-tale signs of bewitchment (enacted by the sender).

Gell's study presents a useful and analogous approach to the study of artifact agency. In an effort to explain art cross-culturally, Gell developed a more general argument for study of artifact agency. He recognized that what is often described as art in various cultures turns out in many cases to be objects that people engage in anthropomorphized social relationships. Recalling Schiffer and Miller's receiver model of inference, Gell argues that we can recognize or measure art objects by focusing on the practices of the human side of the social interaction. Similarly, he transforms these practices into a communication process drawing on an inference-based model that derives from Charles Sanders Peirce's semiotics. In this model, actors are what Schiffer and Miller call communicative interactors; art objects—Schiffer and Miller's artifacts—become indexical signs; artists are senders; receivers are recipients; and information is a prototype (sensu Peirce 1931–1958).

Although sources cited by Schiffer and Miller and Gell betray their different backgrounds—Schiffer and Miller are scientists and Gell is a cultural anthropologist—the terminology and rhetorical styles of both approaches share virtually identical interpretative structures involving three bodies. Schiffer and Miller refer to people making "inferences," emphasizing similarities to the scientific process of archaeological thinking. Gell prefers "abductions," a term coined by Peirce emphasizing a more speculative form of inference common in everyday life, ironically in contradistinction to science-based practices of induction and deduction. Gell notes that the "inferential schemes (abductions) we bring to 'indexical signs' are frequently very like, if not actually identical to, the ones we bring to bear on social others" (Gell 1998:15). The reason for this is that "the index is itself seen as the outcome, and/or the instrument of, social agency" (Gell 1998:15). In both of these models, art and archaeology appear to imitate the practice of life.

Both models blur analytical lines between people and artifacts. They highlight indirect inference and argue that the material patterning of animated objects is the same as that of animated people because objects at times are perceived as being like people. The perception is enhanced by the performance characteristics of both kinds of material objects. People and objects can emit sounds; they can be dressed in clothes, paint, or other materials. Some artifacts move on their own, like people, and a great many—such as animated masks—move through the energy of the people wearing them. The important lesson for archaeologists interested in applying these methods is to not get caught up in the "reality" of these perceptions but instead focus on the materiality of the chains of activity they measure.

The methodological implications of these interpretive models of practice and artifact agency are twofold for the archaeology of war and religion. On one hand, they suggest that inferring the activities of nonhuman agents follows the same structure of inference involved in inferences about human activities. On the other hand, they will likely produce what might seem at times surreal histories because they necessarily include human and nonhuman practices. It is, however, the identification and exploration of just such culturally contextualized histories that unites the contributors of this volume.

Schiffer's (1987) pathway model of inference assumes that patterns in the archaeological record were causally produced by earlier events in the life histories of the artifacts it contains. He argues that, given a detailed knowledge of correlations between activities and artifact patterns in different stages in an object's life (e.g., manufacture, use, reuse, discard) identified in verifiable ethnographic (especially ethnoarchaeological) settings, one can construct analogical arguments about past object histories and the traces they leave in the archaeological record. In effect, the pathway model of inference suggests that any archaeological interpretation is at root a series of inferences about past artifact life histories.

Schiffer and Gell's models of communication and agency suggest that to explore the roles and histories of nonhuman agents in prehistoric warfare, one needs to redefine the problem as one of artifact agency and the communication between human and nonhuman agents. When reformulated in this manner, the important question becomes: What do the life histories of agential artifacts in known ethnographic contexts suggest about the traces of the life histories of agential artifacts in the archaeological record of religion and war?

An important consideration for identifying such traces would be the assumption that the life histories of artifacts involved in social relationships would parallel those of people themselves. For example, the objects' disposal would be more specialized and similar to a mortuary practice. The purposeful destruction and renewal of temples, for example, reflects the results of interactions between animate buildings and their deposition (Mock 1998; Nielsen and Walker 1999; Walker and Lucero 2000). Similarly, in the American Southwest, it is relevant to understand the relationships between human and nonhuman beings in the burial and burning of buildings and their associated deposits.

In the case study that follows, I summarize the current understanding of warfare and religion in the late prehistoric pueblo Southwest and then draw on ethnographic information to infer a more contextualized history of warfare involving human and nonhuman social actors.

Nonhuman Agents and Puebloan Warfare

Warfare has long been recognized as an important influence on the culture histories of pueblos in the American Southwest (see Wilcox and Haas 1994). Although culture historians did not actively pursue it as a research topic, they generally assumed that large-scale regional abandonments and periods of pueblo social reorganization resulted in part from conflicts from immigrating invaders (Kidder 1924; McGregor 1941). More intensive discussion of the topic occurred in the 1940s and '50s as archaeologists began to move away from interpretations of culture traits and migrations and toward institutional and culture process topics (e.g., Ellis 1951; Linton 1944; Schulman 1950; Woodbury 1959). It has been an increasingly important topic in the studies of social complexity and cultural adaptation associated with neoevolutionary theory since the 1970s (e.g., Haas and Creamer 1993; LeBlanc 1999; Mackey and Holbrook 1978; Watson et al. 1980; Wilcox 1979; Wilcox and Haas 1994). This research has been enhanced by the exceptional preservation of archaeological evidence in the region. In the case study below, I summarize contemporary approaches to the study of warfare in the Southwest and discuss ethnographic evidence of the interaction between people and nonhuman agents. I then apply this ethnographic information to an interpretation of a practice history of the puebloan past.

The Pueblo Region

The pueblo region of the northern American Southwest was defined geographically for most of its history by the Neolithic-scale farming peoples of the Anasazi cultural tradition. The Anasazi were located in the high desert (5500–8500 feet above sea level) of the Colorado Plateau of northern Arizona and New Mexico, as well in as portions of southern Utah and southwestern Colorado. This is an upland environment of pinyon and juniper forests dissected by narrow canyons and steep mesas with geologically active rain- and spring-fed ephemeral streams and washes.

Early farmers emerged during the late archaic period around 1500 BC. There is evidence that by AD 200, there were pit-house farming hamlets and villages grouped into local communities with identifiable pottery traditions. These farming communities typically possessed larger pit-structures that appear to have served as ritual structures shared by one or more communities. Based on ethnographic evidence, ritual activities were also conducted by smaller groups in domestic pit houses.

A transition to surface pueblos occurred between AD 700 and AD 850. The earlier subterranean pit houses were transformed during this change into more specialized religious structures known as kivas. The larger communal pitstructures were also transformed into more formal ritual buildings known as "great kivas." In general, the rise of surface pueblos was associated with gradual increases in population. The size of these pueblos varied in relation to social organization and environmental constraints. During times of drought, communities often aggregated into larger villages and towns.

In the late AD 800s, large portions of the Colorado pueblo aggregated into substantial villages during an environmentally tougher time. In the early 900s, as environmental conditions improved, people dispersed into smaller settlements and the Anasazi tradition reached its maximal spatial extent. In Chaco Canyon, however, aggregation continued. People there built larger pueblos—multistory Chacoan pueblos known as "great houses"—in the canyon, and in the mid–eleventh century, they constructed roadways out to other Anasazi communities and placed new great-house pueblos in their midst. Scholars disagree about the organization and causes of this Chacoan cultural tradition, but they generally believe that its spread incorporated shared religious practices. Great kivas and great houses are diagnostic of Chacoan settlements. The regularity of the roadways and prestige goods found in Chaco Canyon's great houses are considered by some as evidence that pilgrimage served to integrate the Chacoan tradition. Stephen Lekson goes further, arguing that Chaco was an Anasazi ceremonial capital (Lekson 1999). When the pueblos in Canyon Chaco were abandoned in the mid-twelfth century during another period of drought, once again the Anasazi aggregated into larger, often defensive pueblos.

Southwestern Warfare

Steven LeBlanc (1999), Jonathan Haas and Winifred Creamer (1993), and other scholars (Bullock 1998; Mackey and Green 1979; Wilcox and Haas 1994; Woodbury 1959) working in the Southwest tend to agree that the strongest evidence of puebloan warfare occurs after the collapse of Chaco in the AD 1200s and continues through late prehistory until Spanish contact. Most studies of prehistoric southwestern warfare, like warfare research in general, stress variants of neoclassical social evolutionary models of ecological change, resource scarcity, and violence (e.g., Haas and Creamer 1993, 1996; LeBlanc 1999; LeBlanc and Rice 2001; Wilcox and Haas 1994; Woodbury 1959).

Jonathan Haas and David Wilcox's (1994) review of pueblo evidence highlighted defensive architecture and settlement patterns (e.g., aggregation, "no man's land," palisades, towers, alcove, and hilltop settings), aggression (e.g., burned sites, skeletal traumas), and weaponry (axes, arrows, shields, clubs). Buildings in all periods of pueblo prehistory, from the earliest pit-house villages through the late prehistoric period (AD 1300–1540), exhibit varying amounts of burning. Burning was most widespread during the pit-house period, with evidence showing the majority of structures from that time to have been burned, but the practice appears to have been more selective in later periods. Nonetheless, one can find sites early and late that are either entirely burned or unburned.

A small but intriguing percentage of puebloan sites from all periods contain human remains suggestive of violence that do not appear to have undergone considerate burials. These occur in a pattern suggestive of ritual violence (Walker 1998, 2006). Earlier instances typically occur in pit houses, while later examples are recovered from kivas. Many of these cases also have been the focus of studies of prehistoric cannibalism (Turner and Turner 1999).

In LeBlanc's recent study of southwestern warfare, he divided these data into three periods reflecting intensity or scale of violence. There is an early, lower-intensity period of endemic warfare up to about AD 900 followed by a relatively peaceful middle pueblo period (AD 900–1150) associated with the rise to prominence of Chacoan tradition. However, approximately half the cases of cannibalism described by Christy Turner and Jacqueline Turner (1999) occur during this Pax Chaco. They argue that Chacoan leaders enforced their hegemony by attacking and consuming their enemies. LeBlanc's final period (AD 1150–1540) is intensely violent, leading to the formation of more complex puebloan polities characterized by larger aggregated pueblos.

The post-Chacoan period of collapse, violence, and subsequent reorganization correlates with drought, loss of flood-plain farmland, unpredictable weather patterns, and other environmental stressors (LeBlanc 1999:38–41, 198–199; Vivian 1990:486–492). LeBlanc argues that decreases in carrying capacity lead to conflict between these larger communities that eventually resulted in the settlement system in place when the Spanish arrived in the mid-sixteenth century. Following Carol Ember and Melvin Ember (1992, 1994), however, Stephen Lekson (2002) highlights the complimentarity of ecologically and fear-based psychological explanations, noting that the perception of dangers or starvation in this period would have been equally causal.

In the immediate post-Chaco period between AD 1150 and 1275, new forms of architecture arose in the Anasazi region as pueblo communities once integrated in the larger Chaco world were balkanized into smaller warring communities where trade declined and fear and famine increased. Toward the end of this period, between AD 1250 and 1300, numerous cliff dwellings, walled sites, and tower constructions appeared (Schulman 1950).

These site placements facilitated line-of-site communication in addition to physical defense. Haas and Creamer (1993), focusing on the thirteenth-century Kayenta sites on the western frontier of the Chaco world, for example, defined a defensive settlement system characterized by intervisible aggregated sites. These data suggested to Hass and Creamer that the warring parties were outsiders. Complimenting these post-Chacoan architectural changes were increases in the production and innovation of lethal weapons, including self bows, *tcamahias* (nodule clubs), and wooden swords (LeBlanc 1999; Lutonsky 1998). This initial period of late prehistoric violence ended circa AD 1300 with the abandonment of large sections of the Colorado Plateau (Lipe 1995).

Migrants moved south off the Colorado Plateau to areas with more reliable water and potentially irrigable farmlands. In an argument similar to Haas and Creamer's but on a much larger scale, Leblanc (1999) notes that at the end of the thirteenth century, people aggregated into even larger communities, forming defensive confederacies or community clusters. These larger villages, while less defensively located, were hypothetically large enough to serve as defacto fortifications in their own right. Evidence of burning, as well as victims of violence in ceremonial contexts, is still found in these sites. The pueblo groups we recognize today (e.g., Hopi, Zuni, Tewa) congealed at this time.

The spatial orientation of these new towns was radically different from that of earlier Anasazi sites. The long-lived architectural pattern of a midden in front, followed by kivas and then a room block of habitation and storage structures, no longer held. Instead, room blocks were built around kivas and dance plazas, creating an enclosure of public space that has been cited as a defensive attribute of these villages. Kivas tended to grow in size and decrease in relative frequency as larger groups, such as moieties and sodalities, were formed in these communities.

Late Period Nonhuman Agents

These late thirteenth- and early fourteenth-century changes (see Varien et al. 1996:105) in architecture also correlated with the rise of Katsina religion, recognizable by dance plazas, masks, and Katsina deity images found on rock art, kiva murals, and ceramics (Adams 1991; Schaafsma 1991). Other indications of pan-southwestern religious reorganization include the spread of horned serpent and other water-oriented iconography. This religious patterning was exemplified at Casas Grandes, one of the largest pueblos in the ancient Southwest. Polychrome ceramics also replaced earlier black and white bichrome-decorated traditions throughout the pueblo Southwest. In a fascinating study of Salado polychromes, one of the most widely distributed ceramic traditions of the time, Patricia Crown argues that their distribution marked a shared southwestern ceremonial theology focused on fertility and moisture. With the exception of Polly Schaafsma's (2000) rock-art study of warfare iconography, however, these religious changes have been poorly integrated into the contemporary study of war. Warfare scholars have tended to view these religious developments as accommodating or legitimizing changes in social organization and environmental adaptations such as aggregation.

Puebloan People and Nonhuman Agencies

When first contacted by Spanish colonizers, pueblo peoples were organized into politically independent towns, or pueblos. At Zuni, there were seven such pueblos. The Zuni and other pueblo ethnic groups were forced by Spanish and, later, American administrations to present themselves as unified nations. These pueblo nations, on the basis of social organization, fall into three general groups that cross-cut ethnic and linguistic differences: the Western Pueblos, Eastern Kersesans, and Tonoans. Tonoans can be broken down further by geographic and linguistic differences into Tewa, Tiwa, and Towa pueblos. All these southwestern groups possess variable kinship and descent systems and order their ceremonial organizations by sodalities and moieties.

Fred Eggan's (1978:227–228) summary of pueblo social organization describes some of this variability. Western pueblos such as those of the Hopi and Zuni possess matrilineal exogamous clans as well as integrative sodalities, including warrior societies. Their civic and ceremonial leaders are clan leaders. Clans control land and ritual technologies, including knowledge of relations with nonhuman agencies, as well as many but not all artifacts used in such interactions (e.g., kivas, idols, altar material). Individual sodality members may control certain important objects such as their Katsina masks. The largest sodality is the Katsina Society. Other pueblo groups, such as Keresans, have matrilineal exogamos clans and a dual-division ceremonial system utilizing two great kivas. Membership in these kivas is patrilineal. Still other pueblo groups, like the Tewa, have bilateral kinship and nonexagmous patrilineal membership in dual divisions or moieties.

Every pueblo had warrior fraternities of one form or another that had interactions with nonhuman agencies. Each town had a war priest and two war captain assistants who, following the lessons of their origin stories, impersonated or represented twin war deities (Parsons 1924). These war chiefs had multiple tasks, including organizing communal labor; investing villages' chiefs; performing in new-year or winter solstice ceremonials; performing prayer rites; and caring for scalps, war-god idols, and associated altar weapons such stone arrow points, tcamahias, and stone axes. Their costumes included warrior decorations composed of weapons, paint, and enemy clothing. When fighting, these leaders went into battle impersonating twin war deities or sent representatives along to assist in the battle. Indeed, when the Hopi challenged the U.S. military, they met it with human warriors and humans in masks that were the temporary embodiment of the warrior twins and spider grandmother, an earth-mother deity. Pueblo war chiefs, idols, their human impersonators, their homes, and their responsibilities were all bundled together in ceremonial calendars that integrated them with rites involving the production of rain, collective harmony, and crop fertility. They were merged with Katsina ceremonials in variable and complicated ways in step with the different social organization of particular puebloan groups.

In these societies, causes of conflict and environmental problems were perceived as the results of immoral actions of people and witches. Witchcraft attacks occurred within pueblos and between them, leading to internal repression and warfare (Ellis 1951). Witches were assumed to kill people directly through hidden arrows, as well as indirectly by changing the weather to cause droughts, floods, and subsequent fam-

ines. Environmental problems were also believed to result from anger-
ing nonhuman agencies controlling moisture. Not surprisingly, witchcraft as well as jealousy, stinginess, and quarrelling were suppressed by
war chiefs.

Warfare was also integrated proactively into fertility and rain cults.
Pueblo warriors took the scalps and clothing of their enemies in battle
in order to harness their animate powers in subsequent rain-making
rites (Ellis 1951). The life force contained in scalps could be used in
dance ceremonials to bring the rain (Parsons 1924). This relationship
with the scalps, however, was not simply an impersonal technology.
The scalps had to be treated respectfully, and to some degree, kindly.
Pueblo peoples talked to them as well as to similar stone animal-spirit
fetishes. Scalps, idols, masks, and other agential objects were fed frequently (daily or weekly). Scalps were stored in sanctified places often
described as their homes. These might be shrine features, kivas, or a
specialized corner in a war chief's home. These human and nonhuman
beings and their organization have histories described in pueblo creation narratives.

Pueblo Emergence Narratives

Pueblo oral traditions assume that the past was a progressive process that entailed a journey through several earlier worlds. Similar to
archaeologists and geologists, pueblo peoples infer and recognize time's
passage in the ruins, distinctive rock outcrops, and fossils beds of their
desert landscape. They also believe that their current world is spatially
located above several previous worlds (Parsons [1939] 1996:210–266) to
which they have access from this one through subterranean features in
the landscape such as caves, springs, lakes, and the kivas and shrines
they have constructed to open portals between the worlds. Although
histories vary from group to group, they share some common themes.

In earlier worlds, anti-social acts—such as greed and jealousy—and
immoral acts—including witchcraft, adultery, and excessive gambling—
were rampant. These misbehaviors displeased other beings in the world,
leading to environmental problems (floods, droughts) and subsequent
human tragedies, including famine and warfare.

In response to these moral/environmental crises, leaders under the
guidance of twin warrior deities (a mythological motif widespread in

North and Central America) led the people out of that last world and into this one. In myth, the people usually climb up some sort of tree or vine through a hole in the first world's sky that leads to a cave or lake or some other entrance into this world. In this process, they do not completely free themselves from their past problems. In some cases, witches follow them into this world. The warrior twins help to combat these witches, new monsters, and other forms of danger. After entrance into this world, pueblo people migrate across the landscape until they reach their present locations, leaving in their wake the many ruins that archaeologists study.

In pueblo peoples' struggles from that time until they settled where they are today, they learned many skills such as agriculture, hunting, and fighting, as well as more esoteric knowledge about the animate nature of life and its transitions. They literally abducted evidence of these earlier events in this and earlier worlds in the landscape of the current one, for example by recognizing that fossils and stone formations were transformed spirits, killed monsters, and other forces of earlier times.

When confronting questions about their understanding of the world around them, these peoples will draw on this history to explain and reason through answers. Their ongoing struggles with witches, for example, while distressing, also explains their understanding of the afterlife. In their creation narratives, the killing of one individual by a witch soon after the people's emergence from the lower world is a common motif. The witch regularly appeals for mercy by having the community look down through the hole of their emergence to see the soul of the deceased alive and well below. The lesson learned is that despite death and sickness, the souls of the dead return to the underworld and live on much like they do in the present world (Cushing 1923:163–171).

There are levels of specificity in these narratives that are more localized. Each community has stories about the founding of pueblos that archaeologists date to the late prehistoric period (Pueblo IV, AD 1300–1540) and less-specific stories about their prior lives, travels, and interactions with nonhuman beings. These stories sometimes incorporate older (Pueblo III, AD 1150–1300 or Pueblo II, AD 900–1150) ruins. It is also not unusual for contemporary puebloans to incorporate more distant ruins (e.g., Teotihuacan) to account for evidence revealed by archaeological study.

The narratives betray a complex dynamic of mythology, prophecy, and community strife (see Parsons 1922) of interest to archaeologists of religion and violence. Although the historical accuracy of these stories does not measure up to scientific standards, they do have ordering principles that affect not just the way people remember the past but also how they act in the present to create the future. In Hopi stories, leaders unable to abate the jealousy, adultery, malicious use of magic by humans, and supernatural activities of witches (nonhumans) routinely invited other groups to attack and destroy their villages. As a testament to their sincerity, these leaders then sacrificed themselves, perishing with their fellow villagers (Malotki 1993). In these and other stories, spirits were active participants. It is common in these narratives for Katsina beings and other deities, like the sun and the spider grandmother, to help them and culture heroes such as the twins to overcome their enemies.

In a longitudinal study of Hopi oral history, Eggan (1967) tracked the transformation of a relatively minor Navajo/Hopi skirmish in the 1860s to a myth that chartered Navaho/Hopi territorial boundaries. As this oral tradition developed, Eggan found, its supernatural and prophetic elements increased with each telling, and the dispositions of daily life helped to reshape the recalled event. After seventy years, the Hopi victims of the skirmish, like leaders in their village-abandonment stories, had been transformed into proactive agents, described as having purposely sacrificed themselves to the Navajo with the assistance of the warrior twins. Tradition in Hopi is never static in part because inferences about communication and interaction with nonhuman agencies are constantly shaping and reshaping it. As a consequence, they are also shaping archaeological evidence.

A critical facet of pueblo peoples' interactions with supernaturals involves their interpretation and anticipation of prophecy (Geertz 1994). They fear and respect spiritual power and want to harness it in their rituals in order to live well and fulfill their destinies. The pursuit of traditions of prophecy allows Hopi peoples to establish seeming continuities in the face of discontinuity. Ethnographers (Geertz 1994; Whiteley 1988 but see Levy 1992) now argue that such agency accounts for the fissioning of the Hopi pueblo of Oraibi at the beginning of the twentieth century. Leaders following prophecy and emulating oral descriptions of earlier leaders manipulated village divisions, sacrificed their own power, and destroyed the community. They did this in the midst of U.S. government aggression, environmental pressures, epidemics, and other social problems. These variables, which are usually the focus of processual study, were not causes as much as context. They were the evidence leaders marshaled in their inferences about relationships with the spirits guided by traditional prophecy. Armin Geertz's description of tradition captures this dynamism:

> The term "traditionalist" does not automatically imply the continuity of tradition. It implies the opposite in fact. It implies the pursuit of tradition, which again implies the traditionalist neither stands in nor accepts the tradition of contemporaries. In other words the traditionalist is an innovator, one who promotes sacred matters that are perceived, as having been displaced by contemporary tradition. (Geertz 1994:5–6)

Oral traditions are not simple records of the past; they are active sources of change.

Southwestern archaeologists interested in warfare sometimes appeal to select passages in these oral histories (e.g., Haas and Creamer 1997; Lightfoot et al. 2001) to create analogical arguments confirming their prehistoric inferences about warfare. They seldom, however, consider the link between human and supernatural practice in these stories. Consider the historic Hopi Pueblo of Awatovi, located on Antelope Mesa. This pueblo was abandoned around 1700 as the result of conflict with neighboring communities from nearby First and Third Mesas. The story describing this abandonment, often cited by archaeologists interested in prehistoric war, conforms thematically with other Hopi oral traditions. All the different versions of the story suggest that the villagers were corrupt and practiced witchcraft (in this case, the witches are the Christian converts) and other dangerous activities. The village chief invited other Hopi villages to attack his village. They accepted and killed most of the men and boys in the kivas, and the Awatovi village chief sacrificed himself in the ensuing battle. Archaeological evidence at Awatovi does not conform completely to this abandonment story (Malotki 1993). While the Awatovi expedition did find that the mission church had been razed, it found little evidence of kiva burning. It found no context resembling a scenario of a surprise attack focused on the kivas (Montgomery 1949).

In excavations at Awatovi in the late nineteenth century, Jesse Walter Fewkes (1893) found one burned kiva containing a body but did not complete the kiva's excavation because his workmen refused to continue. It is usually this reference to violence and not the Peabody Awatovi report that is cited by archaeologists interested in this story (Montgomery et al. 1949). Fewkes refers to an area with burning and bodies in a nearby plaza, but his description is so sparse that one cannot definitively associate it with the site's abandonment. He also notes that large roofing beams from the mission church could still be seen in other villages, where they had been reused as trophies in the houses of important ceremonial leaders. These appear to have been viewed as some form of supernaturally charged objects. The Awatovi case actually exemplifies the ambiguities of warfare evidence found throughout the archaeological record of the Southwest.

This particular story has strong supernatural overtones. The archaeological evidence indicates both violence and ceremonial power as evidenced by the plaza, kiva deposit, burned church, and reuse of the church vigas as trophies. It does not, however, support a more generalized battle as described in the oral traditions and it does not conform in a pragmatic way to a description of battle as an adaptation to environmental change, conflict over resources, or fear of scarcity. Instead, the natural and social environments become the context for Hopi inferences and abductions about nonhuman agencies, human agencies, and the past. These inferences inform practices that then lead to more inferences and practices, so that no static models driven by anthropocentric categories such as oral history, environmental change, economic rationality, or cultural beliefs can corral the resulting complexity. We must instead find a way to integrate perspectives by building nonanthropocentric interpretations.

Puebloan History and Warfare

In a general sense, events in the late prehistoric period after AD 1200 correlate with the structure of known origin stories and the abandonment narratives they contain. It seems that the patterns of warfare, burning, and human remains in kivas found in archaeological sites make sense based on the shared logic of social relations still detectable in ethnohistoric practices, including oral traditions—even though in the ongoing process of history making in the pueblos, discrepancies have arisen between the actual events creating the deposits and the narratives explaining them. We know from archaeological evidence that pueblo peoples' interactions with Katsinas began in the late 1200s and early 1300s during a time of intense environmental stress and the abandonment of the Colorado Plateau. Warrior twin iconography appears earlier in the Southwest, circa AD 1000 (Carr 1979). It seems reasonable that the warrior twins were interacting with the puebloan peoples long before the Katsinas revealed themselves. They may have an even longer history, given that these warrior brothers were culture heroes in the American Southwest and Mesoamerica. Teasing apart that history, as well as more localized human-nonhuman interactions, is a challenge for future research. For now, I will concentrate on the late prehistoric nonhuman agencies.

The best documented evidence of religion and violence are deposits of human remains in kivas and pit houses. These relatively rare deposits appear to be the end result of inter- and intracommunity conflict in a pattern that does not fit the practical explanations of processual archaeologists, and which suggests a form of ritual violence I call kratophany (Walker 1998). These structured deposits appear to be the variable result of applications of a long-lived oral tradition about witches coming from the underworld at the time of creation. In pueblo traditions, fighting malignant spirit people has its own methods and logic. It was not enough to simply kill them. The kiva is a recreation of the hole linking this world with the one beneath. The process of killing and disposal seems to involve an attempt to push witches back into the underworld through the portal of the kiva.

Of course, as Sahlins noted, when such a general cultural worldview is applied in each instance of activity, it leads to new social structures and the transformation of culture through time. Therefore, there is no reason to suppose that such violence would be identical throughout pueblo history, yet like subterranean pit houses transforming into variable kivas, there would be continuity. What constitutes witchery, the scale of witchcraft, and those responsible for punishing it would vary with the changing constellations of past cultures. In the more egalitarian early pit-house communities, this might entail the execution of a shaman suspected of using his powers maliciously, while in a complex organization like that associated with Chaco Canyon, this might entail more intercommunity strife. Perhaps in a large, late-prehistoric pueblo town, there would be a mix of intra- and intercommunity examples.

In addition to violence, of course, there were undoubtedly also nonviolent ritual abandonment traditions that may have involved the destruction of particular buildings or even entire sites. If so, these traditions could have become entwined with inferences about the agency of spirits in later periods just as they did in Eggans's example of the Hopi-Navajo skirmish.

Nonviolent Ritual Abandonment. In many cases, in this long sequence of pithouse and pueblo abandonment, there is contextual evidence demonstrating that the burning of many buildings did not result from surprise attacks (Walker 1998, 2002). These buildings were burned in what appears to be a slower, more deliberate process. Their largest primary weight-bearing beams were removed and presumably reused elsewhere, and floor features—pits, vents, *sipapus*—were sealed. Sometimes structures were burned after they were no longer in use, their roofing material resting in superposition over floors covered by eolian sand. In other cases, burning was controlled and appears to have been purposely smothered soon after ignition. In still other cases, objects such as whole vessels were placed with the burned roofing or in the strata above it. These are not the strata of battles, but instead ritually structured deposits (Walker 2002). Processual inferences of war are completely silent on the contradictions such stratigraphic evidence raises. However, within an expanded understanding of society as comprised of human and non-human beings, such deposits would make sense. The abandonment of particular structures and even entire sites could have been conceptualized as the proper way to transition animate beings from this world to another. Alternatively, such disposal activities might have neutralized the power of these buildings lest they fall into enemy hands, like scalps.

Perhaps the archaeological evidence is telling us that despite fighting and witchcraft persecution, ritual abandonments were also aspects of practice that have been reinterpreted in the later practice of oral traditions. The practice of abandoning a settlement after twenty, fifty, or seventy years should not be taken lightly. Such activities would have had economic, political, and religious consequences. When archaeologists apply analogical arguments drawing on contemporary oral traditions, their arguments benefit from the differences—as well as the similarities—between past and present. By accounting for the differences, they learn as much about the past as from affirming similarities.

One of the most significant longer-term changes introduced by Spanish colonialism was the curtailing of pueblo abandonment and migration activities that had been going on for thousands of years as attested by archaeological and oral traditions. In prehistoric times, members of a village would remember previous abandonments and might even have first-hand knowledge of them that would differ dramatically from that of pueblo peoples living in historic times. Rituals of abandonment no longer in use were forgotten within a few generations. Abandonment became a novel circumstance to be explained with still-current dispositions. The destruction associated with such rituals seems to have become mingled with other memories of warfare and violence of the Pueblo III, Pueblo IV, and earlier periods.

Forgetting and Remembering. But what about those aspects lost to memory altogether? In other words, what about practices that no longer occur and are forgotten (Küchler 1999)? How did spirits contribute to war in Pueblo III and earlier times? Ruth Van Dyke (2004) and others (Bradley 1992, Crown and Wills 2003) have been working on the relationship between memories of Chaco and its importance in archaeological interpretations. It is clear that the D-shaped great ritual structures and coffee mug-style ceramics found in some early post-Chacoan aggregated sites (AD 1200–1300) derive from earlier D-shaped great houses and cylinder vessels found in ceremonial contexts in Chaco Canyon. Perhaps these micro-traditions of practice (sensu Pauketat 2001) reflect attempts to revitalize communication or interaction with earlier Chacoan spirits (Bradley 1992) that did not survive the coming of the Katsinas in the AD 1300s. One can also see similar effects on the defensive towers and lineage kivas. They both disappeared as new relationships were established with Katsinas.

Although defendable, not all thirteenth-century tower attributes are explained well by secular warfare interpretations (Ferguson and Rohn 1987:138). Some towers are associated with kivas. In some cases, they are connected by tunnels. Contrary to LeBlanc's assertion (1999:62), these tunnels do not resemble the escape hatches sometimes found associated with New Guinea men's houses. In some subtle cases, towers are constructed near or over kivas in purposely structured stratigraphic sequences (see Site 16 Mesa Verde, Ferguson and Rohn 1987). These towers connected ceremonial traditions through time and presumably provided a measure of prehistoric nonhuman agency to places or the structures erected in those places (Bradley 2000). Towers are seldom dismantled by archaeologists. When they are, perhaps we will find more communicative ritual objects (such as prayer sticks) in their walls, like those in the tower in Mummy Cave (Morris 1940). Such findings might support my contention that these buildings were central to interaction with nonhuman agencies.

How else can we account for towers at Hovenweep that in some cases were put so close to springs that they lost the tactical advantage of nearby higher ground? They might protect springs, but it is also possible to imagine, based on analogies with ethnohistoric pueblo cultures, that springs, like kivas, served as entrances to the underworld and were often the locus of water spirits of various sorts. As such, towers would have been critical points of contact for interaction with nonhuman agents (e.g., Bunzel 1932:487; Dumarest 1919:190,192; Parsons 1996:213–214; White 1935:91).

Towers were not simply pragmatic constructions, yet in a practice framework, we are not forced to choose between ritual and war or ritual and practicality, but can consider their integration. Perhaps warriors, kivas, and towers as beings all had practical purposes in the social relationships created between people and nonhuman agencies. Warriors on top of towers guarded springs, and towers as spirit warriors also afforded some kind of protection for warriors.

In the contemporary pueblo world and in many other cultures, buildings as well as artifacts are animated. Kivas, for example, have souls (Walker 1999) and names. Masks that facilitate the transformation of pueblo peoples into Katsinas are also animated and called "spirit friends" (Barnett 1990:103). Why not thirteenth-century prehistoric towers? In Pueblo cultures, water contains in its purest form the animating power of life. Therefore, scalps—because they retain this essence from a fallen warrior—were valued tools for interacting with spiritual forces controlling rain and fertility (Parsons 1996:31). At Zuni, rain priests were members of a warrior society entitled the priesthood of the bow. They called scalps "water and seed beings" (Stephen 1936). If towers, like kivas, were also animate beings, they might have defended and propagated water by placating spirits or warding off malevolent human and nonhuman beings.

It is telling that many warfare scholars (e.g., LeBlanc 1998; Lekson 2002; Schaafsma 2000) agree that pueblo war reached its apogee in the subsequent period between 1300 and 1540 (Pueblo IV), yet there are no towers in this period. From a pragmatic perspective, towers would have worked well in a large Pueblo IV pueblo. Indeed, many of these sites are located in more vulnerable open-air settings where a tower would have been more useful than in the alcove and spring contexts of many Pueblo III examples. Towers could have offered at least the same protection to field houses and smaller pueblos that they offered in the preceding period (Pueblo III, AD 1150–1300). They would have also continued to offer line-of-site possibilities. It seems significant that towers disappear along with family or lineage kivas in response to the introduction of a new configuration (see Adams 1991) of nonhuman beings in society (Katsinas).

Although the large-scale environmental and demographic problems informing processual histories highlight boundary conditions for action, they fail to explain the details of historical practices. Interpretations that can wrestle with the integration of practice, history, and science have a lot to offer the study of war and other forms of violence.

Clearly, the late prehistoric pueblo period must have been difficult. The spirits, and by extension the environment they controlled, punished the people with drought and famine. Attempting to find ways to make peace with spirits probably involved the punishment of internal and external enemies, leading to evidence we recognize as coming from war, such as weapons, defensive postures in settlement, and human skeletal assemblages indicative of violence. War would have pitted people and their spirit champions against each other until a new order emerged.

The thirteenth century witnessed a range of social experiments after the fall of Chaco. Fourteenth-century pueblo peoples literally created a new world (the fourth world) of practice under the guidance of Katsinas. This is clear in the radical architectural changes that came to typify many late prehistoric villages. A price of this rearrangement, of course, was an apocalyptic or prophetic understanding of the future that continuously pitted leaders with differing understandings of the spirit world against each other. It also tends to overwhelm details about whether enemies literally forced abandonments such as that of Awatovi or whether supernaturals, through human agency, chose other ways to confront the violence of others and vicissitudes of nature.

It is easy to claim that practice overcomes the limitations imposed by analytical dichotomies such as those between the ideal and real, the mind and body, and belief and action, but it is harder to put such theory into play. We must do more than assert that beliefs are real or that the mind is the body and then continue to do archaeology as usual. Otherwise, we will simply rephrase these terms in a more obscure language that hides their anthropocentric blind spots. To say that artifacts are meaningfully constituted reinforces these types of dichotomies as much as does saying that the meaning of material culture is irrelevant. Rather than playing out the recursive possibilities of these dichotomies, let us rebuild the house.

If we can conceive of nonhuman agencies as social actors rather than the symbols or beliefs of people, then we can also redefine society. Allowing the nonhuman agents in the door counterintuitively has the benefit of helping us reconceptualize the causal consequences of material objects. Ian Hodder (1986) once argued that artifacts are meaningfully constituted. I prefer to say that some artifacts and buildings are alive. They act on their own as well as interact with people. Envisioning them in this way leads to new questions about social organization, the processes that change it, and the study of its history.

PART II

Warfare in Precolonial Central Amazonia

When Carneiro Meets Clastres

Eduardo Góes Neves

> What wonders me the most in these wars and cruelties of them is
> that I could not know from them why do they launch war, since
> they do not have possessions of their own, neither Lordships, King-
> doms or Empires, and they don't know what thing is greed, that is
> property, or selfishness of ruling, what seems to me to be the cause
> of wars and of each unordered act. When we asked them to tell us
> the reason, they did not know how to give another reason, except
> that they said that this curse started before among them and that
> they wanted to avenge the death of their ancestors.
>
> —Amerigo Vespucci, second letter to Lorenzo de Pier Francesco
> de Medici

Few subjects have attracted as much attention in lowland South American ethnology as warfare, and yet disputes about this topic in the literature remain far from settled (Carneiro 1970; Chagnon 1993; Fausto
1998; Ferguson 1995; Ferguson and Whitehead 1992; Redmond 1994).
Many disagreements center on the reasons for and the historical context underlying warfare: Can it be explained in purely biological, materialist, or ideological terms? Was it prevalent in lowland South America before European colonization, or did it develop in the colonial era?
What was the role of warfare in the rise of complex societies? To answer
these questions, it is useful to look at the archaeological record. This
may help us understand the extent to which warfare was common in
the area in the precolonial past, and it will also help us verify whether
such factors as population pressure or competition for resources could
be positively correlated to it.

This chapter has several goals. First, it aims to briefly present archaeological evidence of warfare and conflict from different areas in the

Amazon basin, with a focus on the central Amazon. This evidence relates mostly to defensive structures recently identified at archaeological sites regionally. Second, the data will be evaluated under a wider contextual perspective—including regional chronologies and settlement patterns—to verify whether it matches the expectations that correlate warfare with the emergence of complex societies. This evaluation will show that the match, if present at all, is ambiguous at most. Based on that, a hypothesis will be presented to explain the dynamics of warfare in precolonial Amazonia. Drawing from ethnographic examples and the iconographic of archaeological objects, this hypothesis turns itself away from a social evolutionist perspective and suggests that precolonial warfare can be understood as part of a process of personal identity-building among Amazonian Indians that can be identified in the present as well.

Was There Warfare in the Amazon Before the Sixteenth Century AD?

An understanding of warfare in precolonial Amazonia needs to be sought in the light of a wider debate concerning the changes brought by the European conquest on native patterns of social and political organization. Although latent since the late-1940s publication of *The Handbook of South American Indians* (Lowie 1948; Steward 1948), this debate became stronger around the late 1980s and early 1990s, thanks to contributions both in archaeology and cultural anthropology (see Viveiros de Castro 1996 for an overview). Briefly speaking, the debate relates to the possibility of employing data from contemporary indigenous societies in the explanation of the precolonial archaeological record of the region. To put the matter as a simple question: To what extent are contemporary patterns of social and political organization of Amazonian indigenous societies similar to those of their precolonial ancestors?

The understanding of this general question has generated a large amount of research during the last twenty years, either aiming to show that the changes brought by European colonization were major or aiming to show they were minor (Erickson 2000; McEwan et al. 2001; Meggers 1993–1995; Heckenberger 2003; Heckenberger et al. 1999; Neves 2001; Neves et al. 2003, 2004; Neves and Petersen 2005; Petersen et al. 2001; Roosevelt 1991, 1999; Schaan 2004; Stahl 2002). That so much disagreement can exist over such a basic issue is a direct consequence of the paucity of the archaeological data available for the whole Amazon basin.

At the heart of debate are questions related to the demography of precolonial Amazonia. Those who work under the assumption that the colonial changes were minor also assume that precolonial indigenous population was not much larger than the contemporary population (Meggers 1993–1995). Those that work under the other premise—that the colonial changes were major—accept William Denevan's (1992) general idea that the size of the native Amazonian population in the late fifteenth century could have been around 5.5 million people. The question of population size is also far from understood. In the early twentieth century, when the data that compose *The Handbook of South American Indians* was compiled, the size of indigenous population in the Amazon was much smaller than it is today. In the particular case of lowland South America, it is today clear that some of the patterns recorded ethnographically in the late nineteenth century or early twentieth century result directly or indirectly from the colonial historical past. Such patterns cannot be projected into the past as examples of century-old forms of economic, social, or political organization (Roosevelt 1989).

One of these patterns may be warfare. Napoleon Chagnon proposed in the 1980s that warfare plays a pivotal role among the Yanomami society, allowing successful warriors preferential access to resources and women—which eventually leads to a higher rate of biological reproduction (Chagnon 1983, 1988). Later, in 1995, an alternative hypothesis was presented, in which Yanomami warfare was understood not as a result of competition for reproductive success, but as the outcome of pressures such as disease and competition for trade goods brought up with the European colonization of the New World (Ferguson 1995). Currently, the debate about Yanomami warfare is far from settled (Albert 1989, 1990), having escaped the small circle of lowland South American specialists and entered into the wider circus of American anthropology (Tierney 2000).

Brian Ferguson's argument on Yanomami warfare was previously given in the volume *War on the Tribal Zone: Expanding States and Indigenous Warfare* (Ferguson and Whitehead 1992), in which studies were presented to show how indigenous warfare could be understood as local manifestations, in peripheral contexts, of the general historical process of capitalist expansion outside of Europe from the sixteenth century onwards. In this particular context, the concept of tribal zones was proposed as a way to explain how, in lowland South America, competition and conflict among the European metropolis reproduced itself as conflicts among local polities, leading to the emergence of powerful regional leadership and the reification of patterns of conflict that were previously much more fluid (Whitehead 1990a, 1990b). Neil Whitehead's own example of sixteenth-century conflict between Arawak- and Carib-speaking polities in the lower Orinoco and the lower Antilles is a case in point (Whitehead 1990b). He has demonstrated how allegiances established between Spaniards and Arawaks helped to create the image of savagery that became later associated with the Caribs, from whom indeed the word "cannibal" was coined (Dreyfuss 1983–1984).

Whitehead's argument is pertinent because it fits other known cases of indigenous warfare recorded in the Colonial period in the lowlands. In the Rio Negro basin in north-central Amazonia, the Portuguese launched war against the Manao Indians in the 1720s, having as a pretext their supposed alliance with the Dutch of the Guyanese coast (Rodrigues Ferreira 1983). The Manao, however, were middlemen in an intricate trade network, probably of precolonial origin, that connected—by the Negro, Branco, and Essequibo rivers—the central Amazon to the Guyanese coast (Porro 1985). As the demand for Indian slave labor increased in the Dutch plantations of Guiana, the Manao became slave raiders, launching wars against other indigenous groups placed at the fringes of the trade network (Farage 1991). Such engagement created their reputation as warriors, and competition with the Portuguese— slave raiders themselves—led to their demise in the mid-eighteenth century. A similar process happened among the Tupinambá of the Atlantic coast of what today is southeastern Brazil. Literature from the sixteenth century (Fausto 1992) attests to how the Tupinambá were model warriors, known by a sophisticated cannibalistic complex in which captives were taken and ritually killed after several months of residence among their captors. Tupinambá warfare was also caught up with the conflicts among the countries aiming to control that part of the New World—in this case, France and Portugal. As in the case with the Caribs, different Tupinambá groups sided with either the French or the Portuguese in armed conflicts that lasted during a good part of the sixteenth century, by the end of which these groups were heading towards extinction.

Summarizing the discussion, there is good evidence to support the hypothesis that patterns of warfare and conflict recorded historically and ethnographically among indigenous groups of lowland South America retain some kind of causal association with the European colonization of the continent. The reasons accounting for that association vary and are not mutually exclusive. Among them are competition for foreign trade goods, alignment with different colonial powers, engagement in the slave trade, and conflict over land among populations pushed away from their former homeland. It is important, however, to turn to the archaeological record to check the prevalence of warfare or armed conflicts before the onset of the European colonization and, more importantly, to try to understand the historical milieu in which warfare unfolded in such contexts.

Warfare and Chiefdoms in the Archaeological Record of the Amazon

Despite a century-long tradition of research, the archaeological record of the Amazon is still poorly known (Neves 1999). Thus, any discussions on subjects such as warfare will necessarily be based on scanty data. This limitation notwithstanding, there has been a remarkable increase in research done in the Amazon over the last twenty years. Most of this research was aimed at answering the question of how much change has happened among precolonial indigenous societies as a result of the European colonization (Neves 1999). One particular way to address this question can be seen in efforts done on mapping and excavating large archaeological sites spread across the Amazon basin. These sites are several hectares large and normally multi-componential, containing deep deposits usually associated with anthropic dark soils known in Portuguese as *terras pretas*. The archaeologists and soil scientists working with these sites aim to show that they correspond to large settlements in the past, sometimes occupied by several hundred individuals (DeBoer et al. 1998; Heckenberger et al. 1999, 2003; Petersen et al. 2001; Neves et al. 2003, 2004; Neves and Petersen 2006; Roosevelt 1999; Woods and McCann 1999). This hypothesis suggests that the Amazon was densely occupied in the past and that the people who lived there left a visible imprint in the land-scape in the form of anthropic soils and other features (Denevan 2001). However, the same kind of evidence has been interpreted as the archaeo-logical record of several re-occupations of the same or adjacent spots by small-scale groups, with the successive re-occupations in the end forming large archaeological sites (Meggers 1990; Meggers et al. 1988).

This debate is ongoing but, although grounded in apparently antag-onistic perspectives, many of the scholars involved in it share the same neo-evolutionist theoretical background. One part of neo-evolutionist thinking—in its various forms—is the attempted classification of social formations within general evolutionary groups in order to allow for cross-cultural comparisons (Flannery and Marcus 2000; Trigger 1989). On the theoretical level, the processual approach has never actually achieved a full grasp of Amazonian archaeology (Roosevelt 1991: ch. 2). As in many places of the world where research is mostly exploratory, the major theoretical orientation of archaeology in the Amazon is culture historical. Curiously, however, neo-evolutionism and cultural history blended together in Amazonian archaeology in the development of the debate addressed in the beginning of this chapter. In its most simple form, this debate contrasted two distinct evolutionary stages to frame the discussion of the historical changes brought about by the European conquest—specifically, the conquest's effect on the opposition between tribes and chiefdoms (Myers 1992; Roosevelt 1987, 1991). In this opposi-tion, chiefdoms would have been the prevailing social formation across the floodplains of the Amazon and its major tributaries in the precolo-nial past, while their contemporary descendants—the indigenous peo-ple now living in the Amazon—would have a tribal pattern of politi-cal organization (Beckerman 1979; Roosevelt 1995). The identification of chiefdoms in the archaeological record has traditionally occurred through a sort of checklist approach, namely through identifying those features that supposedly correlate with the emergence of more central-ized forms of political organization (Flannery and Marcus 2000). In the Amazon, such an approach followed the identification of monu-mental structures such as raised mounds, causeways, roads, and central plazas (Erickson 2000; Roosevelt 1991, 1999; Heckenberger et al. 1999, 2003; Pärsinnen et al. 2003). The assumption is that these structures were built through the mobilization of labor, indicating some kind of institutionalized social inequality.

The identification of warfare in the archaeological record has played an important role in the checklist approach. Since the pioneering work of Robert Carneiro (1970, 1987), warfare has been causally associated with the emergence of formalized social inequality of the type characteristic of chiefdoms or states (Earle 1997; Flannery and Marcus 2004). In the case of the Amazon, Carneiro suggested that effective adaptation along the area's fertile floodplains would have generated population growth in geographically restricted areas—those places immediately adjacent to the floodplains, known as *várzeas*. The combination of population growth and the relatively small availability of land in the várzeas would have generated demographic pressure and competition for better access to aquatic resources and fertile soils among local communities, a process he called "geographic circumscription" (Carneiro 1970). In Carneiro's classic formulation, warfare would have been the mechanism allowing for the subjugation and control of local communities in the emergence of regional, centralized policies under contexts of geographic or social circumscription (Carneiro 1970).

Turning back to the archaeological record of the Amazon, then, the identification of warfare in the precolonial past can be seen as a material correlate of the kinds of social-political changes normally associated with the emergence and functioning of chiefdoms. This reasoning would apply even better in those settings where geographic and social circumscription would have been expected to happen in the past, such as the floodplains of the Amazon and its major tributaries. Some evidence for this will be reviewed here.

Case Studies

The case studies presented in this chapter come from different parts of the Amazon basin. Some of them are from first-hand research (Fordred-Green et al. 2003; Heckenberger et al. 1999; Neves 2001, 2003; Neves and Petersen 2005; Neves et al. 2003; Petersen et al. 2001), while some result from research done by others (Arellano López 2002; Heckenberger et al. 2003; Pärsinnen and Korpisaari 2003).

The first example comes from the central Amazon, where we have been conducting systematic fieldwork since 1995 in a research area roughly nine hundred square kilometers wide (fig. 5.1). The area is bound by two of the largest rivers of the Amazon basin: the Amazon

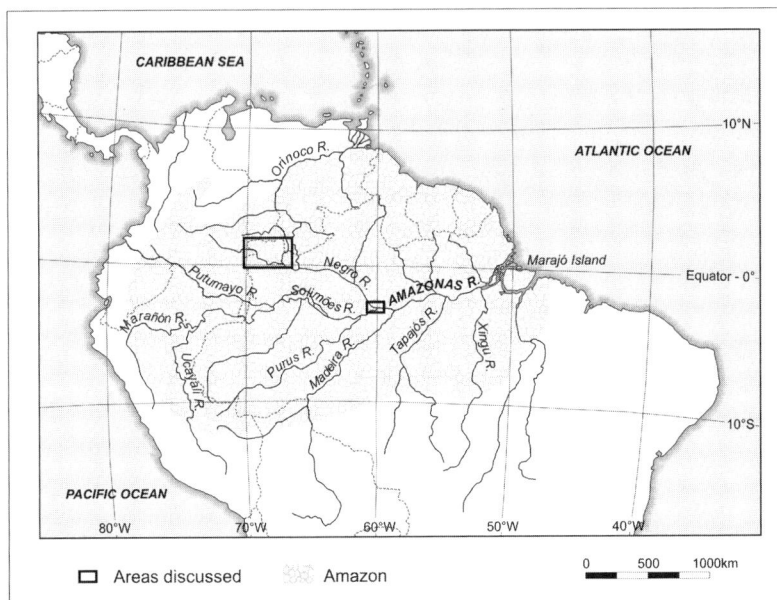

5.1. Areas of northern South America discussed in the text.

itself and the Negro. The Amazon is a whitewater river, rich with nutrients brought from its catchment's area in the Andes. Those nutrients yearly fertilize the floodplains' soils in the wet season. The Amazon is a dynamic river, constantly building new channels and abandoning old ones. As a consequence, its floodplain forms a landscape mosaic composed of oxbow lakes under different stages of activity, swamps, flooded forests, and other landscape formations. The Negro is a much less dynamic river: having its catchment's area in the geologically ancient Guiana plateau, it is a nutrient-poor river whose floods do not fertilize the floodplains. Although work is still in progress in the area, the available data already offer an adequate benchmark for the testing of the correlation between warfare and the emergence of regional political centralization as suggested by Carneiro's model.

Regional survey and stratigraphic excavations allowed for the identification of several dozen archaeological sites and the systematic excavation of eight of them (fig. 5.2). A series of radiocarbon dates provide the backbone of a regional chronology that starts in the early Holocene and goes to the Colonial period (Lima 2003; Lima et al. 2006; Neves

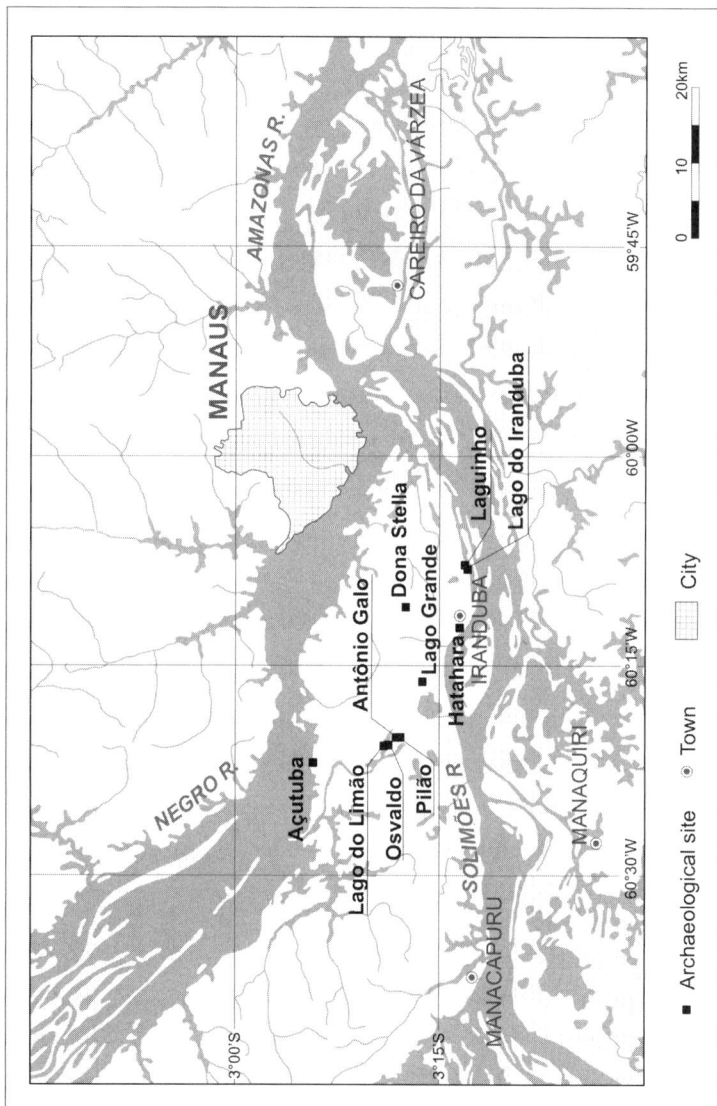

5.2. The central Amazon.

2003; Neves and Petersen 2006; Petersen et al. 2004). The chronology shows that the history of occupation of the area was marked by gaps, mostly during the mid-Holocene (Neves 2006). From circa 2300 year BP, ceramics appear for the first time in the area, and between that time and the beginning of the European colonization eighteen hundred years later, there appeared at least four distinct ceramic complexes now visible there, although the historical relationships among them are not clear (Petersen et al. 2004; Lima 2004). The pottery decoration and technology of the four complexes can be very sophisticated. In all cases, different manufacturing techniques are indicated, including coiling and modeling. Decoration includes different plastic techniques such as incision, excision, grooving, and the application of zoomorphic and anthropomorphic appliqués. Painting is also very common, characterized by the use of white and red slipping and the polychrome combination of white, red, and black, among other techniques. Funerary urns of large size are also common, mostly from the Paredão and Guarita phases, the phases of the two most recent complexes in the sequence.

The size and density of the ceramic sites in the area peaked in the period between the sixth and twelfth centuries AD (Neves 2006). Such a peak in occupation is characterized by dense occupations that produced the anthropic, organic-rich, fertile soils known as terras pretas. Some of these sites are quite large. Açutuba, the largest one excavated and mapped so far, is more than ninety hectares in area. Most of the sites are also multi-componential. In some cases, ceramics of previous occupations were evidently recycled as raw material for the construction of ceremonial structures such as burial mounds (Machado 2005).

The available chronology shows that the length and pace of occupation of these sites varied greatly. Some seem to have been occupied for only a few generations, while others were occupied through much longer periods of several centuries. At this moment, it is not possible to establish the functional relationship among ceramic sites in the area during the different periods of occupation. The presence of intrusive ware in some of the sites indicates that some form of connection via exchange or marriage was prevalent (Donatti 2003).

Potential indicators of centralization and the emergence of hierarchies in settlement patterns could be seen in terms of labor investment in the building of monumental structures. In the central Amazon, these structures are characterized by artificial mounds, made by the accumulation of soils, laterite (iron oxide), and mostly ceramic shards (Machado 2005). One of these mounds was partially excavated at the Hatahara site, next to the floodplain of the Amazon. It dates to the eleventh century AD and is formed by a 140-centimeter-thick accumulation of anthropic dark soil and hundreds of shards over at least eleven human burials (Machado 2005). The burials are individual and direct, except for one collective burial placed at the center of the mound. Artificial mounds of the types identified at Hatahara are, however, widespread over sites of different size and length of occupation found across the area, and they are correlated with occupations during both the Paredão and Guarita phases, from the eighth to the thirteenth centuries AD (Donatti 2003; Heckenberger et al. 1999; Neves 2003; Machado 2005; Morais 2005). This evidence suggests that even if large amounts of labor were recruited to build these structures, these were short-term mobilizations, associated with the events of mound building themselves and not the result of the control of a permanent available labor force (Neves and Petersen 2006).

A final possible indicator of centralization and the emergence of hierarchies was found in site layout. It has been initially proposed that the presence of public areas, in the form of ring villages with central plazas, might be seen as a characteristic feature of those settlements occupying central places in regional hierarchies (Heckenberger et al. 1999). Further work done in the area, however, shows that the ring shape is not particular to large sites with long-term occupation spans; it is also visible in quite smaller settlements with shorter occupation spans (Donatti 2003; Machado 2005; Morais 2005).

Summing up the argument for the central Amazon, then, data on topics such as chronology of occupation, settlement patterns, and site layout show a pattern of regional political decentralization, indicated by the absence of visible hierarchies among the settlements during eighteen hundred or so years of occupation of the area. The theoretical challenge of understanding this pattern arises because the data show that, in some cases, site occupation was quite long, lasting for a few hundred years. Such occupation spans suggest some sort of territorial behavior that could be conducive to conflict or warfare. The available evidence on the topic shows that, in the central Amazon, although defensive structures are found in some of the excavated sites, this does not suggest that they were associated with territorial conquest or competition for the occupation of ecologically more productive spots.

As with most of the data produced so far for the central Amazon, evidence of warfare is still preliminary. Out of the excavated and mapped sites in area, two have artificially built ditches interpreted as defensive structures: the Lago Grande and Açutuba sites. Lago Grande is a Paredão-phase site currently covered by high-growth secondary forest and a garden. It was occupied continuously—according to eighteen radiocarbon dates—from the end of the seventh to the end of the tenth century AD (Donatti 2003; Neves 2003). Over the sequence, Manacapuru-phase ceramics are found in small but constant amounts, evidence that suggests regular contacts between Lago Grande and contemporary Manacapuru-phase sites in the area occupied in the eighth century AD (Abreu 2001; Donatti 2003; Neves 2000).

Lago Grande is located on the top of a high peninsula, thirty meters above a large floodplain (várzea) lake on the north bank of the Amazon River. This is a prime spot because the height of the bluff prevents the site from getting flooded, while at the same time, its location guarantees access to the rich soils and rich faunal resources of the floodplains—especially in the summer, when the lakes dry almost completely. The site was a horseshoe-shaped village about two hectares wide with a group of middens placed around a central plaza.

On the isthmus that connects the peninsula to the adjacent high ground, a ditch was excavated and dated (fig. 5.3). The excavations showed that the ditch is artificial, which can be seen from the inversion in the stratigraphy. The ditch itself is marked by a depression, while at both sides of it one sees the original anthropic A horizon recovered by earth dug for the opening of the ditch. The ditch is roughly 30 meters long, 6 meters wide, and 1.5 meters deep, interrupted at its center by a passage formed by ground that was not dug. The excavation of the trench involved the removal of about 270 cubic meters of soil, a task probably accomplished with digging sticks and baskets. A radiocarbon date from a charcoal sample from the trench's east wall shows that it was not dug before about 1100 BP, roughly 150 years after the beginning of the occupation of the site and another 150 years or so before its abandonment (Donatti 2003; Neves 2003). No visible signs of palisades,

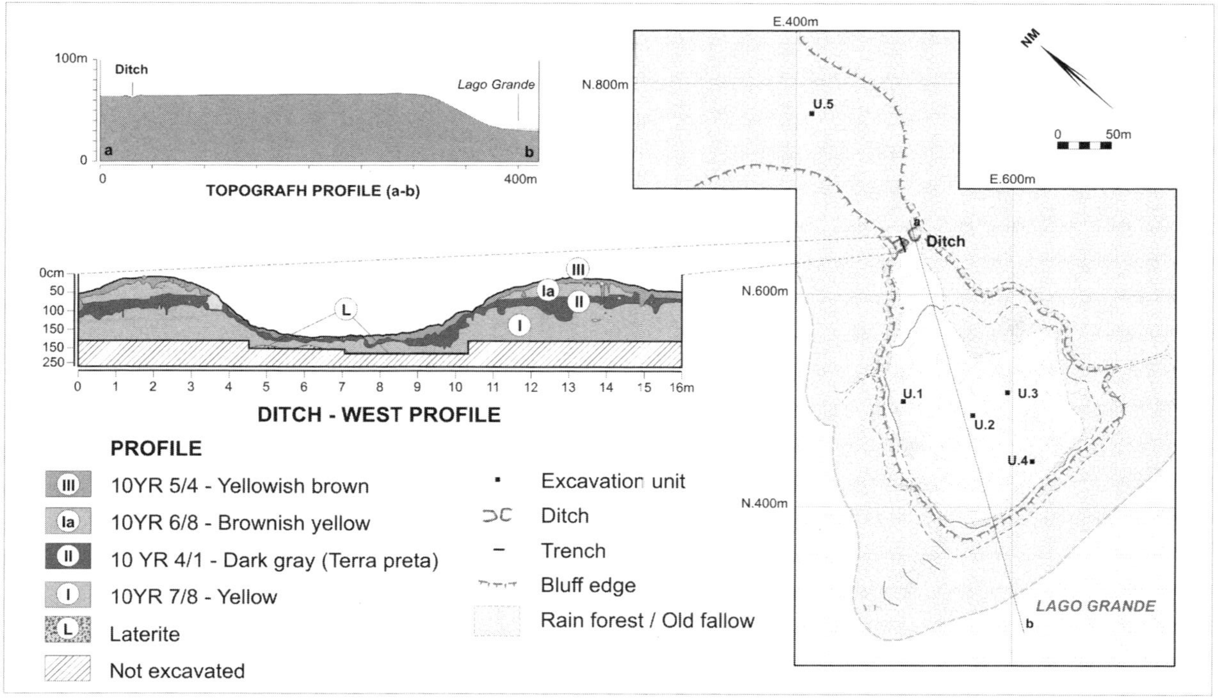
100m
Ditch
Lago Grande
0
a
b
0
400m
TOPOGRAFH PROFILE (a-b)
0cm
50
100
150
150
250
L
III
Ia
II
I
0
1
2
3
4
5
6
7
8
9
10
11
12
13
14
15
16m
DITCH - WEST PROFILE
PROFILE
10YR 5/4 - Yellowish brown
10YR 6/8 - Brownish yellow
10 YR 4/1 - Dark gray (Terra preta)
10YR 7/8 - Yellow
Laterite
Not excavated
Excavation unit
Ditch
Trench
Bluff edge
Rain forest / Old fallow
E.400m
N.800m
U.5
NM
0
50m
E.600m
a
Ditch
N.600m
U.1
U.3
U.2
U.4
N.400m
LAGO GRANDE
b

5.3. Digital map with the isthmus of Lago Grande.

such as post molds, can be seen inside or outside the ditch, which indicates that it worked as a moat.

The location of the Lago Grande site, high on a bluff, and the placement of the ditch on the isthmus—the only straight connection between the peninsula and the adjacent high ground—indicate that it was defensive structure. An alternative hypothesis—that the ditch was an irrigation channel or a water reservoir—seems unlikely due to the location of the site at the edge of a floodplain lake, as well as due to the fact that the ditch does not connect to any sources of water. Lago Grande was abandoned around the end to the tenth century AD. Although the reasons for its abandonment are not clear, armed conflict could have had a part in it. Unfortunately, there is no available information that can help us understand the life history of the moat during the 150 years that spanned between its construction and the abandonment of the village. It is not clear whether the moat was actively used as a defensive structure over that period. If we had that information in hand, it might be possible to understand the extent to which warfare was prevalent in the central Amazon around the end of the first millennium AD. On the other hand, the fact that the site was not occupied after the eleventh century AD is in itself interesting. Going back to Carneiro's general formulation, one would expect a strong degree of competition for settlement in places such as Lago Grande, located next to the highly productive floodplain of the Amazon. This, however, wasn't the case. Aside from what looks like a quite ephemeral Guarita-phase occupation, there are no signs of permanent, sedentary occupations at the site well into the Colonial period. This trend was confirmed by Patrícia Donatti's (2003) survey of the whole north bank of Lago Grande Lake, from which it can be concluded that, for whatever reason, access to more productive areas probably wasn't a primary cause for warfare in the Lago Grande area.

A second case study from the central Amazon comes from the Açu-tuba site. Açutuba was occupied for a longer period than Lago Grande: from the third century AD to the fifteenth century DC, with at least four distinct occupations. This is a large site, with an area of roughly ninety hectares. It has been undergoing excavation and mapping since 1995 (Heckenberger et al. 1999). A likely scenario for the history of occupation of the site is that it was initially a small village settled around 300

BC and that it has been discontinuously reoccupied until the fifteenth century AD, when it was a large Guarita-phase village. As with Lago Grande, Açutuba sits on the top of a bluff, in this case overlooking the Rio Negro. However, unlike Lago Grande, Açutuba is not on a peninsula, the site's main axis being parallel to the river (fig. 5.4).

During mapping, a 150-meter-long ditch running in the back of the site, parallel to the river, was identified. The ditch was cross-sectioned with an excavation trench. Its stratigraphy shows that it is artificial, as it can be seen by the break in the laterite on the profile. Different from Lago Grande, the ditch at Açutuba was filled by a double palisade, as indicated by two post molds visible in the profile. Due to its artificial construction and the post molds in its interior, the ditch is also interpreted as a moat. This interpretation is further corroborated by its placement at the back of the site, the area most vulnerable to attack, since the front end of the site sits on the top of a high bluff. As with Lago Grande, an alternative hypothesis that the ditch was an irrigation channel is discarded because it does not connect to any sources of water (Neves 2000).

The moat at Açutuba is much larger than the one at Lago Grande, although a bit narrower. Its construction involved some planning and mobilization of labor, largely to complete the felling of trees with stone axes to build a double palisade filling the long ditch. The ditch has not been dated yet, but it surrounds the area of the Paredão-phase occupation. Since there is no chronological information on the building of the ditch, it is hard to present any hypothesis on the history of the use of the moat and palisade, but the fact that the post molds were preserved in the archaeological record shows that these were permanent structures. If that is correct, one can presume that warfare was chronic in the area during the Paredão-phase occupation in the early second millennium AD.

A third example comes from the upper Rio Negro basin in the northwest Amazon. The northwest Amazon is a vast area encompassing parts of Colombia, Venezuela, and Brazil. It is occupied today by indigenous societies that speak languages from five different linguistic families but that are more or less integrated by marriage, trade, and/or labor into one large regional system (see, for instance, Hugh-Jones 1979, 1995). The goal of research was to understand whether the origins of the

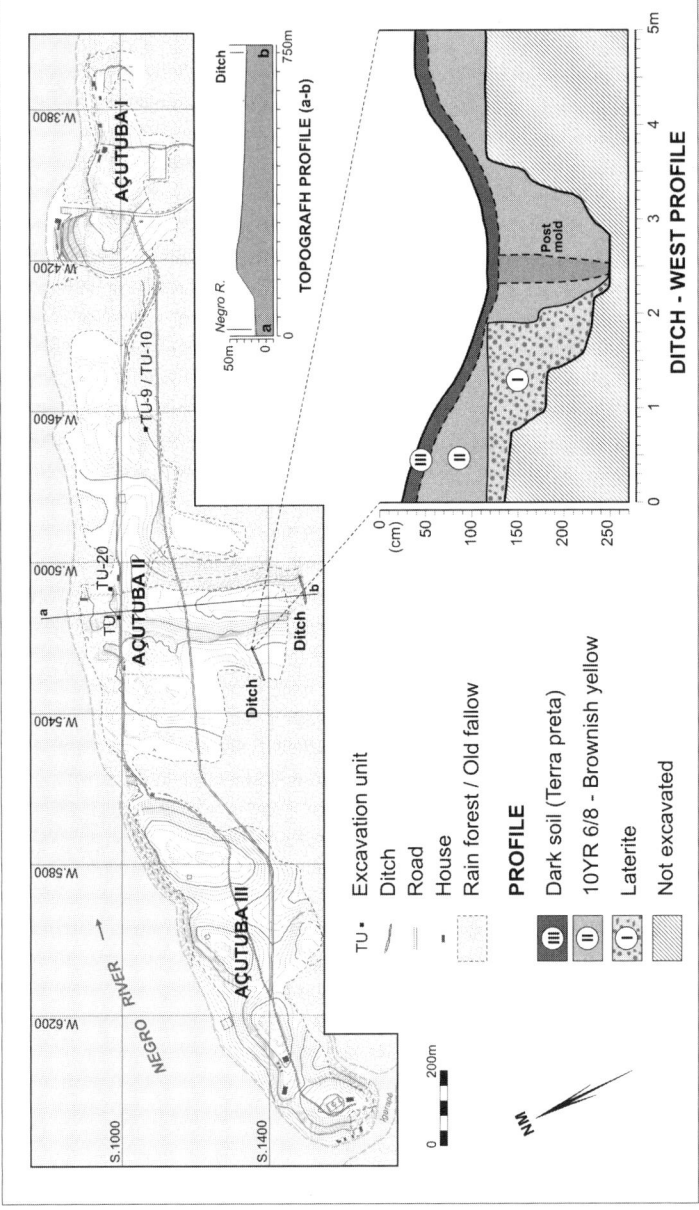

5.4. Digital map of Açutuba.

system were precolonial or whether it developed as a consequence of the European occupation of northern South America, including parts of the Amazon, Guianas, and the Caribbean (Neves 1998, 2001). The addressing of this question ties into the more general research problem presented at the beginning of this chapter, since the presence of regional multilingual systems seemed a contradiction to the tropical forest model presented in *The Handbook of South American Indians*. To address the question of the origins of the northwest Amazon regional system, archaeological and ethnographic fieldwork was done among the Tariana, Arapaço, and other Tukanoan-speaking groups settled in villages along the Uaupés River.

Indigenous peoples of the northwest Amazon have a rich body of oral tradition that has been recorded by the Indians themselves and by anthropologists since the nineteenth century (Amorim 1926; Kumu and Kenhíri 1980; Hugh-Jones 1979; Moreira and Moreira 1994). This body of oral tradition refers both to events that took place in what feels like a remote past—in episodes related to the creation of the world proper—but also to events that happened more recently. Among the Tariana, one of the groups settled in the area, for instance, oral tradition includes narratives about warfare around the time of their arrival at what is today their territory. Such oral tradition was already recorded in the late 1800s, although it wasn't published until the early twentieth century (Amorim 1926). The same narratives were recorded recently by Indians and anthropologists; comparison between nineteenth- and twentieth-century narratives shows striking similarities in features such as place names and names of rivers, creeks, rapids, and hills (Moreira and Moreira 1994; Neves 1998).

There is a consensus among indigenous people in the northwest Amazon that the Tariana were the last indigenous society to settle in the area. Their current territory encompasses twenty-odd villages spread along the Uaupés and one of its tributaries, the lower Papuri (Neves 1998: fig. 5). Oral tradition says that the Tariana migrated from a place in the north and that their arrival in the Uaupés resulted in a war against the other groups already settled there. Episodes of this war are narrated in a series of histories that connect events in the past to actual places scattered over their territory. In the events described in the narratives of the Tariana wars, there is constant reference to their leader, Buopé.

"Buopé," in the northwest Amazon, is a name with different meanings: it is the name of the mythical leader of the Tariana in their wars against other groups, it is the name of a now-extinct indigenous group, and it is also the name of the Uaupés River, one of the main rivers of the area (Wright 1992).

One of the places described in detail in the narratives is a village from which the Tariana launched their last attacks against their enemies, and at which they eventually settled. Located on the top of a hill, in the Jurupari ("devil's") range, this was a village surrounded by ditches and spikes dug and placed as defensive structures. Spiked trenches were apparently very common in the northwest Amazon in the past (Wright 1990). Irving Goldman (1963) says that Cubeo oral tradition refers to them. Janet Chernela (1993:23) states that "as reported by Wanano informants, raiding and warfare were so severe that numerous villages on the Aiarí, affluent of the Içana and the Uaupés rivers, were surrounded by spiked trenches." Although it is likely that raiding increased in the eighteenth century due to the demand for slaves, it may well be that some of these structures were precolonial.

The site of the village is currently under forest cover, more than five kilometers away from the main river, an atypical settlement for the contemporary villages and archaeological sites of the area. In fact, the village has no obvious archaeological visibility—no shards could be seen at the surface whatsoever—and it could not be found without the help of the Tariana (fig. 5.5). Excavation found evidence that matched the oral tradition, showing that two semicircular defensive ditches were dug around a village full of buried ceramic remains that were not visible on the surface (see fig. 5.6). Radiocarbon and ceramic samples collected from the site show that it was occupied around late fourteenth or early fifteenth century AD, indicating that the conflicts narrated in the oral tradition happened before the beginning of the European conquest.

The size and shape of the site show that the ditches surrounded a longhouse, roughly fifteen by fifty meters, that was occupied for a short interval. Oral tradition states that the immediate reasons for conflict between the Tariana and other groups in the area were fights over women. (Bouapés ordered the killing of Arara women who insisted on seeing secret, all-male Jurupari rites.) These narratives indicate that in the beginning of the fifteenth century AD, at the time the spiked-trench

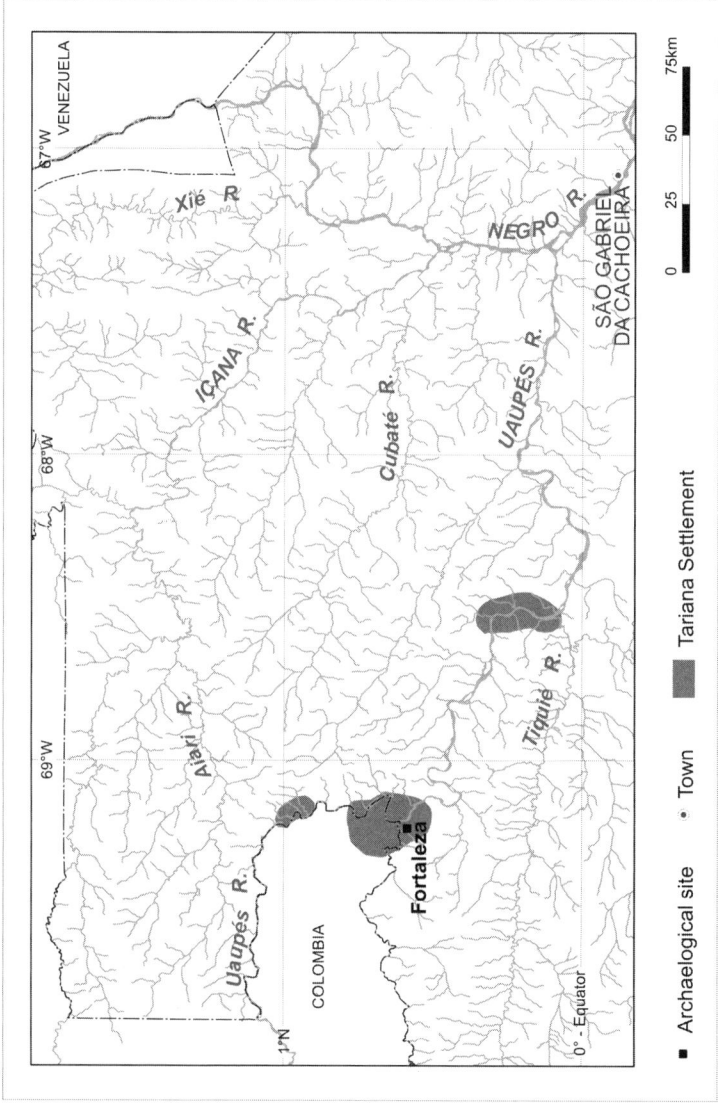

5.5. Map of the Tariana area.

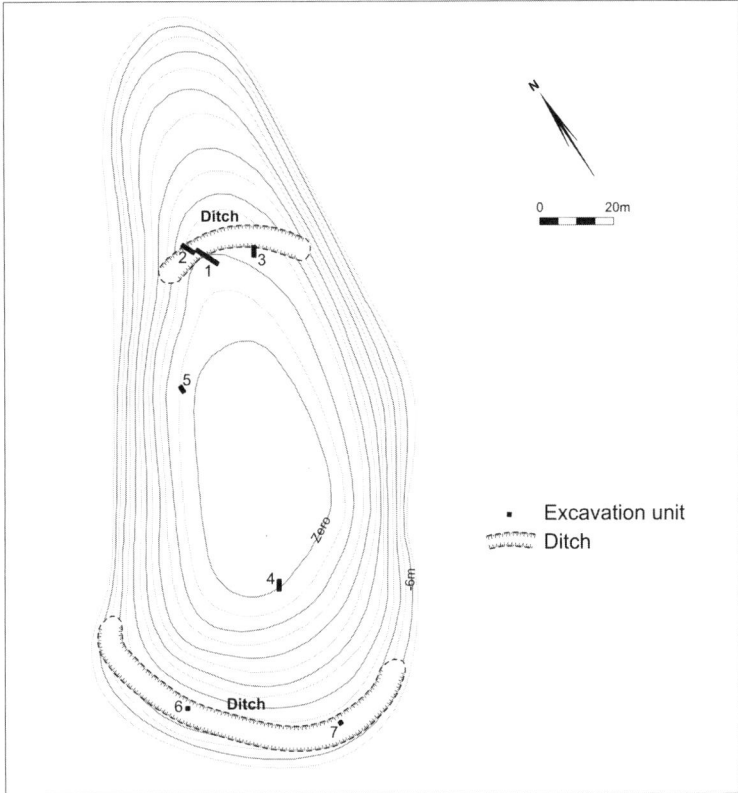

5.6. View of the excavation and site plan at Fortaleza.

fortress was built on the Juruparí hills at the Fortaleza site, the Tariana were already incorporated into the regional system, since they were intermarrying with Arara women. Eventually, with the end of hostilities, the Tariana resettled near the Uaupés River and started again to marry Uanana and Tukano women, among others.

Interpretive Discussion

The archaeological evidence presented above attests to the incidence of armed conflict—in a scale demanding investment in the construction of defensive structures—in different parts of the Amazon basin during precolonial times. Adding to the examples discussed here, there are other known cases worth briefly mentioning. Work in the upper Xingu—in the southern tip of the Amazon—confirms the same trend evidenced in the upper Rio Negro and the central Amazon, as is attested by the recent excavation of large earthworks resembling defensive structures surrounding ring villages dating back to the thirteenth century AD (Heckenberger et al. 2003). Recent fieldwork at the confluence of the Beni and Madre de Dios rivers in Bolivia has also uncovered fortified structures—dating to the second half of the fifteenth century and the early sixteenth century AD—that a Finnish-Brazilian research group suggests could be associated with an Inka foray well into Amazonia (Pärssinen et al. 2003a:71). The same research group has mapped walled geometric earthworks in the Acre state, also in southwestern Amazonia, but it is not clear whether or not these were defensive structures (Pärssinen et al. 2003b:97–133). In the same area, on the Bolivian side of the border, several other sites with moats were also identified (Arellano López 2002:68). Finally, in the Uaçá-Oiapoque area, on the border of Brazil and French Guiana and north of the mouth of the Amazon, recent fieldwork among the Palikur Indians has uncovered defensive structures. The Arawak-speaking Palikur say these are related to their skirmishes with the Carib-speaking Galibi, both groups being currently neighbors in the Uaçá reservation (Fordred-Green et al. 2003). The structures have not been dated, however, and there is a chance that they were built after the beginning of the Colonial period.

Hence, although the establishment of "tribal zones" in the Colonial period (sensu Ferguson and Whitehead 1992; Whitehead 1990b) intensified and somewhat changed the logic of warfare in the Amazon and elsewhere in lowland South America, there was armed conflict there well before the sixteenth century AD. It remains, however, to explain the causes, mechanisms, and consequences of indigenous warfare in precolonial Amazonia. One potential explanation might be found in hypotheses proposed to explain the origins, development, and dynamics of chiefdoms. As already mentioned, Carneiro (1990:192; 1995) has suggested that, in areas such as the Amazon floodplain, warfare would emerge as a consequence of conflicts for access to more productive areas, in addition to as a means to exert control over local communities.

Elsa Redmond (1994:123) has also called attention to the role of warfare in the development of political centralization. From her perspective, warfare has a key role in the consolidation of regional hierarchical
social formations (Carneiro 1987:246; Earle 1997; Flannery and Marcus
2000).

The evidence presented here, however, does not seem to fit this general model. In the case of the Lago Grande site, for instance, the construction of the defensive ditch during the Paredão-phase occupation
preceded site abandonment. Interestingly enough, though, the site was
never to be occupied again after abandonment by the Paredão, even if
it is located at a highly productive spot next to a floodplain lake of the
Amazon. Moreover, after the abandonment of the Lago Grande site,
the central Amazon remained occupied for almost another five hundred years before the arrival of the Europeans. Conversely, at the Açutuba site, conflict seems to have been much more intensive, as attested
to by the construction of a 150-meter-long moat with a double palisade. Surprisingly, however, Açutuba is located at a much less productive spot, next to the nutrient-poor Rio Negro. Indeed, what one knows
so far about the precolonial history of the central Amazon challenges
utilitarian thought, since several sites located next to highly productive
spots were occupied more intermittently and abandoned earlier than
sites such as Açutuba (Neves and Petersen 2006). Warfare in the central
Amazon, therefore, cannot be explained as conflict resulting from the
desire for control of more productive areas.

In the Tariana case, there could be a stronger drive that might
explain the wars they launched against the other groups settled in the
area: their own movement into the Uaupés from a northern homeland.
What is interesting in this case, though, is that after successfully establishing themselves by force in the area, the Tariana rapidly melted into
the regional social systems following the principle of exogamy, taking
wives from other groups. Indeed, oral tradition suggests that exogamy
was already going on even before the conflicts began. Moreover, there
is no historical or archaeological evidence that military success entitled
them to a hegemonic political standing in the Uaupés, or even the control of multi-village polities. The same pattern was recorded among the
Hohodene Baniwa, also in the northwest Amazon (Wright 1990).

It appears, therefore, that practical reason, economic rationalism, or arguments on scarcity do not provide good explanations for the examples of warfare discussed here. First, there seems to be no single reason for warfare. This is as valid among contemporary indigenous Amazonian societies, such as the Parakanã (Fausto 2001:261), as it was for the Carib from 1500 to 1820 (Whitehead 1990:168). Second, both contemporary and precolonial warfare in the Amazon seem not to have been related to conquest, strictly speaking. Third, both presently and in the past, there are no signs of captive taking for the use of slave labor. There are few clear signs of intensive mobilization of labor in the productive activities of precolonial Amazonia (see Clark Erickson [2000] and Michael Heckenberger et al. [1999, 2003] for possible exceptions).

Thus, the explanation for the reasons, mechanics, and outcomes of Amazonian warfare need to be sought in the political realm, more specifically in the internal politics of the different social formations, where persons, more than groups, may have been key players. To develop this argument, it may be useful to turn to the ethnography of the contemporary indigenous societies of Amazonia. There is a theoretical trend in lowland South American ethnology that focuses on the body as the main locus of identity among indigenous societies (Viveiros de Castro 2002a:387). For Carlos Fausto (1999:933), "Amazonian societies are primarily oriented towards the production of persons, not material goods; that is, their focus is not the fabrication of objects through labor, but of persons through ritual and symbolic work." From this perspective, worldviews are marked by a sort of "multinaturalism," an ontology that sees animals and human beings as constituted by an identical and basic form of "culture." In this sense, "animals are people or see themselves as people. Such conception is always associated to the idea that the manifested form of a species is an envelope ('a piece of cloth') hiding an internal human form, normally visible only to the eyes of the own species or of certain trans-specific beings such as shamans" (Viveiros de Castro 2002:351). In other words, these are identities built up on notions of flow and transformation, not on permanence. Thus warfare, as well as all the ritual and logistical activities associated with it—captive taking, killing, and cannibalism—entail the production and reproduction of identities through the consumption—literal or not—of bodies (Fausto 1998). These new identities manifest themselves in one way through new names, including changes in status and leadership. Among the Parakanã and the Yanomami, for instance—placed widely apart in both eastern and northern Amazonia—it is common for more than one person to receive the credit for killing a single enemy in battle, even though that enemy may have been well dead when the shooting actually happened (Albert 1989; Fausto 2001). Warfare therefore allows for the social reproduction of these societies (Fausto 2001:326).

Going back to precolonial Amazonia, the notion of the body as a stage for constant flow and transformation is also visible in the iconography of artifacts. In both pottery and lithics from different areas, figures are represented in half-human, half-animal form, typically with human bodies and animals' heads. In some cases, representations are metonymic, animal features being represented by, for instance, fangs or stingers. In other cases, such as among Tapajonic pottery, transformations manifest themselves through so-called "dual figures," which change form from human to bird, and vice-versa, depending on the angle from which they are seen (Gomes 2001). Representations of flow and cycle are also present in other groups of artifacts, whose forms and iconography relate to themes associated with human reproduction, most often pregnant women and genitalia. Some of the best-known examples of this can be found in Marajoara pottery, in which anthropomorphic funerary urns explicitly portray pregnant women—as indicated, for instance, by protruding bellies—or by anthropomorphic statuettes with clearly phallic forms representing women with genitalia exposed. The idea of a cycle is then brought about by the conjunction, in the same objects, of representations of death (funerary urns) and birth (pregnancy), femaleness (genitalia) and maleness (phallic forms), and so forth (Neves and Dias 2003; Schaan 2001).

Parallelisms of notions of identity between precolonial and contemporary Amazonian Indians warrant the use of ethnographic analogies. The logic of precolonial warfare can be examined using an understanding of the identity-building process discussed here. In the past, as in the present, causes and pretexts for precolonial warfare in Amazonia resulted from the actions of persons who—assuming important roles in the organization and leadership of warfare—could benefit from the symbolic capital attained by such a system of productive consumption. This would explain the apparent lack of material benefits or wealth accumulation in the contexts examined.

Conclusions

More than thirty years ago, Carneiro (1970) called attention to the role of warfare in the emergence of hierarchical social formations such as chiefdoms. It is likely that this could have happened in the Amazonian past. However, Carneiro himself has noticed that, in some instances, warfare could have had a centrifugal role, leading to the pulverization of social formations. Expressed through that statement, his ideas mirror those of Pierre Clastres, who in books such as *Society against the State* (Clastres [1974] 1978) proposed that the very atomization of lowland South American social formations created conditions that precluded the constitution of long-term, supralocal socio-political units (chiefdoms in the sense used by Flannery and Marcus 2000) and, beyond that, the emergence of the state. In precolonial Amazonia, warfare may have reified hierarchies and helped to establish leadership because persons successful in leading and/or defending against aggression would have gained access to broader roles. In such cases, warfare would act as a centripetal force. Warfare, however, could also have had a centrifugal role, leading to decentralization and preventing the establishment of long-term hierarchical social formations. The archaeological data presented here support this argument when showing that there is weak evidence for the establishment of long-term stratified supralocal political formations in the cases studied.

Summing up my argument, the different dimensions of warfare could have had an important role in the consolidation and dissolution of power in precolonial Amazonia. The fluidity in the consolidation and dissolution of political centralization seen in the archaeological record follows conceptions of self and identity among indigenous societies of the Amazon, verified both ethnographically and through the iconography of precolonial artifacts.

This hypothesis mirrors statements from the ethnography of contemporary indigenous societies in Amazonia, through which one can verify the co-existence and alternation of two complementary models of social organization: one hierarchical and other egalitarian (Fausto 2001:533; Hugh-Jones 1995; Viveiros de Castro 2002b). At the root of this constant alternation lies the pulverized structure of productive economies in precolonial Amazonia (Neves and Petersen 2006). Turning back to the problem posed at the beginning of this chapter, then, it can be proposed that despite the large demographic, social, cultural, political, and historical changes brought about by the European conquest, there is something essentially similar between what one sees in the archaeological record and the ethnography of Amazonian Indians. Such similarity has its foundations in a basic worldview in which identity is based on change, not stability.

Acknowledgments

I would like to thank Axel Nielsen and William Walker for the invitation to participate in this symposium. Fieldwork in the central Amazon, upper Rio Negro, and Uaçá has been funded by grants awarded by the Fundação de Amparo à Pesquisa do Estado de São Paulo (FAPESP) (grants 99/02150-0 and 02/02953-7), the Wenner-Green Foundation for Anthropological Research, the universities of Maine and São Paulo, the William Hillman Foundation, and the National Science Foundation (DBS 9223763). Thanks to the owners of the lands where we directly worked—people from the village of Santa Maria, notably Mr. Pedro de Jesus Gomes, Mr. Yodi Ideta, Mr. José Ricardo, Mr. Adilson, and his wife Elaine. Alessandro Barghini kindly provided the source for Vespucci's quote. A lot of the ideas presented here have been discussed over the years with Jim Petersen. Thanks to the staff of the Amerind Foundation for a warm welcome and wonderful stay and to the group of colleagues attending the seminar. Carlos Fausto, Axel Nielsen, William Walker, and two anonymous reviewers provided insightful criticism that contributed to improving this text.

Warfare and Political Complexity
in an Egalitarian Society

An Ethnohistorical Example

Polly Wiessner

Warfare might be minimally defined as socially organized aggression. However, in egalitarian societies in which every man is his own master, organizing war is a General's nightmare. Decisions to go to war must be made by consensus, for no man has authority over others. The actions of all men—right or wrong—must be defended and their deaths avenged, regardless of whether or not such actions derail the original objectives of the war. When the spoils of war are secured, they must be divided among warriors. Finally, warriors are free to pursue their own personal agendas in the arena of warfare, even when these agendas depart from group interest. Individual motives are diverse: to restore honor, to make a name in warfare, to settle old grudges in somebody else's war, to bond with fellow warriors, to win a small piece of land to loot, to profit from ensuing exchanges with allies or enemies, or simply to experience the excitement of battle. As a war unfolds, emotions boil and individual goals may be altered with every turn of events. Because the emotional glue that binds warriors together is composed of fear, brotherhood, and honor, material objectives quickly take second place. The result is often socially disorganized aggression bearing long-term negative repercussions. As an Enga proverb cautions: "The blood of a man does not wash off easily."[1]

The discounting of the cultural context and historical forces that shape warfare has permitted explanations used to account for war in "complex societies" to be applied to "simple" societies—population pressure, attempts at centralization, subjugation of opponents, or coordinated efforts to procure scarce resources. A classic example is Mervyn Meggitt's (1977) thesis that the Mae (Mai) Enga fight over land in response to population pressure. Using population pressure on land as an independent variable, Meggitt (1972:116) positions efforts in ritual and exchange as part of the larger project of defense: "The basic preoccupation of the Mae is, it seems to me, with the possession and defense of clan land. Participation in the Te and other ceremonial exchange is but a means to this end."[2]

However, Meggitt himself expresses reservations in his epilogue, where he admits that land is not really short in Enga, but only perceived to be. And despite the primacy that he places on warfare, Meggitt (1972) acknowledges that military prowess does not help a leader achieve and maintain position, for the heart of Enga interest lies in ceremonial exchange. He then goes on to aptly title his book *Blood Is Their Argument*, not *Land Is Their Argument*. Subsequent studies in central Enga indicate that Meggitt's inclination was correct: land is not their argument (Lakau 1994; Wormsley 1985; Young 2002; Wiessner and Tumu 1998; Wiessner 2006).

In this chapter, I will first revisit Meggitt's explanations of warfare using a practice approach. The strength of the practice approach is that it situates warfare within a historically constituted field of action where actors are guided by cultural institutions and their corresponding incentive structures. It switches the focus of analysis from population pressures to the strategies of agents working within cultural and historical contexts. The result is an understanding of Enga warfare that conforms much more closely to subjective explanations of warfare given by the Enga and to quantitative data on the changes in the causes, the courses, and the outcomes of wars over time.

But a practice approach can only take us so far, for in addition to the operation of historically specific processes, some general principles should be applicable when agents work within certain cultural structures or institutions that shape their agendas and actions. After all, as Margaret Archer (2000) has pointed out, practice is rooted in the body's encounter with the social and natural world. One weakness of a practice approach is that it does not provide a framework for comparing the impacts of widely found cultural institutions on warfare across cultures. For intercultural comparisons, it is necessary to have some understanding of the advantages and drawbacks of certain cultural institutions, in this case egalitarian systems, in order to grasp when and why actors chose to stick to certain practices or to engage in actions that transformed those practices. If such an understanding cannot be achieved, then studies of war will provide little more than historically unique examples. For this reason, I will draw on insights from institutional economics (North 1990) to understand egalitarian institutions and to explore their impact on warfare, focusing on three questions: When and why did actors continue to practice and enforce egalitarian relations? Did egalitarian "rules of the game," coupled with basic human strategic interests, have a predictable impact on the potential of warfare as a prime mover towards political complexity? Finally, under what conditions do actors abandon egalitarian practice in the context of warfare in favor of hierarchy?

Institutional economics and a practice approach have many points in common. They both assume that agents act pragmatically, rationally, and in a self-interested manner within a culturally constructed incentive structure. They both see institutions or "the rules of the game" as having benefits in increasing the predictability of human action and interaction; however, neither sees institutions as maximally efficient. They both propose that history matters and context counts because the decisions of the present are made within institutions or structures built in the past. They both recognize that the rules of the game are not the game and see change as the cumulative result of the actions taken by agents at the margins who bend the rules or play them differently. Institutional economics differs from a practice approach in that it provides a theoretical framework to understand the role of social institutions and generate hypotheses concerning when actors will seek to and succeed in changing the rules of the game. A practice approach situates the dynamics of institutional change in a specific cultural and historical context. Here I will draw on both.

Background

The material presented here comes from a ten-year study carried out by myself, Akii Tumu, and Nitze Pupu among the Enga of Papua New Guinea. The Enga are a highland horticultural population who live at altitudes of 1500 to 2500 meters, cultivating the sweet potato and a wide variety of other crops (Feil 1984; Gordon and Meggitt 1985; Meggitt 1965, 1972, 1977; Talyaga 1982; Waddell 1972). Their staple crop, the sweet potato, which was introduced along local trade routes some 250 to 350 years ago, is cultivated intensively to feed large human and pig populations. Enga homesteads, comprised of a men's house and one or more women's houses, are scattered widely over the landscape and nestled between gardens and stands of casuarinas trees.

The Enga population, which numbers approximately 350,000 today, is divided into a segmentary lineage system composed of phratries or tribes, clans, subclans, and lineages (Meggitt 1965). At the time of Meggitt's studies in the 1950s and 1960s, tribes were composed of between 900 and 5,400 people divided into an average of 7.8 clans with a range of 100 to 1,000 members. As the population grows, clans fission into independent political units, usually after intraclan warfare. Patrilineally inherited clan membership furnishes a pool of people who cooperate in agricultural enterprises, defense, procurement of spouses, raising wealth for a variety of payments, and in the past, holding ceremonies for spirits and ancestors. Affinal and maternal ties, established by exogamous marriage and maintained by reciprocal exchange, provide access to resources and assistance outside the clan. Clan members see the wide range of affinal and maternal ties as enhancing the clan's strength. Except in times of warfare, when affines may be on the enemy side, there is little conflict between loyalties to affinal and agnatic kin.

Enga women devote themselves primarily to family, gardening, and pig husbandry. During warfare, they retreat with their children and pigs to the clans of relatives, where they are immune to violence. Men also invest effort in subsistence production, but this is in addition to engaging intensively in the politics of warfare and exchange and the pursuit of "name" or reputation. Ceremonial Tee exchange involving group distributions of pigs, shells, and other goods and valuables on the clan's ceremonial ground is the propelling force in Enga society; success in Tee exchange is the culturally defined road to status. Because labor shortages constrain the potential of households to amass large amounts of wealth for distribution, families depend heavily on affinal and maternal kin outside the clan to finance exchange.

In contrast to exchange, warfare is deplored by the Enga and seen as a last resort to solve problems because it takes the lives of many men; destroys houses, land, and gardens; and disrupts affinal and maternal kin ties. Meggitt (1977) estimates that 25 percent of Enga men died in warfare prior to contact with Europeans. According to Enga historical traditions, warfare has always been prevalent; there was never

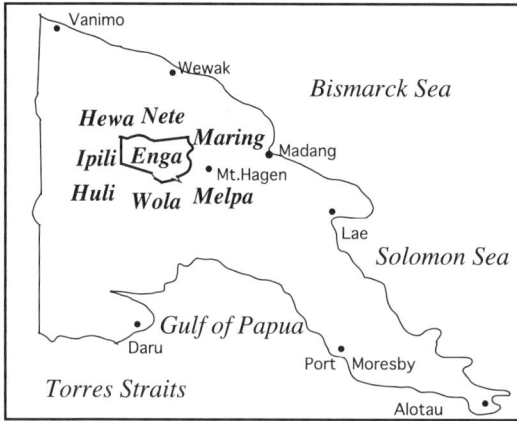

6.1. Location of the Enga province in Papua New Guinea.

a "time before warfare." However, the Enga were not perpetually at war. In the decades prior to contact with Europeans, central Enga clans fought approximately once every five to ten years (calculated from Meggitt 1977), with wars lasting anywhere from days to weeks before participants sought peace settlements. Some clans fought much more frequently than others.

The data used here come from the oral historical traditions of 110 Enga tribes (fig. 6.1) (Wiessner and Tumu 1998). The Enga have rich and detailed oral historical traditions, held distinct from myth, that are said to have originated in eyewitness accounts. Oral historical traditions contain information on subsistence, wars, migrations, agriculture, the development of cults and ceremonial exchange networks, leadership, trade, environmental disasters, and fashions in song and dress. They cover a period that begins just prior to the introduction of the sweet potato (250 to 350 years ago) and continues until the present. Accompanying genealogies allow events to be placed in a chronological framework (for descriptions of our methodology, see Wiessner and Tumu 1998; Wiessner 2002).

The period covered by oral traditions was one of rapid change when the sweet potato, imported through local trade routes, released constraints on production and made substantial surplus production possible for the first time. After the sweet potato took root and pigs were produced in surplus, there were rapid developments in ceremonial exchange, religious cults, and warfare to take advantage of new opportunities and deal with the problems of a changing social and economic landscape. Three large ceremonial exchange systems arose: the Kepele cult network of western Enga, the Great Ceremonial Wars of Central Enga, and the Tee Cycle of the eastern Enga. Gradually, they merged to culminate in the Tee Cycle. By the mid-twentieth century, the Tee Cycle involved some forty thousand participants and the exchange of tens of thousands of pigs and valuables (Wiessner and Tumu 1998, 1999; Wiessner 2002: fig. 2).

Egalitarian Relations, Exchange, and Warfare

Egalitarian institutions are found in many societies. They are characterized by equal access to natural resources, social assets, and status positions, as well as the autonomy of individual decision making. Following Morton Fried's (1960) classic definition, there are as many positions of leadership in egalitarian societies as there are qualified individuals to fill them. Egalitarian relations foster autonomy of decision making, recognition of moral equality, an attitude of respect for the abilities of others, and appreciation of what each person has to contribute. They are usually found in less complex societies, though complex "egalitarian" societies do exist—for example, the Igbo of Nigeria (McIntosh 1999; Uchendu 1965) or the Iroquois of the northeastern United States (Trigger 1990).

Egalitarianism is not the blank slate on which complexity is written; inequality has deep roots in human phylogeny (Boehm 1999) and egalitarian relations must be actively and constantly maintained. All egalitarian systems offer people potential equality—that is, equal access to social and natural resources at the starting line. However, because they are the product of social institutions, they vary greatly in their scope, duration, and nature, just as hierarchical institutions do. For instance, they may encompass people of all ages and both sexes or they may operate within age and sex groups. They may have a temporal dimension, with inequalities being recognized during limited periods or specific events. In some egalitarian societies, individuals are given equal access to resources and encouraged to use these to achieve status by providing benefits for the group (Flanagan 1989; Robbins 1994). In others, primarily foraging societies, people do not give formal status recognition to those who excel, even when they do provide group benefits.

Whether or not achieved status is socially recognized, in no egalitarian societies can the more capable infringe upon the autonomy of others, appropriate their labor, or tell them what to do. Power does not reside in the individual or in a position, but rather in the flow of goods and services. In societies where status can be achieved, the demise of leaders who no longer provide group benefits is rapid. There are a number of ways in which status and influence are achieved in egalitarian societies (Chowning 1979; Roscoe 2000): (1) economic enterprise in production (large-game hunting, gardening, and raising livestock); (2) mediation; (3) social competence in organizing events; (4) knowledge of land tenure, genealogy, history, or ritual; (5) skill in warfare; and (6) control of trade and exchange. Each of these ways has its limits. Broader and more durable differentials in power and influence take shape through the application of integrative skills by leaders who conjoin several areas of leadership to construct and validate more complex institutions or social movements.

The costs of maintaining egalitarian institutions are high and require constant vigilance and repressive leveling measures, such as verbal or physical punishment, witchcraft, withdrawal from reciprocal relations, or ostracism (Boehm 1999; Lee 1993; Kelly 1993; Wiessner 1996). Moreover, individual and economic initiatives may be curbed by excessive leveling and group decision making hamstrung by an inability to reach consensus. Why do actors engage in practices that reproduce egalitarian institutions?

Institutional economics (North 1990) provides some insights into why actors might choose to reproduce egalitarian institutions. Institutions following Douglas North (1990:3) are "the rules of the game" in a society, or, more formally, the humanly devised constraints that shape human interaction. Exchange, whether social or economic, is costly because of the many uncertainties of human interaction that generate transaction costs—for example, assessing what somebody has to give and will indeed give, protecting rights for reciprocation, and enforcing agreements. Institutions contribute to setting the incentive structure of a society by determining the opportunities that are open to individuals and by influencing costs and benefits. Institutions vary widely in their efficiency and consequences for economic performance.

Egalitarian institutions and their accompanying ideologies do much to reduce the transactions costs of exchange. For societies in which the social and economic relations are intertwined, egalitarian relations standardize important information: that partners are social equals. As equals, partners can ask and receive when in need with the trust that assistance given will not be used to dominate. Equality permits all individual group members, as equals, to defend their rights and be defended by kin, reducing the costs of punishment for any single person. Equality fosters the trust required for delayed or long-distance exchange and it facilitates mobility required to organize intergroup enterprises because hierarchies do not mesh easily.

Enga Egalitarian Institutions and Their Impact on Warfare and Exchange

Among the Enga, egalitarian institutions stipulate potential equality within the sexes but not between the sexes. All married Enga men hold equal rights to be granted land, to allocate household labor and its products as they see fit, to receive support from group members in procuring spouses, to be protected by the clan, to have a voice in decision making, and to pursue status. Women enjoy equal status and rights relative to other women. Egalitarian relations and ideals among the Enga apply to potential equality but not to equality of outcome (Robbins 1994). Although initially defined as equals, men are encouraged to excel and achieve status via actions that are perceived to benefit the clan. Given the tension between potential equality and achieved status, potential equality has to be continually maintained. People first tackle affronts to their equality, and when they cannot manage alone, the lineage, clan, or subclan members lend support. All people are thus heavily dependent on the group to help maintain their rights, producing conflicting moralities. The one espouses individual assertiveness, while the other espouses sociality, solidarity, loyalty, and willingness to sacrifice for the group. Serious insult to or injury against one clan member is thus regarded as an offense against all and a challenge to clan honor.

Egalitarian structures differentially impact Enga warfare and exchange. They greatly facilitate cooperation in social and economic exchange for a number of reasons. Enga offer assistance to both agnatic and affinal kin in agricultural enterprises, raising wealth for ceremonial exchange, defense, and other community activities knowing that, as equals, they will be able to request and receive assistance when in need. Equality alleviates fears that any assistance given will be used by others to build position and subordinate the person being helped. For Tee ceremonial exchange, the assurance that a person will be received as an equal facilitates the mobility that is so crucial to the organization of the flow of wealth. Tee exchange involves temporary imbalance and significant time delays; equality between exchange partners is crucial to foster the trust that wealth given will be repaid. It is unlikely that the great exchange networks that emerged in Enga history could have been constructed outside of the matrix of balance and trust fostered by equality. There is ample evidence in historical traditions that Enga leaders were aware of the importance of equality for the development of exchange, because as real economic inequalities increased through time, leaders made efforts to develop or import bachelors and ancestral cults that underwrote equality of men (Wiessner and Tumu 1999). While egalitarian institutions effectively reduced the transactions costs of interindividual and intergroup exchange, they also made decision making cumbersome and fostered runaway competition for leadership that disrupted clan unity.

The very same egalitarian structures that facilitated the expansion of ceremonial exchange inhibited the use of warfare as an effective tool for achieving broader political goals. Testimonies of Enga elders indicate that their grandfathers were aware of the potential of warfare to bind larger political units, strengthen their leadership, and forge exchange ties with allies that would give them greater access to wealth (see also Sillitoe 1978). According to historical traditions, big-men (*kamongo*) sometimes tried to manipulate wars to pursue personal agendas, but these attempts were only successful in isolated cases. The overall rejection of authoritative leadership, the diverse interests of fighters, the unanticipated events of battle, and the desire for revenge to achieve parity often propelled wars into vicious revenge cycles with unintended outcomes (see also Sackschewsky et al. 1970; Young 2002).

The opposing effects of egalitarian relations on warfare and exchange raise interesting questions concerning the development of complexity: Given the new opportunities and disruptions after the introduction of the sweet potato, did the chaos caused by weakly organized warfare lead to hierarchical developments in the organization of warfare itself, or did egalitarian relations become insufficient for organizing large-scale exchange, resulting in hierarchical organization within some arenas of exchange?

A Brief History of Enga Warfare and Exchange

Wars, both large and small, are reported from the very beginning of Enga historical traditions some three hundred years ago. Enga historical traditions do not valorize war or war heroes. They tell of failures as well as successes, and both the winners and losers recount similar events. The Enga deplore war as a last resort to solve problems. War is depicted in both oral historical traditions and current accounts as a way to restore balance and honor after insult or injury. For this analysis, eighty-four Enga wars recalled in some detail in oral historical traditions were ordered by generation and coded for triggering incidents, size, degree of viciousness, approximate duration, and outcome.[3] This sample can be seen as representative of serious wars that had lasting political impact; minor wars that ended in standoffs are unlikely to be remembered in oral historical traditions. I will divide precontact Enga warfare into three periods:

Period I: Eight to Twelve Generations Ago

In the generations just prior to the introduction of the sweet potato and shortly after, Enga historical traditions describe the population as sparse, life as lonely, and spouses as hard to find. Residents of higher altitudes were dependent on hunting and gathering to make a living, and those at middle altitudes and lower altitudes relied on a mixed economy of horticulture and hunting. At the center of historical traditions was the trade of salt, axes, stone, cosmetic oil, and items for ceremonial dress. Small-scale ceremonial exchanges organized by lineages, subclans, or clans were carried out, including modest bridewealth exchanges, funeral feasts, and war reparation payments to allies.

Both skirmishes and full-blown tribal wars are reported for this period. Land was plentiful during this time, and the losing party usually settled on nearby vacant land or was welcomed by relatives in other areas who sought to increase their numbers. Tribal fights appear to be primarily the result of mounting tensions in communities that had grown too large to cooperate. When disputes subsequently broke out over hunting rights, meat sharing, or organizing events, conflicts escalated into wars that split communities into manageable sizes. Conflicts leading to migration of the losing party were also reported at this time. More often than not, these conflicts were described as the unintentional outcome of runaway violence, with the victor expressing regret at losing a "brother" group after a seemingly trivial conflict.

Period II: Six to Seven Generations Ago

The introduction of the sweet potato released constraints on production, made agriculture possible in high altitudes, and allowed people to produce a surplus of pigs. Subsequent generations saw substantial shifts in population as people sought to take advantage of the new crop. Groups that formerly relied upon hunting and gathering in higher altitudes moved down into the valleys, where they were hosted by relatives and were able to settle down to agriculture. Within one or two generations of these moves, the immigrants were often involved in war with their hosts following conflicts triggered by theft, homicide, political disputes, rape, and gardens.

The large wars of this period appear to have been some of the most vicious in Enga history. Though they began with seemingly trivial incidents, they soon escalated into runaway violence, indicating histories of underlying tension that are not well documented in oral historical traditions. For the largest wars, tracts of land covering as many as 160 square kilometers were vacated. The wars of the sixth and seventh generations greatly altered the social landscape of the Enga. Defeated Enga clans fled deep into the Kompiam district to Maramuni and to Porgera, where they occupied empty land or displaced non-Enga groups and prospered (fig. 6.2). It may be at this point in Enga history that the association between warfare and loss of land developed. Land, which represents both pride and independence in Enga, was taken at the end of vicious wars to humiliate the losers. Because most wars were not instigated with the intention of land acquisition, the victors were hard pressed to fill land gained and had to invite relatives and allies to help occupy the vast areas of land gained in some wars.

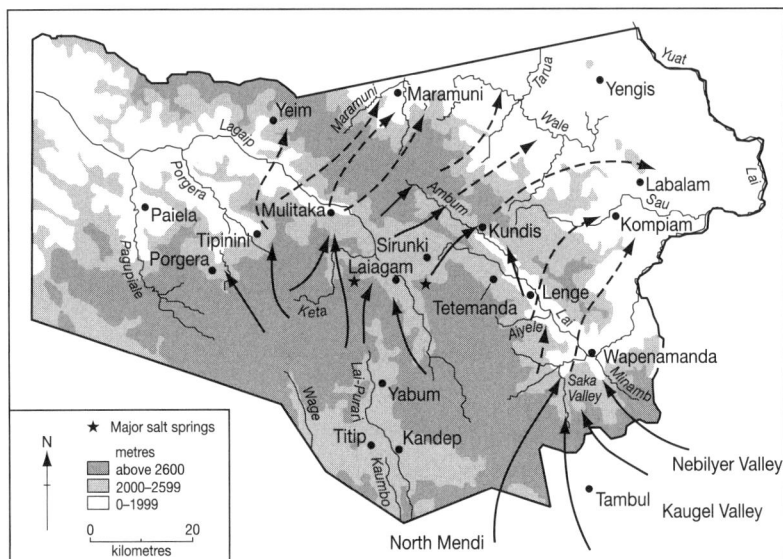

6.2. Schematic representation of major migrations in Enga after the introduction of the sweet potato. The situation is much more complex than illustrated here. From the earliest generations of historical traditions until approximately the fourth generation before the present, we recorded 270 migrations of entire clans or large segments of clans.

During this period, as in the proceeding one, warfare was not directly integrated with production, ritual systems (beyond specific rituals for warfare), or ceremonial exchange. However, it was out of the chaos of the wars of the sixth and seventh generations that the three great ceremonial exchange systems emerged out of attempts to put order to chaos: the Tee Ceremonial Exchange Cycle, the Great Ceremonial Wars, and the Kepele Cult Network (Wiessner and Tumu 1998; Wiessner 2002).

Period III: Four to Five Generations Ago
(ca. 1885–1915)

Around the fourth to fifth generation before the present, much of the land in the major valleys had filled and the large networks for ceremonial exchange flourished, drawing entire clans and tribes into coordinated exchange. At this time, homicide, political disputes, and quarrels over gardens became the predominant triggering incidents for wars. While causes and initial courses of wars of this period appear to have shared many characteristics of preceding ones, the outcomes did not. Traditions tell of wars contained by rules, the return of land taken by opponents, and the establishment of peace through the exchange of pigs, goods, and valuables.

With the institution of peace procedures, clans could fight, exchange wealth, and stay put. There appear to be two motivations for the development of peacemaking efforts. First, the land was filling up and relatives were not as eager to absorb displaced clans. Second and most importantly, opponents did not want to expel neighbors, often brother clans, who provided valuable partners in Tee exchange. It was labor, not land, that was short, and wealth to finance ceremonial exchange had to be obtained on credit from neighbors.

During the fourth to fifth generation, restoring balance, harmony, and clan honor in the face of insult or injury became a central goal in Enga warfare. If a clan could not maintain its "name," chances of attracting investment from other clans declined. The importance of restoring clan honor is expressed in the following quote:

> Now I will talk about warfare. This is what our forefathers said: When a man was killed, the clan of the killers sang songs of bravery and victory. They would shout, "Auu" ("Hurray" or "Well done") to announce the death of an enemy. Then their land would be like a high mountain (manda singi) and that is how it was down through the generations. The members of the deceased's clan would become small (koo injingi). They would be nothing. But when they had avenged the death of their clansman then they would be all right. Their hearts would be open (mona lyangenge). In other words, when one fights and takes revenge for the death of a fellow clansman, then one gets even and back on equal footing. (Tengene Teyao, Yakani Kalia clan of Wakumale, Wabag)

Warfare thus reestablished balance and mutual respect after insult or injury and created a social matrix of equality in which ceremonial exchange could flow. In this sense, warfare had become the handmaiden of exchange.

The impact of peacemaking is reflected in figures on migration. In the sixth and seventh generations before the present, seventy-one clans of central and eastern Enga migrated into other areas after warfare with an average migration distance of twenty-six kilometers per clan. In the fourth to fifth generation, there were only twenty-nine migrations of entire clans to other areas after warfare, with an average migration distance of seventeen kilometers per clan. For the second to third generations, we recorded only five migrations of entire clans out of their homelands to other areas after warfare, with an average migration distance of fifteen kilometers per clan (Wiessner 2006). In all periods, numerous families or entire lineages (*akalyanda*) left their clans of origin to settle with relatives in other places.

We did not collect systematic data on warfare for the period between ca. 1915 and 1945 so as to not make people concerned that the goal of our studies was to provide evidence for land litigation. In the information that we do have for the period spanning 1915 to 1940, there is little indication that triggering incidents, courses of war, and outcomes differed greatly from those of the preceding two generations. Meggitt (1977) presents data on causes of fights in central Enga between ca. 1900 and 1950, but he did not use genealogies to systematically date or sequence Enga wars. Meggitt destroyed his field notes, so it is not possible to reanalyze his entire data set; however, we found that the wars in his sample that we could trace—for example, the five wars in which entire clans were permanently evicted from their land—occurred ca. 1820, not between 1900 and 1950. The wars in his sample can thus only be considered to represent "wars of the past." Otherwise, Meggitt's (1977) description of Enga warfare is excellent and probably applies to most wars fought from approximately the fifth generation until the early Colonial period.[4]

Meggitt's Thesis Revisited

Meggitt's thesis, derived from cultural ecology, contends that Enga warfare is driven by land shortage. A practice approach that situates warfare within the context of history, institutions of exchange, and their incentives structures gives quite another perspective. Amongst other things, it supports the Enga explanation for warfare: that warfare is a last resort to solve problems and restore balance of power between groups so that enchained exchange can flow.[5] The balance of power hypothesis better fits available data than Meggitt's hypothesis on a number of accounts. First, it is labor, not land that is short in Enga. Low production due to labor shortage is offset by wealth received on credit from neighboring tribes to finance ceremonial exchange. Warfare disrupts essential ties with neighbors, reducing the amount of wealth available for distribution. Second, warfare is not valorized in Enga as one would expect if warfare were over resources critical to survival. Third, when land is taken at the end of vicious wars, the intent is usually to punish the loser. It is left as barren no-man's-land for years or even decades and thus is of little immediate value to the victors. Land taken in warfare is disputed for generations. Fourth, the response to the havoc created in the sixth and seventh generations by warfare was the institution of peacemaking processes so that clans could fight, make peace, stay put, and resume profitable exchange. Had the goal been to obtain the land of neighbors, one might expect that developments would have taken another course—for example, the formation of larger social units to procure and hold land. Instead, Enga warfare continued to split large clans into smaller cooperative "brother" clans, who fought, but only occasionally. Fifth, if land shortage were the cause of warfare, then one would expect conflicts over land to increase through time. This is not the case. For the fourth to fifth generation before the present, 28 percent of wars reported in oral historical traditions were ignited by land disputes. Between 1961 to 1975, when the Colonial regime instituted land laws, the percentage of wars triggered by land disputes rose to 67 percent of all Enga wars recorded at the time. Between 1976 and 1990, the percentage of reported wars triggered by land disputes dropped to 30 percent of all wars, and from 1991 to 2004, it dropped to 23 percent (Wiessner 2006). Between 1961 to 2004, the population more than doubled.[6] Thus, the percentage of wars triggered by land disputes has decreased rather than increased with rapid population growth, even though the cultivation of cash crops put additional pressure on land during this period, and the ability of the national and provincial governments to contain warfare has decreased radically (Wiessner 2006).

The high percentage of wars over land recorded by Meggitt can be attributed to two factors. First, Meggitt's thesis was supported by testimonies of Enga big-men who framed explanations for fighting in terms of struggles over land because they had found that land was the only explanation Europeans accepted as "rational" (Ambone Mati, Itapuni Nemani clan, Kopena [Wabag] 1991). Second, conflicts over land were generated as a result of the colonial administration trying to fix land boundaries through land courts and thereby inciting plots on the part of the Enga to win land from neighbors through court decisions.

Peacemaking and Egalitarian Practice

The marriage of warfare and exchange to produce peace was a complex process. Long-standing relationships with kin outside the clan were activated for preliminary negotiations. Then, leaders with skills in oration had to soothe feelings and build bridges. The victim was praised and his death was portrayed as an unfortunate act of fate that should not go uncompensated. Food was offered to the clan of the deceased, and the clan of the killer promised modestly to give what it could in war reparations at a later date, little though that might be. In such a way, anger was quelled and feelings of balance reestablished. Once initial ties had been restored, if the losers had been driven off part or all of their land, they made a payment to be allowed to return to their land after some weeks. As they rebuilt their houses and gardens, contact with their adversaries slowly resumed. Deaths of allies were compensated, and war reparation exchanges between enemies were initiated. These exchanges involved (1) initiatory gifts given by the clan of the deceased to the clan of the killer, (2) a major distribution of pork given by the clan of the killer to the clan of the deceased on the host's ceremonial grounds, (3) a second round of initiatory gifts, and (4) a final payment of live pigs and valuables by the clan of the killer to the clan of the deceased. Over the two- to four-year period of the exchanges, peace was maintained through the promise of material gain. In all phases of compensation, families in one clan gave to families who were friends and relatives in the other to achieve peace and restore a multitude of individual ties. Once peace was established, marriages were sometimes arranged between enemy clans to further strengthen bonds. Pig production was accelerated to meet the demands of compensation payments to the enemy and to fuel the growing Tee Cycle.

When peace with—rather than expulsion of—the enemy was desired, rules were established to contain warfare. Before going to fight in the morning, clansmen met and reviewed these rules: (1) one should not kill a person on the land of another clan; (2) bodies should not be mutilated; (3) women and children should not be killed; (4) if a man is wounded, he should not be pursued and killed; (5) one should avoid killing leaders, for they are the ones with the wealth and political skills to make peace; and (6) wars should be fought on one front. Rules were backed by proverbs, such as *Akali taiyoko ongo kunao napenge* ("The blood of a man does not wash off easily") and *Mena lenge ongo katao londenge, endakali yati lenge ongo katao londala naenge* ("You live long if you plan the death of a pig, but not if you plan the death of a person").

Developments in peacemaking altered the egalitarian structure of Enga society. Two developments occurred in this context. First, according to the Enga oral historical traditions, strong leadership emerged hand in hand with peacemaking. Big-men gained substantial influence for their role in conjoining their economic skills with those of mediation and negotiation for peacemaking. With increasing influence assigned to the role of big-man, interindividual competition also mounted, brewing factions within clans. Second, young men were placed ever more firmly under the control of their elders as bachelors' cults (*sangai* and *sandalu*) arose and spread widely throughout the Enga. Bachelors' cults kept young men out of marriage until they were twenty-five to thirty years of age; imbued young men with values; honed their skills for political analysis; and encouraged agricultural enterprise, skills in public oration, and success in Tee exchange. Prowess in warfare was not valorized in the bachelors' cults. Through the bachelors' cults, an age hierarchy was constructed that allowed older men to contain the enthusiasm of younger, unmarried men for battle, amongst other things.

The role of women also changed with the advent of peacemaking. Throughout Enga history, women deplored warfare and fled the fight zone with children, household possessions, and pigs to live with relatives. They had no say about the course of the war. However, with the institution of peacemaking, women became important private emissaries in negotiating peace, particularly if they were from the enemy clan.

The Great Ceremonial Wars

The general constraints that egalitarian structures put on Enga warfare as a productive force are similar to those described for many other highland New Guinea societies (Feil 1987). However, unlike in most Papua

New Guinea highland societies, Enga big-men were able to circumvent structural constraints and harness warfare towards their economic and political ends at one critical point in history via the Great Ceremonial Wars.

The seeds for the Great Ceremonial Wars are said to have been sown in approximately the sixth to seventh generation before the present (Wiessner and Tumu 1998); by the fifth generation before the present, they were well underway. The early development of the Great Ceremonial Wars is not well documented in oral historical traditions; however, we do know that four great wars were fought between pairs of tribes of central Enga in repeated episodes at ten- to thirty-year intervals beginning in the early to mid-1800s until 1940 (fig. 6.3). The Great Wars developed out of large, vicious wars between entire tribes or pairs of tribes shortly after the introduction of the sweet potato. This was a time when inhabitants from high altitudes in central Enga were moving into the valleys and beginning to switch their subsistence base from hunting and gathering to agriculture. With the large-scale movements of people and a new subsistence base, former exchanges of forest products for agricultural products became obsolete and routes for long-distance trade were disrupted. When these major wars ceased, big-men sought to alter the rules of conventional warfare in order to preserve some of its benefits without the destruction. Benefits included the formation of larger cooperative units between instigators of the fight and their allies, consolidation of leadership, and the development of far-flung pathways of exchange. However, the Great Ceremonial Wars did not replace conventional warfare. Smaller conventional interclan wars continued in the long intervals between ceremonial wars, and as far as we can determine, they had no connection to the Great Ceremonial Wars.[7]

Participants describe the goals of the Great Wars as follows:

> The Great Wars were planned and planted like a garden for the exchange that would follow. They were arranged when goods and valuables were plentiful and when there were so many pigs that women complained about their workloads. Everybody knew what they were in for, how reparations were to be paid for deaths, and what the results would be. They were designed to open up new areas, further existing exchange relations, foster tribal unity, and provide a competitive, but structured environment in which young men could strive for leadership. These

6.3. Spheres of the Tee Cycle, the Great Ceremonial Wars, and Kepele Cult networks.

qualities of the Great Wars made them differ from conventional wars, which disrupted relationships of trade and exchange, causing havoc and sometimes-irreparable damage. The distributions of wealth that took place after the Great Wars brought trade goods from outlying areas into the Wabag area on the trade paths initially established by the salt trade. (Ambone Mati, Itapuni Nemani clan, Kopena [Wabag])

The underlying purpose of these wars was to bring people together— they were formal and ceremonial. They were fought to show the numerical strength and solidarity of a tribe and the physical build and wealth of the warriors; figuratively, it is said that in the wars, "They exposed themselves to the sun." The Great Wars were events for socializing. After getting to know each other, they would kill many pigs and hold feasts (Great War exchanges).

(Depoane of the Yakani Timali tribe, Lenge [Wabag])

The Great Wars were fought between entire tribes or pairs of tribes who were "the owners of the war" and intermediary tribes who hosted men from the respective sides. The hosts provided their guests with food, water, entertainment, and front-line fighters. The timing and location of Great War episodes was first negotiated by fight leaders (*watenge*), big-men chosen from participant clans for their ability to plan the Great Wars, to put on spectacular public performances, and to organize the ensuing exchanges. For weeks before the battles began, people from the hosting tribes on both sides received warriors in their own houses and began to make preparations, sing, dance, and brew the fighting spirit. Meanwhile, fight leaders drew up plans for battle and, most importantly, for the exchanges that would follow.

On an appointed day, hundreds of warriors—or in later generations, some two thousand warriors—appeared on the battlefield in full ceremonial regalia. Fight leaders engaged in flamboyant ritualized competition and announced a formal beginning of battle. Fight leaders were considered fair targets for humiliation—for example, warriors sought to capture or to steal their plumes—but it was considered foul play to kill them because they were the men who would orchestrate the Great War exchanges at the end of battle. Fighting took place in a designated zone on the land of the hosts, so that no land could be gained or lost. By day, warriors fought in front of hundreds or thousands of spectators, while the women sang and danced on the sidelines. By night, they ate, drank, talked with their hosts, and courted the hosts' daughters. Death rates were generally low (three to ten deaths for several weeks or months of battles), owing to the formal arrangement of the battle and the avoidance of lethal tactics such as night raids and ambushes. The battles continued until fight leaders decided to hold a closing ceremony and cast their arms into the river. In these "fights without anger," there were no winners of losers, no land could be gained or lost, no damage was inflicted on property, and the men who died were said to have given their lives for a worthy cause. Their deaths were not avenged.

Following the battle, a series of massive and festive reciprocal exchanges of pigs, cassowaries, goods, and valuables was initiated between owners of the fight, hosts, and allies. These exchanges continued for two to four years and transformed the newly formed relationships between hosts and hosted into strong exchange partnerships. In the last great fight of ca. 1940, some twenty participant clans slaughtered some one thousand to two thousand pigs on one day. Hosts and allies awoke at dawn and traveled over hill and dale from clan to clan collecting pork from families they had hosted. Food was so abundant that it was said that even the dogs could eat no more. The Great War courtship parties generated post-war marriages, further strengthening ties between groups. Interaction between opposing sides could be resumed soon after the closing feast, and former opponents became desirable exchange partners.

The Great Wars were ingeniously constructed to circumvent disruptive forces of conventional warfare on at least four accounts:

1. The Great Wars united all warriors for a common cause. Their goals to enhance reputation and foster exchange potentially benefited all participants; individual interests and group goals were one and the same.
2. Exchange ties were disrupted with the enemy during the period of fighting only and thus did not differentially affect the networks of participants as conventional wars did. Meanwhile, strong ties were forged with hosts and allies during the war.
3. The possibility for gains and losses in land and property as a result of warfare was removed from the equation, reducing internal conflicts and preventing hostilities from continuing.
4. Battles were formalized to avoid excessive casualties, and deaths were not to be avenged so as not to incite a revenge cycle.

The Great Wars grew to much greater proportions than did any other events in Enga history. Far-flung, enchained exchange networks constructed during Great War exchanges linked western, central, and eastern Enga. Ever more wealth was required to fuel the exchanges. Around the fourth to fifth generation before the present, big-men tapped into the emerging Tee Ceremonial Exchange Cycle of eastern Enga in order to provision the Great War exchanges and invest the wealth flowing out of the exchanges by repaying creditors. With this additional influx of wealth, both the Tee Cycle and the Great Wars expanded greatly, forging networks between three major valley systems (fig. 6.3). The most recent Great Wars involved up to three thousand warriors, the exchange of six thousand to ten thousand pigs, and many trade goods. By contrast, most conventional wars drew only a couple hundred warriors and involved the exchange of some sixty to three hundred pigs.

Despite their lifespan of a mere three to four generations, the Great Wars altered the egalitarian structure of central Enga society. They arose out of the initial need to reorganize the map of cooperation, trade, and exchange in central Enga after the devastating wars of the sixth and seventh generations. While formerly, ceremonial exchange had involved units no larger than clans or subclans, the Great Wars drew in entire tribes, their hosts, and their allies. Successful organization required strong leadership and continuity. While equality continued to facilitate reciprocity between partners within local exchange systems, it hindered the organization of larger interclan networks of enchained exchange. Strict egalitarian practice was relaxed in the context of the Great Wars, affording leaders status and privilege; their names were known throughout Enga. Between Great War episodes, they acted as powerful leaders for organizing other events. According to oral historical traditions, the public called on the capable sons of the Great Wars to replace their fathers in the interest of continuity and predictability. Genealogies indicate that by the fourth to fifth generation before the present, the position of Great War leader was passed from father to son (or to nephew, if a man had no suitable son).

Around 1900, the Great Wars became formidable to organize. Once again, interests in exchange drove developments in warfare as Great War organizers realized that it would be more profitable to discontinue the Great Wars and replace their exchange networks with those of the Tee Cycle (fig. 6.3). One by one, the Great War exchanges were discontinued, and the last episode of the fourth Great War was fought around 1940. Thereafter only smaller, conventional wars continued to be fought as they had been for centuries.

Summary

The course of Enga warfare over the past ten generations was the product of general structural constraints imposed by egalitarian institutions, which can be elucidated by economic theory, as well as culturally and historically specific dynamics, which can be brought to the fore by a practice approach. The egalitarian structures that so facilitated cooperation and mobility for small-scale exchange made warfare ineffective for reaching goals beyond avenging insult and injury, restoring honor, dispersing large communities, and allowing individuals to pursue diverse personal agendas. Aspects characteristic of Enga warfare, such as the rejection of authoritative leadership, weakly disciplined fighting forces, diverse interests of individual warriors, and unanticipated outcomes of battle, are found in many acephalous societies (Boehm 1999; Chagnon 1983; van der Dennen 1995; Keeley 1996; Naroll and Divalle 1974; Meyer 1990; Naroll 1976; Otterbein 1970; Turney-High 1949). Under these conditions, few broader political goals could be realized through warfare, and the losses resulting from warfare often outweighed the gains.

Warfare only became a moving force toward political complexity in Enga society when it created such havoc for ceremonial exchange that measures were necessary for its containment. This was achieved by integrating warfare with co-existing institutions of exchange to bring about peace. The first successful effort to contain conventional warfare was its elaboration into the Great Ceremonial Wars. Here, warfare was structured and oriented towards the exchange that would follow to the benefit of all participants, other than the very few who fell in battle. Egalitarian structures became insufficient for organizing these large-scale events. Apparently, the public recognized this fact and sought to reduce competition for leadership that could tear groups apart; inherited leadership first appears in Enga historical traditions not only as a product of individual initiative, but also upon popular demand for continuity in leadership.[8]

While the Great Wars were restricted to central Enga, the integration of warfare and exchange occurred throughout the rest of Enga around the fifth generation with the conjoining of warfare and exchange to establish peace. Warfare became a means though which balance and respect were constantly reestablished between clans to provide the matrix for large-scale, enchained exchange networks to expand and flourish. As with the Great Ceremonial Wars, the Enga were willing to relinquish some equality in the interest of strong leadership for peacemaking. Meanwhile, bachelors' cults were instituted to form a hierarchy of men who could restrict the enthusiasm of young hotheads for battle.

While egalitarian structures would be expected to have a predictable effect on warfare itself, when warfare is integrated with co-existing cultural institutions of ritual and exchange, far more complex structures will be formed. Configurations will vary greatly from society to society. To give one example, the Ilahita Arapesh of the middle Sepik imported the complex ritual organization of the Tamboran cult from the neighboring Abelam to provide a hierarchical structure in the face of military

threats from neighboring linguistic groups (Tuzin 2001). While relations of equality prevailed in everyday life, hierarchy was activated in
certain contexts in order to counteract tendencies towards fission and
thereby maintain military strength. The upshot was the formation of a
community of 1500 people in a society formerly made up of small hamlets composed of a few extended families (Tuzin 2001).

Implications for Archaeology

What relevance does looking at warfare in ethnographically or historically known societies have for archaeology? Amongst other things, ethnographic models can help archaeologists envision what is not easily
visible in the archaeological record and formulate key questions. First,
ethnographic studies indicate that virtually all societies that permit
achieved status differences in return for delivery of benefits to the group
engage in some form of warfare (Feil 1987; van der Dennen 1995). It
appears that, once competition is permitted, friction is likely to occur
within or between groups, resulting in coalitionary aggression (Kelly
2000). This would suggest that wherever material remains indicate
achieved social inequalities, the presence of warfare should not be ruled
out, even when there is little direct evidence for warfare in the archaeological record. Second, in the absence of evidence for pronounced institutionalized social inequalities, whether and how warfare could be an
independent variable driving change towards hierarchical complexity
is an important question that is not easily answered from archaeological data. Evidence from ethnographic studies suggests that warfare by
itself is unlikely to move egalitarian societies towards hierarchical complexity because the interests and actions among coalitions of fighters
are too diverse to channel wars to achieve concerted goals. This implies
that when evidence for the elaboration of warfare does appear in the
archaeological record of societies that appear egalitarian, developments
in warfare were likely to have been ushered in under the umbrella of
other dominant institutions, such as religious ritual or exchange, or by
a threat from outside. It is often necessary to turn to other dimensions
of a culture in order to understand how warfare could be harnessed to
achieve broader political goals. Third, one of the most important developments driven by warfare may be even less visible than warfare itself—

namely, initiatives in forming alliances or in peacemaking that in turn foster social complexity. The archaeological signatures of peacemaking are rarely explored in studies of warfare, leaving a significant gap in our understanding of intergroup conflict and cooperation. For many of these issues, the answer might be in the archaeological record if ethno-historical studies can identify the right questions to ask.

7

Warfare, Space, and Identity in the South-Central Andes

Constraints and Choices

Elizabeth Arkush

The innovative focus of this volume offers the chance to look at warfare through the lens of practice theory in a cultural context in which the approach is perhaps counterintuitive. In many other contexts, including those discussed by several other contributors—Maya, Mississippian, and Moche, to name a few—there is ample evidence that elaborate ideologies surrounded warfare, death in war, and warrior personae. Archaeologists can draw on a body of material culture freighted with symbolic meaning about warfare: "sacra" (to use Vernon James Knight's [1986] term) proclaiming warrior prowess and valorizing elite status, human trophies, prestige goods circulated in peacemaking or alliances, monuments intended to display the military might of a center or a leader, and iconographies that are richly illustrative of warfare practice and belief. Other material remains testify to ritually charged activities linked to warfare, such as the construction of war temples or the public sacrifice of war captives. These provide a fertile field for practice-based analyses.

In general, these kinds of evidence are lacking in the Titicaca basin of the south-central Andes in the era prior to the Inka conquest, the Late Intermediate Period (LIP, ca. AD 1000–1450). Contact-period documents state there were powerful warring leaders in the region who might be expected to have engaged in militaristic ideologies. Earlier in the archaeological sequence in this same region, rituals of violence and the iconography of violence are highly developed. In the LIP, evidence of fortifications, weapons, and skeletal trauma suggests that warfare was quite intense and probably affected many aspects of life at the time. However, rituals and beliefs about warfare, while they surely existed, did not leave obvious traces. Warfare in the LIP in the Titicaca basin thus lacks the flamboyant, culturally distinctive quality of some other areas and periods.

In consequence, this case highlights some of the most basic theoretical issues raised by this volume. In particular, it poses the question of whether it is appropriate to restrict our ideas of practice in war to warfare behavior that is distinctively shaped by its cultural context—in other words, behavior that does not fit our notions of least-effort functionality or rationality. Such behavior is not particularly evident in the LIP in the Titicaca basin; what are detectable instead are expedient choices about territory and group identity, made in a context of intensifying warfare. I argue that these choices, made within the constraints of severe competitive pressure, can usefully be seen as aspects of practice.

This chapter first discusses the way practice theory is applied to warfare and outlines the ways in which it can inform the differentiation of social groups and group territories in space during wartime. The next section briefly discusses the LIP in the Titicaca basin and the archaeological sequence leading up to it. Finally, I take a practice approach to the development of a sociopolitical landscape of group identities and group boundaries differentiating ally from foe.

Practice and Constraint in War

Practice theory bridges the realms of individual action and "structure," allowing both cultural persistence and culture change to be seen as the cumulative result of many actions—often routinized or habitual—taken by many individual agents. In these actions, individuals reproduce and embody fundamental, shared beliefs about the ordering of the world. As people face new challenges and changing situations, they actively reformulate existing practices to further their interests (social, economic, and political) and to make sense of the world they inhabit. This conscious and creative retooling of practice is what archaeologists usually mean by "agency." Practice theory is thus a powerful tool for explaining how individuals' actions both reproduce and alter larger-scale social patterns, or structure. It rests on the assumption that both individual actions and the larger patterns they create are not determined—that there is "wiggle room" for individual improvisation as well as for cultural variability between societies. Thus, implicit in practice theory, especially in contrast to earlier anthropological theory, is the idea of relaxed constraints on action and outcome. Indeed, Pierre Bourdieu's project, in outlining a theory of practice, was partly to explain social reproduction without recourse to "rules" or other rigid, deterministic ordering structures (Bourdieu 1978, 1980). Those examples with which he illustrates his theory are precisely behaviors and patterns that vary greatly from one society to the next: language, festivals, rituals, gift exchange, etiquette, the spatial organization of houses.

Thus, when locating practice in the archaeological record, archaeologists tend to look for those aspects of behavior that are not solely explicable through practical reason, in the sense explored by William Walker (2002)—that are not universal or obviously utilitarian. Examples of such behavior abound in warfare. Beliefs, rituals, depictions, and displays about warfare are potentially unlimited in their variability, and indeed, they can only be explained with reference to their particular cultural contexts. Signals—the mutilation of a war victim, the wearing of fearsome war paint or dress, the display of group strength through various means—form a centrally important element of warfare in most societies and, like any other form of communication, belong to a culturally specific lexicon. Another example might be conventionalized practices that altered actual combat, such as counting coup among the Plains Indians (Mishkin 1940), Greek hoplite battles (Runciman 1998; Lynn 2003), or places of asylum from violence in Hawaiian warfare (Kolb and Dixon 2002). While these practices were often discarded in confrontations with external groups, they could arise and be maintained (at least for a time) among competing societies that shared a cultural framework. Finally, patterns of warfare that perpetuate themselves without an obvious external cause could be seen as the result of repeated practices that reinforce ideas about war, validate vengeance, and instill a warrior identity in new generations. For instance, Clayton Robarchek and Carole Robarchek (1998) attribute persistent war among the Waorani of Ecuador not to external factors such as population pressure or protein scarcity, but to a variety of cultural norms that are themselves exacerbated by endemic warfare. These aspects of warfare, because they allow room for cultural idiosyncrasy, are at first glance more attractive choices for a practice-based analysis than behaviors that are strongly constrained by external pressures. Particularly attractive are those behaviors that seem to lessen (or at least not increase) the odds of winning. For instance, Walker (2002) specifically identifies evidence of (to the Western observer) "impractical," "irrational," and "nonutilitarian" behavior as a window into the ritual aspects of warfare in the American Southwest.

However, if we focus exclusively on this cultural variability, we risk losing sight of the limits on warfare practice. We fall into the trap, not of pacifying the past (as Lawrence Keeley [1996] asserts), but of failing to recognize the difficult decisions, the fear, and the sacrifices that must often have pervaded life in times of war. Indeed, most scholars consider warfare so hazardous and traumatic that it must be explained as a response to acute needs and powerful incentives (e.g., Ferguson 1984, 1990). In addition, warfare itself creates an environment of strong competitive pressures, wherein a group that does not effectively defend itself risks physical or social extinction. As much as warfare practice is embedded in culture, it is also driven by the desire to avoid defeat. This utilitarian need results in many patterns that are *not* culturally specific—for instance, commonalities in the way fortifications and weapons were designed around the globe. While walls and weapons were indeed produced and reproduced through a framework of culturally transmitted practical knowledge, one can explain their form satisfactorily without recourse to practice theory.

An alternative approach, then, is to consider the practice of warfare as something that takes place *within* external constraints, and within the bounded realm of what is militarily effective in a given social, demographic, and environmental context. For instance, Brian Ferguson (1990) proposes a model of "a nested hierarchy of constraining factors, progressively limiting possibilities," in which material and infrastructural factors are the "hard" limits within which there is play or leeway for social and ultimately ideological structure to influence warfare practice. This model may be too rigid in some ways (for instance, in insisting that warfare must always derive from material causes at root and failing to recognize the way practices themselves alter infrastructure), but it is a useful starting point for thinking about limits on warfare practice and their degree of flexibility.

One could make the objection (as do Axel Nielsen and William Walker, this volume) that, from an emic perspective, such a model is meaningless. Every choice is utilitarian within its cultural framework. Cultural norms and beliefs may constrain action just as much as external factors; indeed, to an actor, the distinction between "internal" and "external" factors may be irrelevant. However, the distinction is relevant to us as observers whenever we take a comparative approach to warfare—as we do in this volume. Contexts of war differ greatly and change greatly over time; we as anthropologists would naturally like to know why (even if the question would not occur to an actor within such a context). One fundamental reason for such differences is different settings of demography, environment, and sociopolitical organization, and recognizing these factors, as Ferguson suggests, allows us to identify the realm of agency and cultural variability where practice becomes useful as a conceptual tool. Because there are remarkable contrasts between the Titicaca basin and other areas covered in this volume, it is clear that the balance of infrastructural constraints and culturally specific practices changes from one place to another or over time in the same place.

Here, I simply restate the obvious: that human choices, including those about warfare (especially those about warfare, we could guess), take place in a world of constraints and pressures. As warfare intensifies, the pressure increases on individuals and groups to do whatever is necessary to improve the chances of survival. These choices and practices made in an environment of intense competitive pressures cannot be said to violate "practical reason"—indeed, they are strongly utilitarian. In this sense, they may not strongly reflect a distinctive cultural context. Nevertheless, they can fruitfully be seen as "practice" in that they create historically specific situations that further inform and constrain action. One of the main ways they do so is in creating and reproducing social identities in space.

Conflict, Cultural Identity, and Space

The concept of social identity, particularly ethnicity, is a dominant theoretical theme in anthropological archaeology. Since at least the 1960s, social scientists have viewed ethnic groups as defined by subjective affiliation or ascription rather than by objective criteria such as genealogy, language, or "race" (Fenton 2003). At the same time, most would agree that ethnic identity must be based on at least a perception of shared culture, history, and ancestry for it to be seen as "ethnic" to begin with. A primary question has been how much conscious awareness and control human actors have in the construction of ethnicity: whether ethnic affiliation or ethnic ascription by outsiders is tied to deep, involuntary loyalties to kin, place of origin, and "our way" of doing things (the "primordialist" view; Shils 1957; Geertz 1963; Gil-White 1999) or is constructed and mobilized strategically and expediently in pursuit of shared political interests, rendering it fluid and contingent (the "instrumentalist" view; Barth 1969; Glazer and Moynihan 1975).

For archaeologists, this question has posed problems for the most basic task of identifying cultural groups—a task that has almost always been based on stylistic differences in material culture. Much archaeological and ethnoarchaeological literature has centered on whether style reflects ethnicity, as well as on finding a palatable explanation for why it should do so (e.g., Wiessner 1983, 1990; Hodder 1982; Sackett 1986, 1990; Shennan 1989; Carr and Neitzel 1995; Emberling 1997; Jones 1997; Dietler and Herbich 1998; Stark 1998). Several scholars have concluded that the problem is best resolved through a practice theory approach (Bentley 1987; Jones 1997; Dietler and Herbich 1998). In this view, the repeated practices of daily life make up a sense of cultural identity and custom that is deeply felt, but that can also be consciously manipulated. Style may express identity and affiliation in both routinized and intentional ways. Indeed, the act of differentiating one's group from others through material culture style is a significant part of the "ethnic process." Recently, archaeologists have also considered the ways cultural identity may be expressed in house structure and other aspects of the use of space, in mortuary ritual, craft production, food choices, and bodily modification; however, they have not closely examined how it is articulated in violence against outsiders.

Nevertheless, there is an obvious relationship between group affiliation and acts of violence against those outside the group. The modern world is rife with tensions that are portrayed by outsiders and conceptualized by insiders as rooted in ethnic and religious identities, although scholars who study these conflicts vigorously reject the idea that ethnic or sectarian difference ultimately *causes* conflict (Allen and Eade 1996; Eller 1999). Rather, they consider other problems to lead to the mobilization and militarization of ethnic groups (see Brubaker and Laitin 1998). For instance, David Turton (1997) stresses the role of "ethnic entrepreneurs" in the former Yugoslavia, Rwanda, and Ethiopia: politicians and intellectuals who consolidated their political power by galvanizing ethnic sentiment and selectively recrafting ethnic histories of pride and grievance. Violence further sharpens ethnic distinctions and plays into the hands of these factional leaders. Thus, warfare can play an active role in the ethnic process.

Meanwhile, anthropologists who study conflict and violence in traditional societies have proposed interesting ideas about how kinship-based social structures direct violence outwards, towards peoples perceived as less genealogically related (e.g., Sahlins 1961; Otterbein 1968, 1970; see Solometo 2006). When violent conflict arises between communities that consider themselves to be related, it is typically more restricted and less brutal than wars with outsiders—for instance, mutilation and trophy taking may be permissible in wars against outsiders, but not in wars between communities with strong social and kinship ties (Solometo 2006). Hence, while collective violence may have its roots in other phenomena—physical and economic insecurity, for instance—ideas of shared ancestry, patterns of social interaction, and alliances built on perceived relatedness channel this violence to the boundaries of the larger social group. In the process, this structured violence reinforces and reproduces the group as a meaningful entity. Thus, regional histories of group identities, allegiances, and hostilities inform each new act of war, even if they may not ultimately "cause" it (Pauketat, this volume).

Ethnic groups are often closely associated with specific territories in space, and the relationship of people to the space they use is also reinforced by warfare. Wars over land or over resources fixed in space reinforce the concept of exclusive territory. They may lead the group to signal its territorial rights with visible, durable markers that become closely connected with the group's identity and history. A group dispossessed of its land is also robbed of its cultural identity and its shared past. Even when wars are not pursued specifically to gain territory, chronic warfare creates a hostile environment in which portions of the landscape are "enemy territory"—too dangerous to venture into in the normal course of life. Warfare results in the creation of no-man's-lands or buffer zones, further demarcating friendly from hostile terrain; it causes dislocations, defensive nucleation, and the building of new communities geared for defense. Fortifications proclaim control of territory and vividly define sociopolitical groups at the level of the fort, as insiders versus outsiders (Adams 1966; Liu and Allen 1999). These effects, while familiar to archaeologists, are not normally examined from a practice approach. However, the fit is a natural one. Since the inception of practice theory, a dominant concern has been the way people reproduce and embody the social and cosmological order in the spatial arrangements of houses or settlements, and the ways these physical structures, in turn, reinforce the social order through daily practice (Bourdieu 1978). Archaeologists have expanded the spatial analysis of practice to treat the social meaning embedded in landscapes (e.g., Deetz 1990; Tilley 1994). Archaeologists who use insights from practice theory treat landscape as a form of materialized and lasting "structure"—something that is shaped by humans, and that in turn durably orders and influences human action. Landscapes of war certainly fit this description. They not only reflect group identities and group boundaries; they reproduce these relationships through proximity with friends, distance from enemies, and built defenses.

While landscapes of group identities and group boundaries may have emerged in varied ways, violent conflict was probably often involved, and endemic warfare can be seen as a crucible for the creation or hardening of identities of Us and Other. For instance, Jonathan Haas (1990) traces the process of tribalization in the American Southwest during a time of environmental degradation and intensified warfare after the Chaco collapse. Archaeologists and historians examine how the Persian Wars were crucial in crystallizing the idea of a pan-Greek ethnic identity vis-à-vis the foreigner (Bovon 1963; Hall 1997). The emergence and persistence of territorially defined groups in conflict-ridden contexts illustrates Frederik Barth's fundamental insight (1969) that ethnic groups are created and maintained through the creation and active maintenance of ethnic *boundaries*.

The Andean region has all the ingredients discussed above: a complex mosaic of ethnic groups, very strong ties between these groups and the lands they inhabited, and a history of endemic warfare. At the time of the Spanish conquest, the Inka empire had only been in existence for about a century, and under a veneer of empire-wide institutions lay a patchwork of native groups whose identities were closely linked to specific places in the Andean geography. (Indeed, the importance of specific lands to Andean groups is indicated by the Inka policy of the forced resettlement of recalcitrant subject populations, sometimes over very long distances.) I believe that these identities took their specific forms prior to Inka conquest in the LIP, and that they did so in the context of frequent warfare, evidenced by widespread defensive settlement patterns. Here, I examine how warfare may have been related to space and group identity in one case, in the northern Titicaca basin of Peru.

Warfare in the Titicaca Basin

In the Titicaca basin, the Late Intermediate Period followed the collapse of the state of Tiwanaku in the southern basin ca. AD 1000 and was succeeded by the Inka conquest of the area around AD 1450. This intervening period was described by Inkas and other native informants as a general time of war in the Andes. The Titicaca basin was said to have been dominated by warlike regional ethnic groups: in the northern basin, the paramount lord of the Collas had politically consolidated a large region through the conquest of many other Colla lords and battled with the lords of the Lupacas and Canas, similarly sized groups to the south and north respectively (e.g., Betanzos [1551–1557] 1996:93; Cobo [1653] 1979:139–140; Cieza [1553] 1984:274, 279; [1553] 1985:15, 22, 110, 121; Sarmiento [1572] 1988:105–106). These accounts portray the Collas and their southern neighbors, the Lupacas, as some of the largest and most politically centralized of the Andes prior to Inka conquest, proto-states led by powerful hereditary warlords who vied for control of the region.

Archaeologically, one of the most noticeable trends of the LIP was a shift to dispersed hillside and hilltop settlements, often fortified. This defensive settlement pattern is evidence of much more intense warfare than at any other point in the prehispanic sequence. The ultimate cause of intensified warfare is not certain, but the timing of the construction and occupation of fortified sites suggests that a long-term series of droughts was a significant factor (Arkush 2008). In the high-altitude, frost-prone altiplano, agriculture is very sensitive to precipitation, and productive agricultural lands are limited.

The pressure of warfare in the LIP manifests itself most strikingly to the archaeologist as a great proliferation of hilltop fortified sites known in the Andes as *pukaras*. From 2000 to 2002, I conducted a research project specifically aimed at clarifying pukara characteristics, use patterns, and regional distribution in the northern and northwestern Titicaca basin, territory associated with the Collas (Arkush 2005; Arkush 2008). Pukaras are very numerous in the Colla area, and in most of the rest of the Titicaca basin as well (Barreda 1958; Bennett 1933, 1950; Frye 1997; Hyslop 1976; Neira 1967; Stanish et al. 1997; Stanish 2003; Tapia Pineda 1978a, b, 1985). Their use varied; some were lightly used outposts or refuges, while others were permanent settlements.

The builders of pukaras paid close attention to the strength and design of defenses. Pukaras follow a canon of one to seven concentric defensive walls, interrupted by cliffs. Walls are thickest and highest on the sides of the hill that are most approachable, and sometimes they peter out on steep ground. Entrances usually consist of several small doorways that could be easily blocked from the inside, and they are often (though not always) staggered from one wall to the next, creating an enclosed "killing alley" that attackers would have to pass through. Walls vary greatly in size; at the largest pukaras, they are truly massive, up to about four meters thick and five meters high. They often have a parapet remaining, especially on the most vulnerable sides. Piles of river cobbles for use as slingstones or throwing stones can be found just inside the defensive walls in some cases. Many pukaras are on very high hills that are difficult to access and exposed to stormy weather, and in more than four-fifths of the forty-four pukaras surveyed in the project, we could locate no year-round water source on site. For thirteen pukaras, including at least eight settlements with substantial domestic architecture or artifacts, water is at least half an hour's walk away, and this is true for at least eight settlements with substantial domestic architecture or artifacts. In other words, the danger of attack was great enough that people made substantial sacrifices for safety; they were willing to live on high hills inconveniently far from water and to invest considerable labor in building defenses. Nevertheless, their preparations were not always effective. At the funerary cave of Molino-Chilacachi in the southwestern basin, which dates to the time of pukara use, 15 percent of the forty-four adult crania from disarticulated skeletons (probable secondary burials) had frontal or parietal fractures, probably from maces or slingstones, including several healed injuries (de la Vega et al. 2002).

In surveying the high, windy, inaccessible peaks where Colla people chose to build, the cumulative subjective impression is of a human landscape shaped powerfully by fear. This settlement pattern suggests that intense warfare placed great constraints on the room for decision, including decisions about warfare itself: where and how to live, how to expend collective labor, how to protect a settlement, how far it was acceptable to travel to fields and water sources. These decisions and the larger patterns they constituted were rigorously driven by necessity.

Warfare and Cultural Context in the LIP

Several contributors to this volume relate warfare to its cultural context by specifically focusing on the rituals and ideologies that surround warfare. But in the LIP, in contrast to earlier periods, there are very few clues in the Titicaca basin archaeological record about war-related ritual and belief. Weapons have been found in some LIP graves. Tombs in the northern Titicaca basin near Ayaviri, excavated by David Bustina Menéndez (1960), contained polished *bola* stones[1]. Among the grave goods at the funerary cave of Molino-Chilacachi were *macanas*, swords made of hardwood from the lowlands east of the basin (de la Vega 2002). Because these, like most LIP burials, were multiple burials, the weapons and other grave goods were not clearly associated with any particular individual (or gender). But the fact that people were buried with weapons suggests that the weapons were an important element of their social identity, and possibly that individuals needed protection after death. Supporting evidence comes from the later contact period. In Nicasio in the middle of Colla territory in the late 1540s, the early Colonial observer Pedro Cieza de León witnessed mourning ceremonies for the funeral of a great lord, in which lamenting women went through the town carrying the lord's arms, headdress, clothing, and seat (1984:279 [I.c]). Given that Cieza describes this as typical of Colla burial customs, the deceased was probably an important local lord rather than an Inka governor. Again, arms, as well as other symbols of status, were closely enough associated with a leader to be displayed in a procession on his death.

Rituals or conventions may have applied during and after actual combat, if early Colonial descriptions of remembered pre-Inka warfare are reliable. Don Pedro Mercado de Peñalosa gives one of the few accounts of warfare practice, for the Pacajes of the southern Titicaca basin ([1586] 1885:59). He states that the Pacajes traditionally fought nude, protected by wooden shields, their limbs and face daubed with colors to appear fierce to their enemies. Cieza mentions war-related rituals, stating that after battle, Andean groups "went triumphantly back to the heights of the hills where they had their castles, and there made sacrifices to the gods they worshipped, pouring before the rocks and idols much blood of humans and animals" (1985:6 [II.iv]). However, this is a statement about the Andes in general, and there is no guarantee it applies to the people of the Titicaca basin.

It is also possible to make some loose speculations based on our knowledge of later Titicaca basin society. At the time of Spanish contact, the Titicaca basin (like much of the rest of the Andes) was populated by a legion of powerful place spirits (associated with mountains, rivers, and springs), and ancestors were present and active in human affairs. Propitiating these powerful spirits and seeking the help of ancestors was probably an important aspect of warfare. Harry Tschopik (1946:563), in his ethnography of the Aymara, noted the importance of divination before any significant undertaking, and we might guess that several centuries before, divination was a prelude to violent action. Nevertheless, these rituals and beliefs are difficult to verify archaeologically and do not have any support at this point.

Indeed, one of the most notable changes of the LIP from earlier periods (along with intensified warfare) is the marked diminution of material expressions of ideology, including the disappearance of long-lived forms of ceremonial architecture such as platform mounds and sunken courts. Even the largest LIP centers had relatively little in the way of civic/ceremonial architecture, elaborate elite residences, or finely crafted items, suggesting a flattening of the social hierarchy and a decreased reliance on ceremony and ideology to legitimize leaders. In particular, the indicators of a link between warfare and elite legitimation that are present at Tiwanaku and earlier centers—such as warlike iconography and large-scale public sacrifices—are not present in the LIP or have not been found. For instance, trophy-head iconography, which was central to earlier traditions, essentially disappears in the LIP (even while actual opportunities for garnering trophy heads must have increased). In fact, nearly all the complex figurative iconography of Tiwanaku and its predecessors disappeared, replaced on ceramics and in petroglyphs with primarily abstract designs. In the Colla area, representational art is restricted to rare depictions of camelids on pottery and in rock art, as well as small, molded ceramic figurines or vessel handles of human or animal bodies or heads (which do not appear to be trophy heads). Meanwhile, new burial traditions were adopted: above-ground circular grave markers (collar or slab-cist tombs) and less commonly, mortuary towers or *chullpas*. These tombs were both the religious and the physical focus of LIP communities: cemeteries are often located at the center and highest point of Colla pukaras.

Despite the early Colonial accounts of powerful pre-Inka warlords, social hierarchies were not elaborately expressed in material culture at this time. There is not much evidence for segregated elite areas, and elite differentiation through house size is apparent more as a continuum than as a distinct category (see also Frye 1997). While chullpas may have signified higher status than other forms of LIP burial (such as cave burials and cist tombs), all of these types are normally multiple burials (sometimes of quite large numbers of individuals),[2] and probably no type of burial was restricted to a single class.

To contextualize, many of the shifts of the LIP affected not only the Titicaca basin but large portions of the Andean highlands. Defensive settlement patterns and (where the information is available) high rates of skeletal trauma are very widespread in the LIP, covering vast stretches of the Andean highlands from Argentina and Chile to Ecuador (Parsons and Hastings 1988; Arkush 2006). The widespread extent of defensive settlement patterns in the LIP poses the question of where to locate agency and cultural variability in the practice of warfare in this time period. How much leeway was there for variability if all across the Andes, people responded to their problems in much the same way?

Colla Groups and Group Identity in Space

One of the behavioral realms in which practice altered large-scale societal patterns in the LIP was in the relationships of groups to each other and to the space they inhabited. Group identities in the LIP were probably very place based, linked to both built and natural aspects of the environment. Prominent chullpas visibly signified ancestral ties to particular places, ties repeatedly renewed by the living in offerings, ceremonies and successive interments. Walls not only protected those inside, they demarcated the local community and reinforced categories of insider and outsider. *Pacarinas*, sacred natural spots on the local landscape, were identified as the origin places or mythical founding ancestors of particular groups. Settlements were enmeshed in a spatial network of powerful place spirits associated with particular mountains. Pukaras and distinctively shaped hills or ranges are visible from great distances, especially from the vantage of the hilltop settlements of the Collas, and form ever-present reference points, concretely locating the observer in space at all times.

Beyond these rather conjectural relationships to the land, a social landscape is visible archaeologically, the result of communities' political actions (violent or otherwise) with allies against enemies. Colla people structured their societies around relationships with allies placed strategically in the local geography, and these relationships formed the seeds of ethnic and political blocs that were later documented in the early Colonial era. While these spatial patterns of allies and enemies were constrained by necessity, they nevertheless represent the accumulated and continual negotiations, affiliations, interactions, and choices that constitute practice.

Patterns of Pukara Distribution

When we turn to the archaeological landscape of the northern basin in the LIP, the picture of political unification given in the early Colonial sources is clearly contradicted. Figure 7.1 shows the distribution of pukaras in the northwest Titicaca basin.[3] The map is centered on Colla territory but includes some portions of Lupaca and Canas territory to the far north and south as indicated by the early colonial literature. Pukaras are hilltop sites by definition, and here they are found not in the plains but on those hills that border the plains, river valleys, and transportation routes, where agriculture can be supported on the terraced hillsides and at the bases of hills. It can be seen that pukaras are very numerous in the study area. Even taking into account the preferential zones where they appear, pukaras are about as

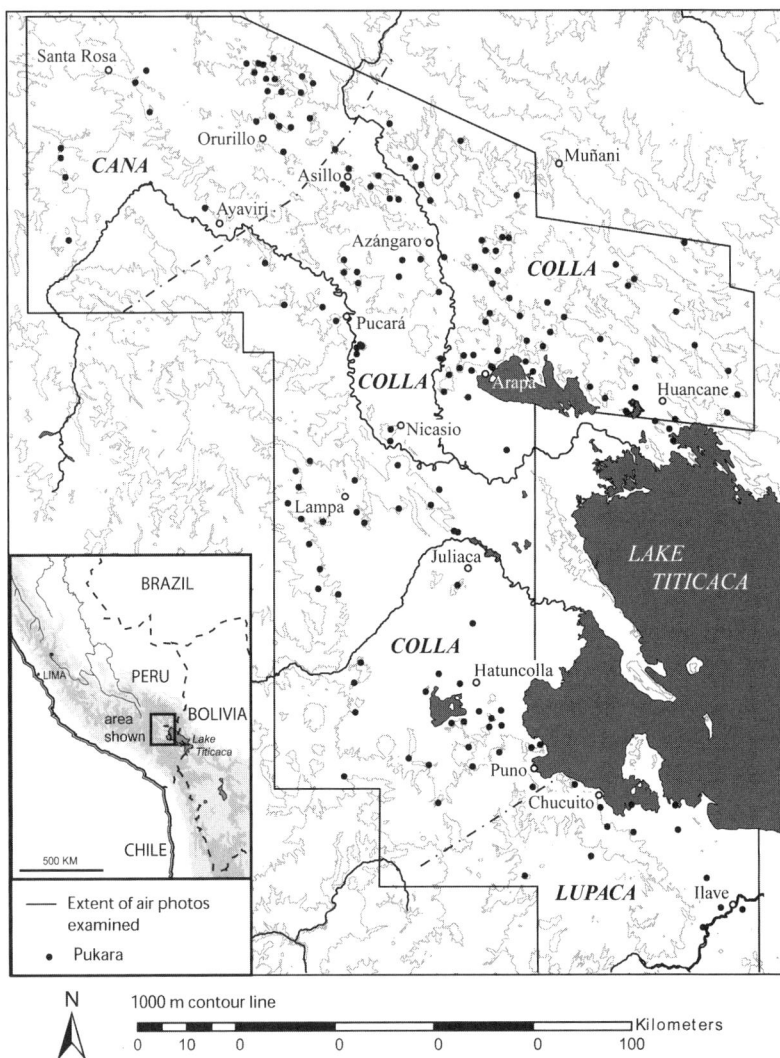

7.1. Known pukaras of the northern Titicaca basin, identified primarily through air photos.

common in the interior of Colla territory as on its edges, a fact that is difficult to reconcile with the ethnohistoric portrayal of the Collas as a centralized kingdom. Furthermore, they are not evenly distributed across these prime zones but clustered in certain areas and largely absent in others, even where the topography is similar. In some cases, two or more pukaras are located quite close together, on the same large hill or range of hills. In other places, gaps of more than ten kilometers with no pukaras separate dense clusters. The resulting picture is one of an extraordinarily fragmented landscape, with few areas of protected or pacified heartland.

However, if pukaras did not defend one unified polity, neither were they wholly autonomous. Most (65 percent of the total of 173 identified) are two to four kilometers away from their nearest neighbor pukara, and almost no pukaras (3 percent) are located ten or more kilometers away from their nearest neighbor. This close spacing in and of itself suggests that many pukaras had friendly rather than hostile relationships; a distance of two to four kilometers is uncomfortably close for an enemy, much closer than the spacing of settlements in societies where conflict is endemic and every village is autonomous (e.g., Chagnon 1968:117). Additionally, pukaras vary greatly in size and defensibility (fig. 7.2). Some were minor outposts or unoccupied refuges; others were large centers with hundreds of densely packed houses. Some had high, thick walls and were located on high, steep hills, while others had minimal defenses and were much easier to approach. In other words, there is a great deal of variation in the scale of attacks that pukaras were intended to resist. I initially assumed, following Charles Stanish (2003), that small and minimally defensive pukaras gradually gave way to larger and better-protected pukaras over the course of the LIP. However, radiocarbon dates from fifteen Colla pukaras indicate that while a few small or minimally defensive pukaras were used earlier, most pukaras of all sizes date to a time frame about 150 years long at the end of the LIP.[4] The best explanation for the variation in pukara size and defensibility in the late LIP is that pukaras, including the small ones, were embedded in larger alliance or hierarchy networks that enabled the populations that used them to take refuge in larger, better-protected pukaras in times of crisis, and to draw on reinforcements from neighboring allies.[5] Thus, rather than existing as a well-integrated polity, Colla territory was most likely characterized by several loose defensive coalitions, each perhaps dominated by a particular center or a skilled war leader.

What were these coalitions? Because the nearest-neighbor spacing of pukaras is so consistent, it is plausible to conclude that any hill zone (that is, terrain suitable for pukaras) wider than ten kilometers with no pukaras can be considered a buffer zone. However, a better approach to understanding spatial patterns of alliance and antagonism is through

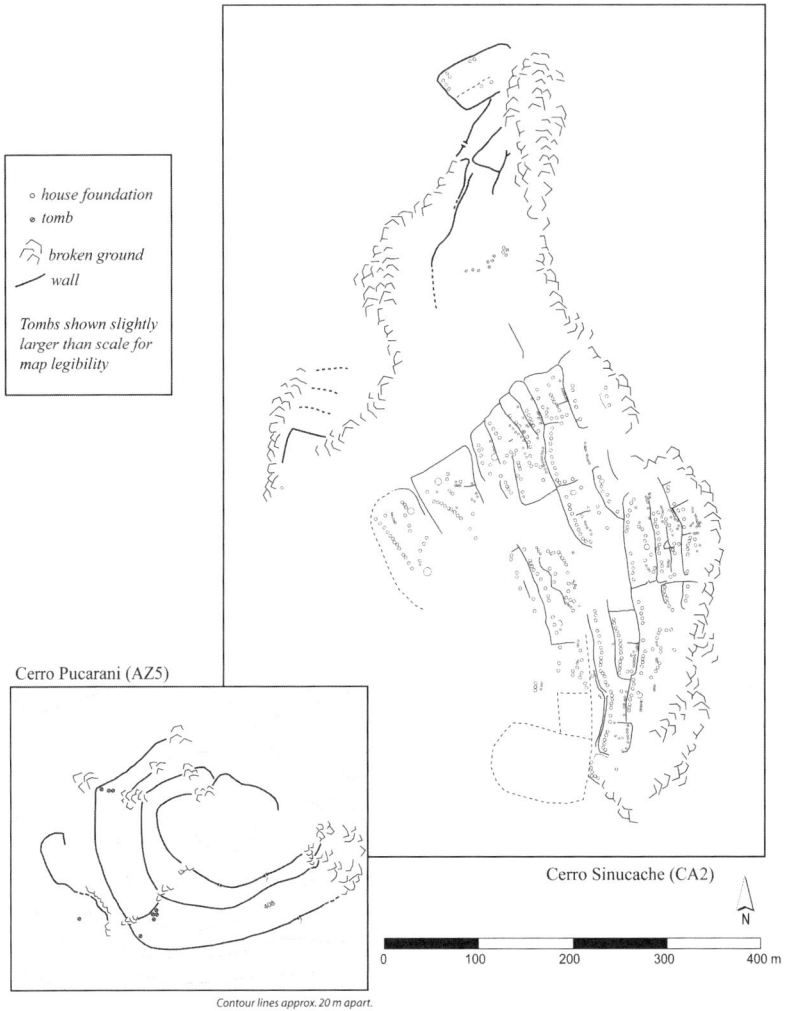

o house foundation
• tomb
broken ground
wall

*Tombs shown slightly
larger than scale for
map legibility*

Cerro Pucarani (AZ5)

Cerro Sinucache (CA2)

N

0   100   200   300   400 m

Contour lines approx. 20 m apart.

7.2. Two Colla pukaras with contrasting sizes and uses.

lines of sight. Lines of sight have been used by several archaeologists, primarily in the American Southwest, to reconstruct hypothetical networks of alliances (e.g., Haas and Creamer 1993; Wilcox et al. 2001), and there is some documentary evidence for the use of smoke or fire signals in wartime in the Titicaca basin (Bandelier 1910:89; Chervin 1913:65; de la Vega [1609:VI.vii] 1966:329; H. Tschopik 1946:548; see Stanish 2003:220). It is expected that pukara sight lines were of great strategic importance in the largely treeless terrain of the Titicaca basin,

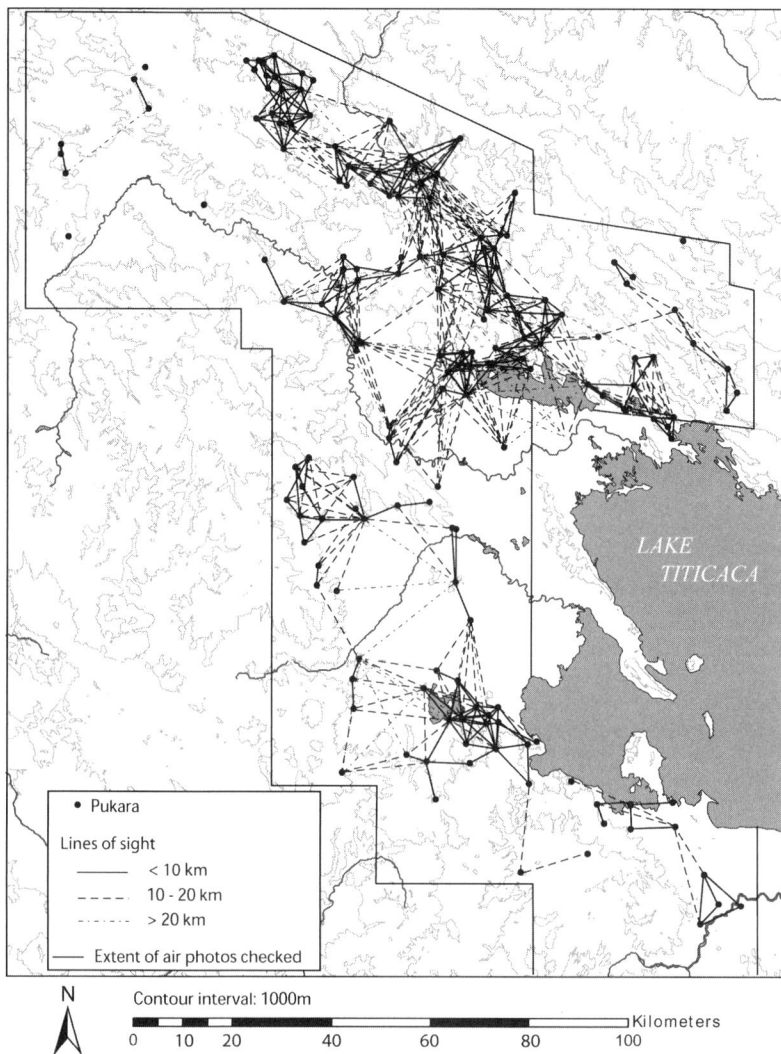

7.3. All unimpeded lines of sight between pukaras. (Where sightlines cross a lake, the distance to detour around the lake is used instead.)

where an observer standing on the peak of a pukara can see a huge area. Pukaras may have used visual signals to summon aid from their allies, and they also probably watched other pukaras closely for signs of battle, wall building, or a departing war party.

Figure 7.3 displays lines of sight between pukaras in the study area, generated through a GIS analysis and confirmed wherever possible in the field. Clearly, pukara visibility is very good in this region. It can be seen that pukaras tend to cluster in mutually visible groupings. Some of these clusters are quite distinct, such as the tight network of pukaras in the far north, around the modern town of Orurillo. Some pukaras have more line-of-sight connections than others, and pukaras with exceptionally good viewsheds may have been used partly as signaling or sentry "hubs." By contrast, some pukaras are arranged in chains that may have been used to pass signals over long distances.

Building on this pattern, possible alliance clusters, defined as networks of pukaras connected by lines of sight less than ten kilometers long, can be tentatively proposed. This is a very conservative interpretation of the size of the groups; for instance, where two larger clusters are connected by only one pukara, they are represented as two overlapping groups rather than one. These groups are about twenty to thirty kilometers in diameter and would take no longer than a day's journey to cross. On the ground, some alliance networks may have incorporated several of the indicated groups or may have grown to include several over time. In other words, the hypothetical political landscape in figure 7.3 is almost certainly wrong in some respects, but it serves as a starting point for evaluation against other evidence. One line of evidence is spatial patterns in ceramic style.

Ceramic Type Distributions

Ceramics in the project were collected from forty-four pukaras spread across a large swath of Colla territory.[6] These ceramics fall into three major types: Collao, Sillustani, and Pucarani. In addition, this analysis was able to identify a distinctive subtype of Collao, termed Asillo. Collao is the most common type and is distributed across the survey area. First defined by Marion Tschopik (1946), it consists of bowls, vertical-sided vessels (*keros*), and both large storage jars and smaller pouring or serving jars, as well as a variety of less-common miniature vessels, figurines, appliqué animals, and other elements. The defining characteristic of Collao wares is a paste with large angular or subangular grit inclusions that often show through on the surface and are visible through slip (if any). Collao vessels may be decorated with

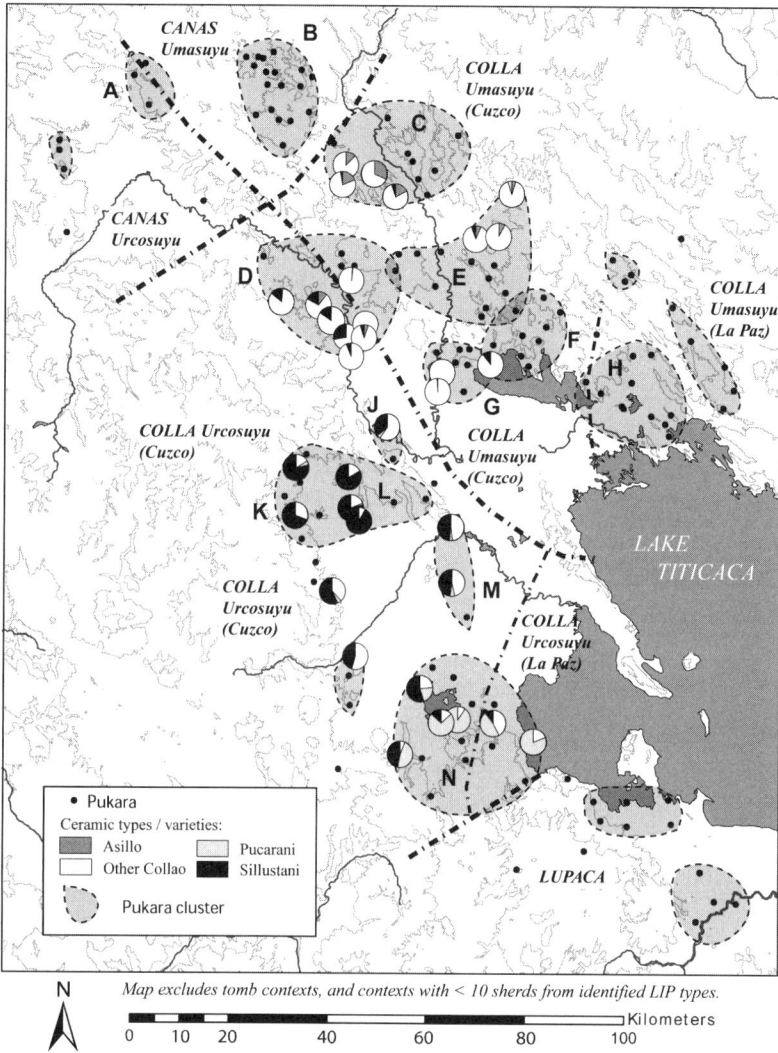

7.4. Spatial comparison of line-of-sight clusters, LIP ceramic types, and later administrative divisions.

rather crudely executed abstract designs in matte black paint (Collao black-on-red); again, paste inclusions show through the black. These motifs are quite variable.

Asillo is a Collao subtype restricted to jars. Most common around the town of the same name, Asillo jars are distinguished by appliqué bands marked by grooves or indentations, placed vertically or horizontally anywhere on the vessel wall. (Collao jars frequently have punctuate bands, but they always appear horizontally around the neck.) In addition, Asillo jars often have a design motif of branching lines unique to this subtype.

Pucarani is a type found in the southwest Titicaca basin (de la Vega 1990) that extends into the southern portion of the study zone. Pucarani pastes are sand-tempered and softer than Collao pastes. This type is composed of bowls, small or medium wide-mouth jars, and a double-stacked large jar form; bowls may be slipped a deep, matte red on the interior and jars are slipped red around the inside rim.

Finally, Sillustani (M. Tschopik 1946) is a type that includes bowls and large jars with quite fine pastes and thick, glossy slip. Sillustani black-on-red bowls are slipped red and often burnished, and they feature a small set of distinctive motifs in black; Sillustani polychrome bowls have red and black fine-line designs on a white ground. Among the ceramic types of this region, Sillustani is the most standardized and skillfully crafted, and it seems the most likely to be associated with special serving events or prestigious uses.

In figure 7.4, the line-of-sight clusters are overlaid with proportions of ceramic types at surveyed pukaras. (This map also includes later territorial divisions, discussed below.) It can be seen that there is very clear regional variation in LIP ceramic types across the survey zone. In general, pukaras within a line-of-sight cluster have a homogenous ceramic assemblage. Collao is distributed throughout. The extent of Pucarani distribution within Colla territory is essentially coterminous with the southern pukara cluster identified here (N); it also extends to the south into Lupaca territory. The Asillo subtype is found within the cluster around the modern town of Asillo, in the north-central part of the study zone (C), and is distributed in smaller amounts on the northeast side of the study zone. The large extent of the Sillustani ceramic style, which spans several clusters in the middle and west of the study area, suggests that these groups may have interacted frequently or perhaps belonged at some points to a larger confederation.

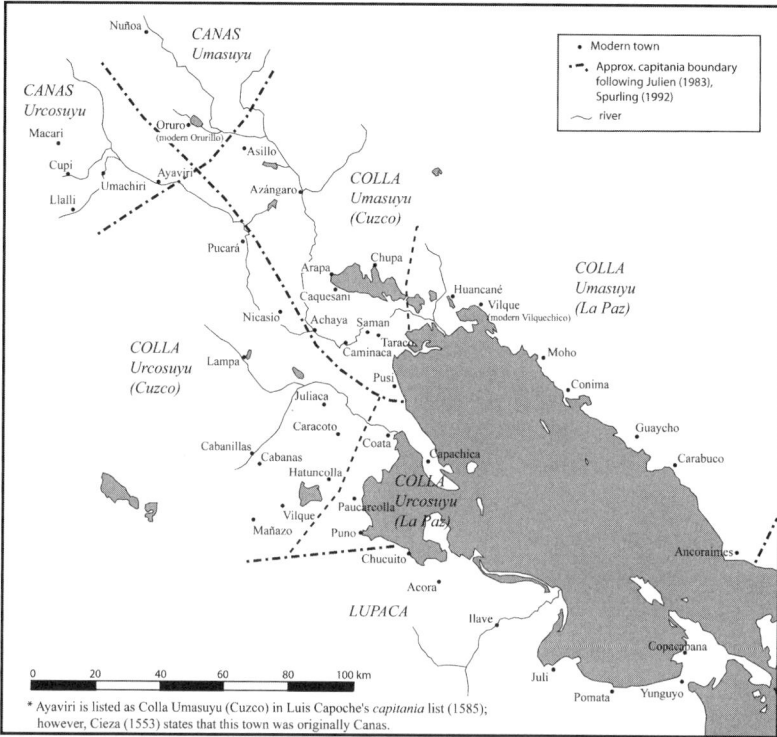

7.5. Ethnic and administrative divisions in the northern basin according to the *capitanía* list of Luis Capoche (1585).

Sixteenth-Century Administrative and Ethnic Divisions

Line-of-sight clusters and spatial ceramic patterns also align in some cases with administrative and ethnic divisions documented in early Colonial sources. These territorial divisions have been explored in some depth by Catherine Julien (1983, 1993) and Geoffrey Spurling (1992), and this section borrows heavily from their research. Both authors have proposed that Spanish Colonial rulers reused Inka-period provincial boundaries in the Titicaca basin instead of redrawing them from scratch, especially since there is evidence for these territorial units from the earliest years of Spanish control; additionally, Spurling (1992) speculates that they originated before Inka times, in LIP sociopolitical boundaries. Here, these territorial divisions are briefly reviewed; then the "fit" is assessed between them and the sociopolitical landscape of the Collas in the LIP as evidenced archaeologically.

Colonial-era documents, in listing the ethnic affiliation of different individual *encomiendas* in the Titicaca basin—land grants to Spaniards of specific Indian communities and their native leaders—allow for the reconstruction of the territorial extent of large ethnic groups such as the Collas, Lupacas, and Canas.[7] In the early Colonial period, a large portion of the northern Titicaca basin was recognized as Colla. This area was carved into a number of smaller pieces for administrative purposes (fig. 7.5). The northern portion of Colla territory was administered from Cuzco, while the southern was governed from La Paz. On at least the eastern side of the lake, the Cuzco/La Paz border followed an Inka-period administrative boundary; a litigation document presented by Spurling (1992) attributes the boundary to Inka times, and the recruitment of *mitima* colonists for Cochabamba under the Inka ruler Wayna Capac followed the boundary (Julien 1993:184). It is duplicated in the early encomienda grant of much of Colla Umasuyu-La Paz to Francisco de Carvajal by the 1540s (Julien 1983:29).

Another major division split the Collas, and other groups, in two, creating the west (Urcosuyu) and east (Umasuyu) sides of Lake Titicaca. The sociopolitical significance of the *suyu* division is supported by several early documents that refer to inhabitants of the two suyus separately, as coming from separate ethnicities, polities, or labor recruitment divisions, rather than lumping them together as Collas (Arkush 2005; Spurling 1992). Urco and Uma sides were also associated with different lifeways and different extra-basin colonies in the documents (Bouysse-Cassagne 1986). For instance, in the Toledo *tasa*, a tax census compiled in 1570, much of Colla Urcosuyu to the west (as well as Cana Urcosuyu) paid no tax on agricultural products, but only on livestock and wool.[8] This area may have been largely pastoral in the sixteenth century. Inhabitants of the Umasuyu side to the east, on the other hand, were universally taxed in agricultural products (always *chuño*—potatoes dehydrated for long-term storage—and often maize), as well as livestock. Most of the maize probably came from the warmer valleys of Larecaja and Carabaya to the east, where nearly every Colla Umasuyu encomienda had mitima colonists, purportedly since the time of the

Inkas (Spurling 1992:95–98; Saignes 1986); meanwhile, no Colla Urco-suyu encomienda was taxed in maize. This division between west and east is also evident in tribute assessments in precious metals. All Colla Umasuyu taxpayers were expected to pay tax in gold as well as silver, demonstrating that they had access to the gold mines in the eastern valleys; no Colla Urcosuyu community paid in gold except for the borderline encomienda of Pucará and Quipa.

The Urcosuyu/Umasuyu division, while it was apparently not an ethnic distinction, did have associations with essentialist personality stereotypes and basic Andean conceptual dichotomies. An early Colonial document explains that inhabitants of the Urcosuyu side of the lake lived on hilltops and were strong, masculine, and held in high esteem, while Umasuyu people were feminine, had lower status, and lived in lowlands near water (Capoche [1585] 1959:140, cited in Bouysse-Cassagne 1986:222). As Thérèse Bouysse-Cassagne (1986) discusses, drawing on the 1612 Aymara dictionary of Ludovico Bertonio, the terms *urco* and *uma* formed a conceptual dichotomy with manifold meanings (male/female, dry/wet, high/low), a dichotomy that corresponded to indigenous Aymara characterizations of the inhabitants of the Urco and Uma suyus of Lake Titicaca, and that was reutilized and reworked for Inka administrative purposes. Finally, the Urcosuyu/Umasuyu division also corresponded to two branches of the Inka road that split at Ayaviri and went around the lake. While there is evidence for smaller territorial blocs, especially on the Umasuyu (eastern) side (Spurling 1992; Julien 1993), their extent is not as clear.

Let us return to the question of how far back in time these territorial divisions went. Figure 7.4 overlays the LIP spatial patterns already identified with these sixteenth-century divisions and largely supports Spurling's (1992) contention of their pre-Inka origin. The correspondence is particularly clear for Colla ethnic boundaries with the Canas and Lupacas. The Colla-Lupaca boundary to the south corresponds to a mountainous but nearly pukara-less zone south of the southern pukara cluster shown on the map (N). The Colla–Canas division lies along pre-Inka buffer zones dividing the Orurillo cluster (B) from those at Pucará and Asillo (D and C). Within Colla territory, the Cuzco–La Paz administrative division on the eastern side corresponds to a range of high hills that blocks most lines of sight. It thus runs between clusters centered near Huancané (H) and near Chupa (F). On the other hand, the Cuzco–La Paz division on the western side crosses a line-of-sight cluster that shared ceramic styles, so this division may have been an Inka or Colonial-period innovation. Finally, the division between Colla Urcosuyu and Umasuyu runs between the clusters centered at Arapa and Nicasio (G and J, respectively), although it bisects the Pucará group (D), following the Río Pucara. In fact, this division is closely paralleled not by one single pukara cluster, but by the extent of the Sillustani pottery style. Overall, several of the administrative/ethnic divisions inherited from the Inka era were probably based on sociopolitical realities of the LIP.

To summarize, patterns of pukara placement and intervisibility suggest that the Colla area in the LIP was broken into smaller territories associated with allied groups of pukaras. The evidence for these groups is supported by ceramic type distributions and duplicated in several cases by territorial divisions attested in later documents. These patterns suggest that LIP ceramic types did indeed have associations with particular group identities in this region. Furthermore, according to later documents, ethnic or essentialist meanings were connected to these groups.

The Case of Pucará

The locality of Pucará is anomalous in a number of ways in this analysis and can be considered the exception that proves the rule. The cluster of pukaras in the vicinity (D) is located more or less on the border that later divided Colla Urcosuyu from Umasuyu. It is separated from the Nicasio cluster to the south (J) by an apparent buffer zone without pukaras, and it is linked to the cluster near Azángaro (E) by short lines of sight, but only tenuously. Its participation in the Sillustani ceramic style is ambiguous—pukaras on the western side of the Río Pucara have Sillustani surface ceramics, but in considerably smaller quantities than pukaras further south. East of the river, there are no Sillustani ceramics among seventy-nine diagnostic sherds collected at the site of Inka Pucara (PKP8).

This ambiguity is reflected in later documents. The communities of Pucará and nearby Quipa (which was later resettled to Pucará) belonged to Colla Urcosuyu (Capoche [1585] 1959; Miranda [1583] 1906), but other documents show they were sometimes grouped instead with communities in Colla Umasuyu. For instance, Tupa Inka owned a royal estate comprising Asillo, Azángaro, a portion of Carabaya to the north-east, and Quipa (near Pucará), according to a list of the Inka ruler's land holdings (Rostworowski 1970:162). Bernabé Cobo likewise recounts that a son of Tupa Inka won from him in a gambling game "the five towns of . . . Nuñoa, Oruro, Asillo, Asangaro, and Pucará" ([1653: bk. 12, ch. 15] 1979:149). All the towns in Tupa Inka's holdings belonged to the Umasuyu side except for Pucará. Pucará, like other communities in the Inka's royal estate, traditionally sent mitima colonists to the east-ern mines in Carabaya, and thus Pucará, alone among Colla Umasuyu towns, paid taxes in gold in the early Colonial period. Finally, in an early list of *tambos*—the way stations that dotted the Inka roads at peri-odic intervals—Cristobal Vaca de Castro ([1543] 1908:457–458) groups the tambo at Pucará not with Nicasio and centers to the south, but with Ayaviri to the north and Pupuja to the east on the Umasuyu side of the division. This grouping, again, may indicate an early separation of some sort between Pucará and the rest of Colla Urcosuyu to the south.

By the 1550s, the encomienda ownership of Pucará and Quipa was split off from any other place in the Colla region, and instead lumped with lands near Cuzco (Covey and Amado Gonzalez 2008:52). Thus, Pucará does not fall neatly into one or another category. Its reduced participation in the Sillustani ceramic style in the LIP and its physical separation from pukaras further to the south suggest that the sociopolit-ical links of Pucará inhabitants with others to the south in the LIP were limited. The locality may have had social ties to groups on both sides of the division, occupying a position that may be reflected in its eventual assignment to Tupa Inka's royal estate, as well as in the ambiguous treat-ment of Pucará in later administrative documents.

Discussion

How does the case presented here relate to a practice-based framework, how can we reinterpret spatial pattern as process, and what does this case tell us about the practice of war?

In the Titicaca basin in the LIP, the necessity of affiliating politi-cally with other nearby communities led to a built landscape in which cooperative and conflictual relationships were reproduced through patterns of proximity, fortification, and visibility. For Collas living in and near pukaras in the LIP, other communities visible nearby were probably allies and confederates, forming a network whose common identity was emphasized through the use of distinctive ceramic types or varieties. Members of these clusters reinforced their relationships through repeated cooperative action in wartime, and perhaps in peace as well. While violent conflict may sometimes have erupted within a cluster, conflict between clusters should have been fiercer and more frequent: buffer zones with no pukaras are visible evidence of these antagonisms over terrain that could instead have supported a network of allied pukaras. In the case of the Sillustani-using people of the area later called Colla Urcosuyu, a larger collection of allied clusters also participated in stylistic exchange and may have recognized a common political identity for at least some portion of the time frame of pukara use. This Sillustani-using group is the most likely candidate for the subjects associated with the Colla lord who is described in the chronicles.

These sociopolitical affiliations of the LIP retained salience for Titicaca basin people two centuries later, long after endemic warfare had ceased. In the sixteenth century, they were partly expressed in ethnic identities of Colla vs. Lupaca vs. Cana, or in essentialist ideas about altiplano inhabitants' qualities: urco (masculine, high) vs. uma (feminine, low). They also affected the relationships of Titicaca basin communities with lowland groups or colonies, relationships later reflected in tribute-payment categories in the Toledo tasa. These categories could be elided at times to suit changing conditions. After the basin had been conquered by the Inkas, its people rebelled at least once, and in this revolt, we are told, many Titicaca basin peoples participated: Collas of both Urcosuyu and Umasuyu sides, Lupacas, and possibly Pacajes—defining themselves as a group against their common foes, the Inkas (Cieza 1985:155 [II.liii]; Rowe 1985:214). However, ethnic identities persisted and continued to affect warmaking in the Titicaca basin. Lupacas attacked Colla settlements in 1538, prompting the Collas to appeal for help to the Spanish conquistadors based in Cuzco (Hemming 1970:242). Hernando Pizarro, with a small force of Spaniards and a much larger one of Andean auxiliaries, routed the Lupacas and quelled the unrest in the altiplano. The eastern portion of Colla lands, from Huancané southward, was also warring with Canas and Canchis to the north shortly after the fall of the Inkas, according to local testimony in the 1580s (Saignes 1986: n. 27). It seems that in the Titicaca basin, while larger coalitions could be built (as in the rebellion against the Inkas), old identities and enmities were entrenched enough to have lasting effects long after the period of pervasive conflict.

This is just one of many archaeological landscapes of groups and group boundaries. That it crystallized in a time of war is not surprising. But approaching it through practice theory allows us to glimpse warfare as a process of cumulative actions—violent attacks, wall building, defensive relocations, the negotiation and renewing of alliances—leading to a durable landscape that reflected and reinforced these social relationships.

Warfare in practice and in representation went through significant transformations in the Titicaca basin over the course of the archaeological sequence. In the LIP, warfare ceased to be mined by elites as a conceptual source of symbol and prestige, even as it intensified to an unprecedented level, forcing communities to relocate and reorganize against the threat of attack. These contrasts illustrate the extent to which an environment of shifting external constraints affected practice over time. In the LIP, a local group's survival depended on its ability to create and cement relationships with allied communities. Within the constraints of geography and defensive pressures, many individual and collective choices about warfare and group affiliation resulted in a specific sociopolitical landscape that in turn participated in the reproduction of enduring identities and animosities. A practice approach is as appropriate for examining these subtle choices made within narrow constraints as for examining the most culturally distinctive behaviors in the repertoire of human violence.

Acknowledgments

Archaeological fieldwork was supported by an International Studies and Overseas Programs fellowship from the University of California, Los Angeles (UCLA), and by research grants from the Department of Anthropology, the Friends of Archaeology, and the Latin American Center at UCLA. Funds for radiocarbon dating of organic materials were provided by a National Science Foundation Dissertation Improvement Grant. This chapter benefited from the comments of fellow participants at the Amerind Advanced Seminar. I thank Axel Nielsen and William Walker for organizing the symposium and book, as well as for inviting me to participate in it.

Ancestors at War

Meaningful Conflict and Social Process in the South Andes

Axel E. Nielsen

War as Practice

It is widely recognized that the time before the Inkas, known as the Regional Developments Period (or RDP, AD 1000–1450), was characterized by endemic warfare throughout the Andes. Archaeological data from the Circumpuna area—the Andes of northwestern Argentina, northern Chile, and southwestern Bolivia (fig. 8.1)—indicate that the conflicts broke out around AD 1200 and continued until the Inka expansion in the 1400s. This phenomenon seems to be temporally correlated with the worst period of a long-term cycle of drought that affected the Andean highlands (Thompson et al. 1985), indicating that a scarcity of water-related resources was one condition underlying these confrontations (Nielsen 2001).

Conflicts must have effected multiple and significant changes in the societies involved. In the Circumpuna area, for example, the RDP is associated with rapid population aggregation, the emergence of well-defined settlement hierarchies, and the development of regionally distinctive material culture forms (e.g., textile and ceramic styles, burial patterns, and domestic architecture), indicating that a new political order was born out of this time of turmoil. Few archaeologists, however, have analyzed in detail how war and sociopolitical change relate to each other, and when they have done such analysis, they have done it mostly on the basis of utilitarian assumptions concerning how armed conflicts universally relate to sociopolitical phenomena, such as prestige accumulation, territorial expansion, or surplus appropriation among "chiefdom" societies (Carneiro 1981; Redmond 1994).

Consider the case of the Central Andean Wanka—people of the upper Mantaro—who are presented as a paradigm of the evolutionary

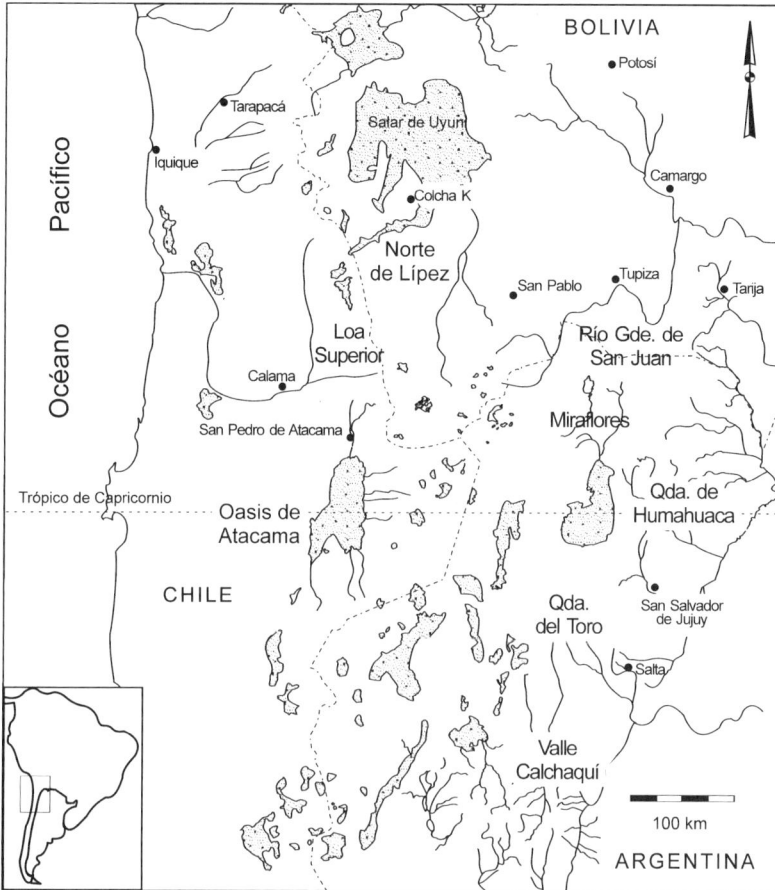

8.1. South Andean regions mentioned in the text.

limitations of military-based chiefdoms in Timothy Earle's (1997) essay *How Chiefs Come to Power*. According to the author, Wanka chiefdoms were "weakly institutionalized" (49), never evolved into large, centralized polities, and lacked a distinct elite class. War chiefs (*cinchekona*) rose to power by offering protection to their communities, but failed to organize stable regional systems because they were unable to develop a strong financial base for their leadership due to environmental constraints on intensive agriculture, as well as their lack of an elaborated ruling ideology "materialized in ways that allowed control through a linkage with the political economy" (196).

Some of Earle's observations regarding the Wanka, like their relatively low levels of political centralization and conspicuous consumption, seem to apply to many pre-Inka societies of the Circumpuna area and of the south Andes in general at this time. Instead of construing these observed characteristics as a political failure of leaders, however, one could interpret them as testimonies of a very successful political project—but clearly different from the Polynesian one—that managed to hold together very large communities (of thousands) during more than a century of endemic violence, with a rather egalitarian distribution of resources and power. Moreover, ethnohistory has demonstrated that the political order that the Europeans found in the sixteenth century rested on a complex and strong institutional base, supported by an elaborated cosmology, whose roots can be archaeologically traced back for millennia in Andean history (e.g., Murra 1975; Platt 1987; Zuidema 1989). Indeed, these roots were so strong that elements of this ideology and institutional framework have resisted more than five centuries of imperial intervention in some rural areas.

We cannot explain the role of armed conflicts in the emergence of this unique political order, however, if we *reduce* warfare to a mechanical response to environmental stress (i.e., an exacerbated form of competition) or an aggrandizing strategy implemented by war leaders alone in a cultural and social vacuum. Practice entails meaningful action and multiple agencies. A practice approach to conflict is necessary—first, because people construe their goals and projects, experience their conditions, and assess alternative courses of action (fighting amongst them) within an inherited framework of representations and dispositions; and second, because war always involves negotiation, not only between enemies, but among a host of knowledgeable actors (human and nonhuman) that were part of those communities (Latour 2005). In the ancient Andes, these included, for example, live people and ancestors; farmers, pastoralists, and craft specialists; warriors, animals, mountain spirits, and other deities; and men and women, authorities, and other members of a people's lineage. Neither war nor its relation to any other phenomena (e.g., weather, politics, or the economy) can be understood without taking into account these two facts. Moreover, since (1) the appropriation of anything as a resource (a necessary condition for action) by someone presupposes some form of (conscious or embodied) interpretation and (2) these frameworks and associated dispositions are always reproduced and contested in practice, it follows that the study of cultural subjectivities and power are inseparable parts of a single project. The goal of such an endeavor would be to understand how cultural, practical logics (sensu Bourdieu 1980) and power relations constitute each other through the practice of war in specific historical settings.

The Semantics of Conflict in Materiality

As argued in the introduction to this volume, archaeological studies of war are currently dominated by a materialist approach. Culturally specific perceptions and dispositions are of little explanatory value when using this approach because they are understood as purely mental and arbitrary phenomena, whose roots and internal workings lie outside of what people really do and the materiality of this doing. This conception is the mirror image of the textual model of culture, based on a Sassurean semiology that treats material culture and actions as signifiers of ideas, according to a relationship or code which is essentially arbitrary. Whether this idealist conception is explicitly presented (e.g., "artifacts as text" [Hodder 1991]) or is dubbed as a form of materialism (e.g., artifacts as materializations of implicitly mental ideologies (DeMarrais et al. 1996]), it is not very promising for archaeology, since it places the causal powers away from what we study (material arrangements and, inferentially, the acts that produced them) and in a purely arbitrary relation with it. It is not surprising that many people resist the notion that something as serious as war needs to be explained by some inaccessible cultural convention (e.g., Ferguson 2001:104).

In the last few years, however, some scholars have pointed out that the semiotics of Charles Peirce is a better heuristic model to treat the meaningful aspects of practice (e.g., Gottdiener 1995; Keane 2005; Parmentier 1997; Preucel 2006). Peirce conceived of semiosis as a triadic process that related a sign (*representamen*), an object being represented, and an interpretation (*interpretant*), or effect of the sign on an interpreter. Anything (a quality, an object, an action, an idea, a law) can be a sign—"something which stands to somebody for something in some respect or capacity" (Nöth 1990:42)—if it becomes part of this process, and indeed most things do (Marafioti 2004:75). Three aspects of Peirce's model are relevant here. First, by conceiving of semiosis as a triad that cannot be reduced to pairs, it situates every act of representation in relation to specific actors. The signification of practice can only be understood with reference to particular interpretants, including various kinds of agents who are affected by them.

Second, the relationship between representamen and object is not always arbitrary and intentional, as in language. Peirce's fundamental classification of these relationships in icons, indexes, and symbols based on shared qualities, actual connection, and convention, respectively, takes into account a wide range of nonarbitrary ways (or *grounds*) in which practices are potentially meaningful, justifying the incorporation of material attributes that archaeologists commonly record (e.g., spatial and formal relationships, artifact function, raw material, and technical traits) when approaching the significance of past practices. It simultaneously allows for a number of pragmatic, embodied, placed, and objectified semantic connections that are not necessarily mediated by ideas, mental codes, or intentionality (cf. Bourdieu 1977). In this paper, the terms "meaning" and "significance" will be used in this practical sense.

Third, Peirce's model takes into account the active role that material culture can assume in the semiotic process, not only as passive recipients of meanings arbitrarily given to them by people, but as motors of signification—Peirce's notion of the "dynamic object" as "the Reality which by some means contrives to determine the Sign or its Representation" (in Nöth 1990:43). When applied to material entities, this idea falls close to the notion of "object agency" that thrives in the current literature on materiality (e.g., Boast 1997; Meskell 2004; Walker, this volume; cf. Gell 1998; Latour 1993). It is also important for situating conflict in a historical context. Materials carry with them a "memory" of the past in which they participated, a memory that shapes the further significance that they or their uses may acquire in new contexts.

Using these ideas as a background, in the first part of this chapter we will explore some meanings and cultural implications that war had for Andean people in the past. We will not use language as our main point of entry, however, as other authors have analyzed the lexical repertoire associated with war and conflict in the Andes (e.g., Platt 1987; Topic and Topic 1997, this volume). Instead, we will focus on a number of war-related artifacts that are frequently found in archaeological contexts of the Regional Developments and Inka periods of the south Andes,[1] looking at their multiple uses, iconography, history, associations, and—when available—ethnohistorical and ethnographic accounts of their significance. The goal of this exercise is to sketch the semantic web that the materiality of these objects wove around conflict in late prehispanic times, as a way of getting closer to what war may have meant for those who suffered it in the past, rather than what it means to us who study it today.

Contesting Power Meaningfully in Times of War

Practice approaches look at people's doings as the locus for reproduction of power structures (Ortner 1984:149). A practice approach to war, then, emphasizes the implications of armed conflict for power distribution, taking into account that every agent—as a subject of practice—has some power and, therefore, social relations are always the unstable result of negotiation. But a practice approach also requires looking at these political processes as emerging property of practice itself, of how war is conducted and understood in specific cases.

The notions of "corporate power" and "heterarchy" have received considerable attention in the recent archaeological literature on social complexity (Blanton et al. 1996; Ehrenreich et al. 1995). This literature is relevant to Andean archaeology because ethnohistory and ethnography suggest that traditional political practices in the area had a strong corporate orientation. A drawback of these models, however, is their typological orientation, in which the collective appropriation of resources is taken as one trait among others that characterize a type of society. One problem with this approach is that the necessary interdependence of the traits is not granted; the other is that it does not explain why one or the other strategy prevails. Both problems can be overcome by focusing on political practice and history rather than typological evolution, i.e., by looking at *how* power is negotiated by agents organized as networks of individuals or as collectives and *how* other practices (such as war and cooperation) affect the outcome of these negotiations in specific cases.

Thinking of these specificities brings back our earlier point regarding the relationship between power and semiosis (see also DeCerteau 1984; Foucault 1999). We can conceive this relationship as twofold. First, socially instituted power presupposes an accepted relationship among agents, resources, and tasks. This is a "sense of the game" (Bourdieu

1980:66), a shared understanding that certain things are fit to some actions by certain individuals under certain circumstances. For example, in a late Andean context, labor tribute (*mit'a*) could be used by a leader (*kuraka*) to finance the cult of community deities (*wak'as*), but not to compensate a follower for a personal benefit (unless it was construed as a collective one), a possibility that could be legitimate in a different culture. When we look at this relationship as a semiotic process, the centrality of interpretation and subjectivities for the constitution of power becomes evident (cf. *doxa* in Bourdieu 1977). Second, these kinds of relationships are unstable and contingent upon previous practices (and powers) that weave new meaningful connections. Intentionally or not, people shape and contest frameworks of meanings in their actions, and in so doing, they transform power relations.

Consequently, the unfolding of semiotic webs—like that of the social relations immanent to them—is a structuration process (Giddens 1984) in which individuals constantly reproduce and transform in their actions the conditions of their own "cultural" existence. As long as this process involves contingency, it has to be explained—at least partially—"with reference to the genealogy of practices or the tradition of negotiations" (Pauketat 2001:80) that brought it about. The second part of this chapter takes this path, tracing the main steps of the history through which the semantic system and its political structure unfolded in changing practices, under conditions of increasing environmental hardship. In this way, we hope to get closer to the interaction among war, culture, and power during the late prehispanic history of the south Andes, highlighting some specificities of the institutional arrangements that emerged from this process.

Metaphors of War

The legendary account of the Age of the Awqaruna ("the warring people") given by the Indian author Felipe Guamán Poma de Ayala (fig. 8.2) offers a good point of entry into Andean perceptions of war and specifically of the era of conflict that preceded the expansion of the Inka:

> From their towns in low ground they moved to live in high [places] and
> mountains and outcrops and in order to defend themselves they began

8.2. The age of the *awqaruna*, or "warring people," according to Felipe Guamán Poma de Ayala (1980:51).

to make fortresses that they call *pucara*. They built walls and fences and inside them houses and fortresses and hide-outs and pits to get water from where they drank.

And they began to fight, and battle, and a lot of war and death with
their lord and king against another lord and king, brave captains and
courageous and spirited men, and they fought with weapons that they
called *chasca chuqui, zachac chuqui* [spears], *sacmana, chanbi* [maces],
uaraca [sling], *conca cuchona, ayri uallcanca* [axes], *pura pura* [metal pec-
toral], *uma chuco* [helmet], *uaylla quepa* [shell trumpet], *antara* [Pan's
flute]. And with these weapons they defeated each other and there was
a lot of death and bloodshed until they subdued each other. (Guamán
Poma de Ayala 1980:52, my translation)

This gives us a first list of items that, in the eyes of an early
seventeenth-century Andean person at least, were directly associated
with warfare. In this chapter, I will focus on five of these items—metal
plaques (often used as chest plates), axes, slings, pukaras, and trumpets—and a sixth one Guamán Poma introduces later in his portrayal
of warriors (fig. 8.3), disembodied heads, using them as windows into
the "pragmatic semantics" of Andean war.

Metal Plaques

Metal plaques with different shapes, functions, and iconography
were used in northwestern Argentina—where they are most frequent—
and adjacent regions of Bolivia and Chile for more than two millennia,
since the Middle Formative Period (500 BC–AD 500) until the Spanish
conquest (González 1992; Latcham 1938:329). RDP plaques are made
of bronze or silver and are rectangular or circular, ranging in diameter between eight and forty centimeters. Some of them were held like
shields using hide straps tied to semicircular metal ears, others were fas-
tened to rigid handles so they could be carried as standards, and others
were used as pectorals, suspended from the neck (González 1979:173–
174). Some of them are plain, others are decorated with one or more of
the following motifs: snakes with two heads (amphisbaenas), anthropomorphic heads (represented without the body), and warriors holding
shields (González 1992:173–180).

In his illustrated "letter to the King," Guamán Poma always depicts
warriors wearing circular plaques on their chests *when they are in battle*,
but without them when they are not (e.g., 1980:51, 128, 134, 136; cf. 123,
130, 138, 168; compare figs. 2 and 3).[2] In the Circumpuna highlands,
rock-art representations of fights sometimes also show individuals wear-

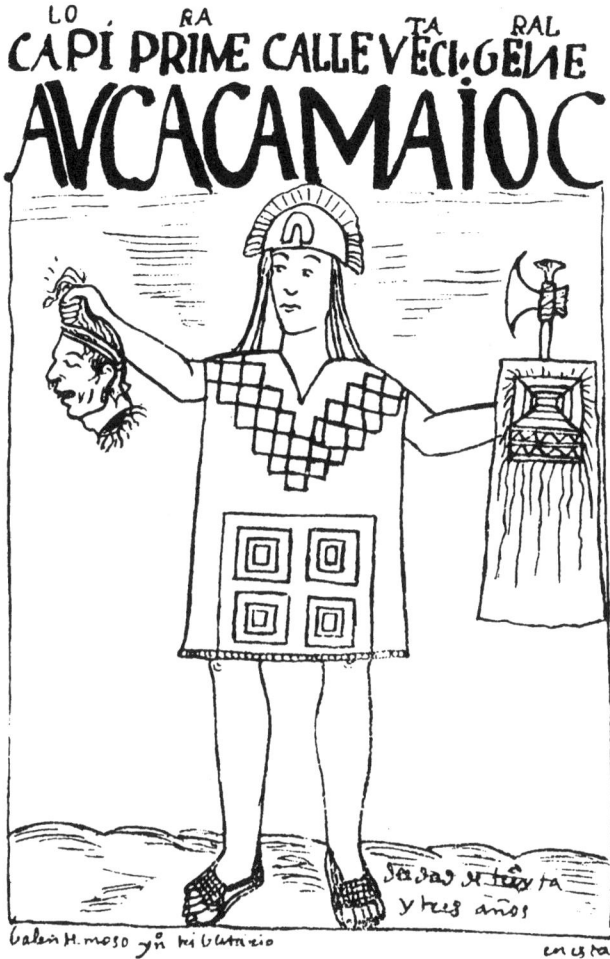

8.3. An *awqa kamayuq*, or "warrior," according to Guamán
Poma (1980:168).

ing these chest plaques (fig. 8.4). This suggests that plaques were specifically associated with battles rather than with the identity of warriors.

A comprehensive study of South American plaques has led Alberto
Rex González (1992) to the conclusion that they were material referents of the Andean sun god—called Punchao by the Inkas—a deity
that iconographically can be traced back to at least the Middle Period.
According to the testimony of an Araucanian chief reported by González

8.4. Rock paintings at Kollpayoc (Quebrada de Humahuaca, Argentina). Note the circular plaques in the warriors' costumes.

(1992:187), the reflective properties of these objects were used in agricultural ceremonies to project sunlight onto the crops, thus enhancing the fertility of the fields. Were the plaques used in a similar way, as mirrors to shed sunlight on the combatants? Was this a way of invoking the power of Punchao to aid Andean warriors in battle? Another Indian author writing in the early seventeenth century, Joan de Santa Cruz Pachacuti Yamqui Salcamaygua, supports this interpretation when stating that the Inkas made "many *pura puras* in silver and gold and silver for the soldiers, to put on their chests and backs, so the arrows and spears would not harm their bodies" (1993:216).[3] The relationship between the solar cult and war implied by this practice is further indicated by another of Guamán Poma's illustrations; it depicts Cuci Uanchire—third commander of the Inka armies—who "first had to drink with the sun his father" before going to battle (1980:126–127).

Axes

Axes with bronze blades of different shapes, plain or decorated, have been found in late prehispanic contexts throughout the Circumpuna area (González 1979; Latcham 1938:318), but like metal plaques, they appear for the first time several centuries before. Although some axes may have been effective for fighting, their relatively small numbers and the exceptional skills invested in the manufacture and decoration of some of them suggests that they served more as special ritual tools or emblems than as ordinary weapons.

The signifying powers of axes are strongly indicated by the iconographic relevance they acquired, as isolated elements (Aschero 1979) or as part of other images like the so-called "sacrificer." This is a complex motif that represents an individual, sometimes showing feline attributes, who holds an ax in one hand and a disembodied head in the other, analogous to what Guamán Poma explicitly refers to as "a warrior" (fig. 8.3). The "head taker"—as we will call it to focus on the icon rather than on its interpretation—has been portrayed on a variety of media, including rock art, metal plaques, ceramic vessels, and wooden snuffing paraphernalia. In the central Andes, this motif goes back at least to the Formative Period (e.g., Paracas, Moche), but only reaches the deserts south of the Titicaca basin during the Middle Horizon, as can be seen in the decoration of imported Tiwanaku artifacts found in San Pedro de Atacama (Mostny 1958; Torres 1987) or in the iconography of La Aguada (González 1998).

It is widely accepted that axes served as emblems of power among Circumpuna peoples and that they had an extraordinary value at the time (Aschero 2000a; Núñez 1987:99). A few specimens have been found in exceptionally rich burials far away from their probable manufacture areas—such as on the Atlantic coast of Patagonia (Gómez Otero and Dahinten 1999)—highlighting the fame of these objects beyond the Andes. In rock art, axes and related motifs (e.g., double axes or anchorlike motifs, ax-shaped head pieces) appear mainly in association with interregional traffic routes (Aschero 2000a; Berenguer 2004; Núñez 1987:98). In Quebrada de Humahuaca, they have also been found on the edge of the agricultural axis of the valley (e.g., in Ucumazo), as if they were marking the frontiers of the Humahuaca realm. This ratifies the emblematic character of axes and their relationship with the definition of the territorial boundaries and trade routes that pre-Inka warfare may have threatened.

Trumpets

Unlike Central Andean trumpets, commonly made of shell, Circumpuna ones are mostly made of animal—exceptionally human—

bone (Gudemos 1998). They are composite instruments made with three fitted pieces sealed with resin. Trumpets have been found mainly in burials and are particularly frequent in Quebrada del Toro and Quebrada de Humahuaca. Some of them have engraved decorations, e.g., geometric designs or "shield" motifs.

The sound of trumpets, like the wind, was considered the voice of supernatural beings or wak'as, so these instruments were necessary in situations that required the intervention of these entities, like initiation or war (Gruszczynska-Ziótkowska 1995:135; Martínez 1995:85–86). As such, trumpets embodied the beneficial as well as the dangerous powers of deities; their ability to both create and destroy was a duality summarized in the perception of these objects as weapons, which bring prosperity to the community by destroying those who threaten it.

Andean armies played trumpets not only in battle, but also during the propitiatory ceremonies held before going to war (Murúa 1962, II:94; Pizarro 1965, f75). Their sound was considered so powerful that, according to some chronicles (Pachacuti Yamqui Salcamaygua 1993:264), Atahualpa—the last Inka—used a special division of deaf-mute soldiers (*yndios mudos*) as a vanguard in his attack against his brother Huascar (Gudemos 1998).

According to the logic outlined before, trumpets were also played to "fight" the elements that could threaten the crops or the herds (i.e., insects, hale, or lightning). Guamán Poma, for example, describes a procession to expel

> hale and ice and lightning which they drive out with weapons and drums and flutes and trumpets and little bells, yelling, saying: . . . [Thieve, spoiler of the people, I will cut your throat, I don't want to see you ever again!] (Guamán Poma de Ayala 1980:259)

José Luis Martínez (1995) has also demonstrated that trumpets— together with stools (*tianas*), biers (*andas*), and feathers—were widely recognized emblems of authority in the Andes, from the Inka to the lesser kurakas. This association is further sustained by the fact that several trumpets found in Quebrada de Humahuaca are engraved with the "shield motif," which some interpret as a highly stylized representation of a shield, another element that would associate trumpets and war. This design, which as noted before, appears also on metal plaques, ceramic vessels, and rock art throughout the Circumpuna area, has been repeatedly interpreted as an emblem of political power (Aschero 2000a:38).

Slings and Pukaras

Slings (*warakas*) have been rarely preserved, except in extremely arid environments like the deserts of northern Chile. However, the piles of selected, fist-sized cobbles commonly found behind the defensive walls of Circumpuna fortresses attest to the importance of these artifacts in pre-Inka warfare.

The double function of slings as weapons and herding tools establishes a first connection between war and pastoral production or related concepts (e.g., life, fertility), as implied in Guamán Poma's depiction of the Coia Raimi or "solemn feast of the queen" in September (fig. 8.5):

> And in this month the Yngas ordered to throw away the diseases and pestilences of all the kingdom. The men, *armed as if they were going to fight a war,* shoot with fire slings, saying, "Get out, diseases and pestilences, from among these people and this town! Leave us!" with a very loud voice. (Guamán Poma de Ayala 1980:227, my emphasis and translation)

Herders use slings to lead their animals and keep them together while grazing, an activity that evokes widely shared concepts of authority. Until recent times, slings served as emblems of ethnic chiefs or *jilakatas* in the southern Bolivian altiplano. The sling was the weapon chosen by the Inka himself when going to battle, as in Guamán Poma's account of Wayna Capac shooting stones of fine gold to his enemy Apo Pinto (a mountain spirit? a mythical ancestor?) to conquer the northern provinces of the empire (1980:304).

Oral accounts throughout the Circumpuna highlands commonly envision mountain spirits, known as Mallkus, using the sling to look after the herds and to fight. These stories typically have two Mallkus quarreling over a female mountain spirit (T´alla), disputes that usually end with one of them beheading the other with a slingshot. Mallkus are described as community chiefs (indeed, the word "mallku" also designates ethnic authorities), as warriors who protect the territory, and as sources of abundance and fertility (*aviadores*), who control the rain

SETIENBRE

COIA RAIMI

quilla

8.5. Men fighting pestilences with fire slings at the Coia Raymi, according to Guamán Poma (1980:226).

and the reproduction of both humans and herds. Ultimately, all these roles derive from the representation of mountains as mythical ancestors, founders of the community or *ayllu* and first conquerors of its land in a mythical time that is commonly associated with the pre-Inka Age of the Awqaruna.

The important role played by ancestors and materialities infused with their agency in Andean warfare was noted by several Spanish witnesses in the sixteenth century (see Rowe 1946:280–281). In the Circumpuna area, this role is further indicated by archaeological evidence from the pukaras of the southern Bolivian altiplano (Nielsen 2002). These fortresses, which were built around AD 1300, are always surrounded by dozens, even hundreds of stone chullpa towers. Chullpas—which appear in great numbers at about the same time as pukaras—served as ancestor monuments (Aldunate and Castro 1981; Isbell 1997), analogous to the *wanka* monoliths (Duviols 1979) or to the "above-ground sepulchers" found in other parts of the Circumpuna highlands. As such, chullpas had multiple functions (as sepulchers, altars, storage chambers, or landmarks)[4] that were coherent within a practical logic that conceived of ancestors as ultimate sources of life, political power, and protection (Nielsen 2006). In the case of fortresses, chullpa towers are aligned to "protect" the most vulnerable sides, distributed along outer walls or, sometimes, inset within these walls, as if they were part of the defensive engineering of the sites.

Disembodied Heads

Skulls specifically prepared as trophies only appear in the Circumpuna highlands (specially northwestern Argentina) during the RDP, around AD 1200 (Nielsen 2001), although beheading has earlier iconographic expression in the area with the "head taker" motif described before. During the RDP, iconographic references to decapitation are less frequent but consistently present, not only on metal plaques and snuffing paraphernalia, but also in rock art (e.g., Berenguer 2004).

Isolated head burials without the characteristic mutilations of trophies, on the other hand, appear even before, in the early formative (e.g., Yacobaccio 2000). If we consider them as part of a more general set of practices involving the manipulation of selected parts of the human skeleton (curation, transport, and reburial), then the genealogy of these practices may go back to the Archaic Period (Aschero 2000b; Standen and Santoro 1994), although they also become particularly frequent in the area during the RDP. Moreover, both trophy heads and caches of selected human parts have been documented at the same sites, suggesting that these two practices may have been quite related at the time.

Disembodied heads and decapitation have received multiple interpretations and have raised considerable debate in the Andean literature. Usually these arguments are confronted in an attempt to decide which one of them captures best the significance of this practice for ancient people in specific cases. The complexity of the evidence—which explains why the debate remains unresolved—suggests that beheading may have been part of different chains of action and may have had multiple meanings in different contexts. If this is true, instead of searching for *the* true interpretation, we could look at disembodied heads as polisemic items that weave all the objects and semantic domains we have considered thus far into a single system. On one hand, they have been associated with war (Rowe 1946:279). The possession of trophy heads allowed warriors to control the spiritual powers of their enemies and served as a form of protection against their agency after death (Vignati 1930). Their public display could be an effective way of commemorating the achievements of warriors in battle, claiming the social recognition they deserved (Redmond 1994; cf. Guamán Poma 1980:131).

On the other hand, John Verano (1995) has interpreted the ritual manipulation of human remains, disembodied heads included, as testimony of ancestor worship. This interpretation is supported by archaeological data from Los Amarillos and Tastil, two large, defensive settlements in northwestern Argentina dating to the 1300s. In the first of these sites, burials of selected human parts have been found in direct association with three "above-ground graves" analogous to chullpas built on an artificial platform overlooking the main public area of the site (Nielsen 2006). In Tastil, 75 percent of the adult skeletons found in the 140 burials that were excavated by Eduardo Cigliano (1973) were missing their skulls, including the two mummy bundles found with wealthy offerings inside an above-ground grave found in the middle of the central plaza.

This affiliation with ancestorship is probably responsible for the frequent use of disembodied heads in fertility rituals noted by several authors (e.g., DeLeonardis 2000:382; González 1992:185; Sawyer 1961). Armando Vivante (1973), for example, cites several cases from the highlands of Peru and Bolivia, in which skulls taken from modern cemeteries or archaeological sites are exposed to stop the rain, whereas in the southern Bolivian altiplano, we have observed the current practice of burying llama and sheep skulls in the summit of "powerful" mountains (Mallkus) in order to call the rain.

When discussing axes, we mentioned the relationship between disembodied heads—as part of the "head taker" motif—and the concept of authority. The frequent representation of the "head taker" on snuff trays and tubes (Mostny 1958; Torres 1987) relates decapitation to hallucinatory experiences, introducing a whole new semantic domain, which lies at the heart of Andean concepts of power and which we may call "transmutation." This relationship has been interpreted as an indication of the importance of both human sacrifice and shamanistic practices for the constitution of political leadership in Middle Period societies like Aguada and Tiwanaku (Berenguer 1998; González 1998; Pérez 2000). Given the absence of significant warfare indicators at this time, it is thought that the "head taker" does not portray a real human warrior but a supernatural entity or mythical character endowed with these attributes.

During the RDP, however, all these elements probably relate to actual armed conflicts. As we mentioned before, when presenting the *awqa kamayuq* or pre-Inka warrior, Guamán Poma shows a man holding a bleeding head in his right hand and an ax and a shield in his left hand (fig. 8.3).[5] Archaeological evidences from the Circumpuna show contextual and iconographic associations between trumpets (a war emblem) and the consumption of hallucinogenic substances (Gudemos 1998:91; Pérez de Arce 1995). One snuff tube found in the late fortified settlement of La Paya, for example, is decorated with the representation of a camelid and a man holding a heavy ax and blowing a trumpet (Ambrosetti 1906:22). These relationships recall yet another set of images associated with warfare, which also fall within the concept of "transmutation." In his account of the Age of the Awqaruna, Guamán Poma describes the metamorphosis of warriors into mythical animals during battle: "Out of their courage, the *aucaruna* or *amqa* became great captains and brave princes. They say that during the battle they turned into lions and tigers and foxes and vultures, sparrows and wildcats" (1980:52; cf. Platt 1987:88–89).

This passage suggests that warriors fought with a strength that was greater than their own; through this transmutation, which was probably aided by the use of masks, special costumes, and—perhaps—hallucinogenic substances, they embodied the powers of mythical animals and other supernatural entities.[6] Perhaps this explains why some of the cuirasses of Circumpuna warriors were made with wild animal skins. One of these cuirasses, purchased by Stig Rydén (1944) was found in Lasana, where it preserved due to the extremely dry conditions that prevail in the Atacama Desert. It is made of cayman (alligator) hide decorated with monkey fur, a combination that would be hard to attribute to utilitarianism, considering that these animals live more than five hundred kilometers east of Atacama as the bird flies, on the other side of the Andes. Similar cuirasses depicted in rock art are dotted, as if they were made of jaguar skins, or have bird designs on them. Could we establish a genealogical connection between the metamorphosis of warriors in battle and the transmutation of earlier shaman/leaders into jaguars and other deities during hallucinatory trances?

On How Ancestors Defeated Warriors

The analysis of war-related objects from the Circumpuna Regional Developments Period has revealed a network of practical, mutually reinforcing metaphors that relate warfare to other semantic domains, such as fertility, authority, transmutation, and ancestorship. Now we will look briefly at the events and processes that brought this system into being.

Let us go back to the pre-war time, to the end of the first millennium AD, when the influence of Tiwanaku on the southern highlands began to wane. This seems to have been a time of relative prosperity. Favored by good environmental conditions, the population was growing and agropastoral economies were expanding, colonizing even the agriculturally marginal habitats of the dry puna. Long-distance caravans and other forms of exchange allowed a relatively fluid circulation of people and resources across regions.

Current data indicate that local autonomy of small communities was the prevailing pattern in the political landscape of the south Andes. It has been proposed on the basis of iconographic affinities and the distribution of trade goods that two "interaction spheres" developed in this area during the Middle Period (AD 500–1000); one of them was related to Tiwanaku and locally centered in San Pedro de Atacama (northern Chile), while the second one was associated with the Aguada culture complex and developed in the temperate valleys of northwestern Argentina. We purposefully use this vague concept because (1) the degree of cultural homogeneity that took place during this period of "integration" has not been sufficiently documented; (2) if these trends toward unification were demonstrated, it is unclear what kinds of practices could account for them or whether they involved some form of subordination between communities; and (3) it is clear that many areas within the Circumpuna Andes were not significantly affected by these supraregional phenomena. These caveats notwithstanding, it seems that during the second half of the first millennium AD, the populations of the south Andes shared—or knew about—a number of general dispositions and representations regarding cosmology and political power, even if they had different importance and practical implications in each region. We want to point out three aspects of this common view. First, political power was associated with religious beliefs that show some homologies with Tiwanaku;[7] these beliefs involved some of the themes we have identified in later cosmologies, like the sun god, the sacrificer, felines, reptiles, and other zoomorphic deities. Second, shamanic experiences induced by the consumption of hallucinogenic substances (*Anadenanthera* sp.) were central to the constitution of political authority, perhaps because they were thought to give certain individuals a special contact with those deities. Third, access to nonlocal goods and some sophisticated crafts (metal artifacts, semiprecious stones, fine pottery, textiles, hallucinogenic substances) was central to the reproduction of both spheres and the political order associated with them.

Between AD 1000 and 1200, Aguada, San Pedro, and Tiwanaku collapsed, a phenomenon that has been related to the onset around AD 1040 of a long-term cycle of droughts that would acquire catastrophic proportions between 1250 and 1310 (Thompson et al. 1985; cf. Binford et al. 1997; González 1998). Whatever the merits of this explanation, the disarticulation of Middle Period interaction networks marks the beginning of the Regional Developments Period. Unfortunately, this early phase of the RDP (AD 1000–1250) is poorly known. Some Middle

Period themes and representations persist, but they seem to be incorporated into new semantic structures.

In the Circumpuna area, climatic deterioration probably affected most severely the populations of the dry puna (the central and southern Bolivian altiplano, the Argentine puna), who depended on a combination of dry farming and pastoralism, two activities that are extremely sensitive to variations in precipitation (Nielsen 2001:247). This situation may have been handled through technological improvements and the intensification of traditional relations of economic "complementarity," such as caravan journeys, verticality, and other arrangements that allowed for multiethnic access to key resources. This would be consistent with archaeological data that show the construction of irrigation systems, the appearance of highland colonies in the temperate valleys of northern Chile (Berenguer 2004; Núñez and Dillehay 1979), and a progressive intensification of interregional trade within the Circumpuna area.

By the late 1200s, when droughts became most severe, violence broke out. Was it due to fear of scarcity? Probably, but how did this fear translate into practice? How did cooperation turn into hostility? How did war transform the cultural framework in which new power arrangements were forged? Good answers to these questions will have to wait for more research; here, we can only point to some facts that are relevant to those answers. First, conflicts crystallized interregional identity contrasts (cf. Arkush, this volume, for the central Andes). Differences in ceramics, architecture, and textiles, as well as in burial customs and the organization of domestic spaces, speak to the materiality of this new cultural mosaic and point to some of the practices that created these collective identities.

Available evidence suggests that armed conflicts did not prevent interregional exchange. On the modern assumption that war and trade are incompatible, south Andean archaeologists have proposed several scenarios to account for the coexistence of war and trade. According to one of them, conflicts happened relatively early in the RDP, then stopped and trade was resumed (Berenguer 2004; Schiappacasse et al. 1989). Conversely, others think that exchange and colonization came first and led to war (Núñez and Dillehay 1979; Núñez Regueiro 1974). The problem is that chronological data from multiple regions (Nielsen

2001, 2002) indicate that the main time of conflict was the fourteenth century, continuing until the expansion of the Inkas in the fifteenth century. Some of the defensive settlements of the RDP were still inhabited at the time of European contact.

The question of how war and trade coexisted is still a matter of speculation. Their co-occurrence could be related to the emergence of non-affiliated caravan drovers (for an example of shifting ethnic labels used by caravanners in the sixteenth century, see Lozano Machuca 1992) or could be part of a purposefully ambiguous policy that allowed for rapidly alternating, even coexisting, forms of interaction (see Wiessner, this volume). The multiple connotations of some of the weapons-emblems previously analyzed, which combine notions of life and danger, seem to materially sustain these ambiguities, and the notion that their skillful manipulation was an important dimension of authority.

The new concern with conflict is revealed by the proliferation of weapons, changes in their design, the adoption of armory (cuirasses and helmets), trophies, and the iconographic protagonism that references to fighting acquired in the rock art of some regions.[8] The best index of the insecurity that hostility produced among people is the radical change in settlement patterns that took place in the late thirteenth and fourteenth centuries throughout the area. People aggregated, left vulnerable positions for defensible ones, fortified their villages or built hilltop refuges close to them (pukaras), and abandoned areas to create buffer zones. The "loss of place" and associated memory implied by this settlement shift was certainly an important condition that facilitated the political transformations of this time. In addition to these, Circumpuna peoples mobilized a number of culturally unique resources to face this period of conflicts; with familiar ritual techniques (mirrors, music, costumes, libations, psychotropic drugs), they invoked the protection and destructive powers of ancient (Middle Period) deities like the sun god and his zoomorphic companions; beheading their rivals, they embodied mythical heroes.

Most significant, however, was the intervention of ancestors. As noted before, archaeological evidence of ancestor worship—that is, the manipulation of human parts or lithic representations of ancestors (García Azcárate 1996; Yacobaccio 2000)—appear sporadically in different parts of the Circumpuna highlands since Preceramic times. During the Middle Period, these expressions are not particularly visible, but in the 1200s, when warfare indicators proliferate in the area, ancestors come back to the scene in monumental forms—such as chullpa towers and *wanka* monoliths—taking control of the landscape, the surrounding communities (*llacta*), and their fields (*chakra*).

Ancestors were also central to the organizational changes of this time. As noted before, the population aggregated rapidly to form settlements of up to one or two thousand people and integrated into multicommunity political structures attested by the emergence of clearly defined settlement hierarchies. This allowed the coordinated action of large numbers of people for defense. The association of chullpas, aboveground graves, and wankas with the plazas of these residential conglomerates, together with indicators of the preparation and consumption of large quantities of food and maize beer (*chicha*) in some of them, suggest that a combination of ancestor worship and feasting played a central role in the integration of the new "polities" (Nielsen 2006).

Unlike Middle Period political systems, which apparently formed through the attraction exerted by certain ceremonial/economic nodes on an "unbound" network of interacting communities, RDP polities crystallized through violent territorial competition. Tristan Platt's model of "segmentary fusion" (1987:95) is probably a good approximation of the processes of socioterritorial integration that took place in the Circumpuna highlands between AD 1200 and 1400.[9] Initially, outstanding war leaders would harness the support of their own communities; through their victories, some of these leaders (*mallkus*) would gain authority over other communities, thus building new, more encompassing levels of political action. During this process, armed confrontations (*ch'ajwa*), which at the beginning may have been ubiquitous, were gradually pushed out toward the newly developed territorial boundaries, turning local tensions into controlled, intragroup competitions, perhaps analogous to the "ritual battles" (*tinku*) documented by ethnohistory and ethnography (Hopkins 1982; Urton 1994).[10] At the time the Inkas began to expand over the south Andes, this process had integrated to some extent—if not as unified polities, at least as confederations or alliances—the populations of every major agricultural region of the Circumpuna area, as indicated by their association with distinctive "archaeological cultures" separated by "buffer zones" (e.g., Nielsen 2001).[11]

The emerging power structures were different from their Middle Period predecessors in other important ways. They integrated larger numbers of people and were able to mobilize a superior labor force (as reflected in the scale of the agricultural infrastructure and other construction projects), with a more limited use of wealth or iconographic display, a fact that has led some authors to talk about an era of "cultural impoverishment" in some regions (Núñez 1991:61) and evokes Timothy Earle's disillusion with the Wankas' lack of elaborate, materialized ideology. What this contrast seems to imply is a shift—in some regions at least—from an exclusionary to a corporate mode of political action that was only superficially masked by the continuity of individual icons, practices, and representations re-signified in the new historical context. Guamán Poma seems to support this interpretation and gives us a hint of the semantic transformation involved by explicitly drawing a connection among warfare, transmutation, ancestorship, and the political order that emerged from the Age of the Awqaruna:

> Out of their courage, they [the *awqaruna* or warriors] became great captains and brave princes. They say that during the battle they turned into lions and tigers and foxes and vultures, sparrows and wildcats. Thus, their descendants up to now are called *poma* [lion], *otorongo* [jaguar], *atoc* [fox], *condor*, *anca* [sparrow], *uaco* [wildcat], wind, *acapana* [sky], bird, *uayanay* [macaw], *machacuay* [snake], *amaro* [serpent]. And in this way their names were called like other animals and weapons carried by their ancestors; they won them in the battles they held. The most prized lord names were *poma*, *guamán* hawk, *anca*, *condor*, *acapana*, *guayanay*, *curi* [gold], *cullque* [silver], as it is until today. (Guamán Poma de Ayala 1980:52; my translation)

Because of the merits they gained in battle, ancient warriors achieved (institutionalized) positions of authority (such as that of captain or prince) and became the founders of lineages (i.e., ancestors). They were not remembered by their individual names, however, but by the names of the guardian animals and other supernatural forces they embodied in war, a point that highlights the fundamentally corporate nature of political power in these social formations. Ancestors (and therefore their descendants as a collectivity) rather than individual war leaders were the ones who harnessed the prestige and political power born out of conflict.

This passage also brings back the association between the ability to transmute and political power, this time applied to warriors who turned into ancestral animals. The appearance of the ancestors in this context should not be surprising, considering the obvious fact that war brings life and death close together, thus facilitating the communication between the living and their forefathers; in the same way that a person may die in battle, the dead may slip back to life (Bouysse-Cassagne 1975:203). But it could also be one aspect of a more general perception of battles as liminal contexts, where a number of extraordinary transformations may occur—like the metamorphosis of warriors into ancestral animals—or of war as a condition that can effect a complete change in society.

Note also that the "weapons carried by their ancestors" (or perhaps some of the materials of which they were made, like the gold or silver of pectorals) became "the most prized lord names." If the political order we have just outlined was born out of war, it is only reasonable that weapons would become important emblems of authority. Although some of them (e.g., axes) may have already carried similar connotations since the Middle Period, others (e.g., slings) may have only acquired such meanings at this time. This practical connection would be later ratified by the use of the same weapons by the Inkas.

Finally, we should consider the semantic connections among warfare, the authorities that emerged from it, and fertility or related concepts (life, prosperity). Based on the analysis of ethnographic and lexical data, it has been demonstrated that (1) for Andeans, there was no sharp distinction between "real" (chaxwa) and "ritual" (tinku) warfare—they were extremes in a continuum; and (2) both belong to a larger semantic field that comprises multiple kinds of relationships among contraries, including reconciliation, sexual intercourse, and marriage (Bouysse-Cassagne and Harris 1987; Platt 1987; Topic and Topic 1997). These semantic connections reveal that armed hostility was viewed as a temporary form of interaction between different and symbolically opposite social units, whose cooperation in other times would be an important source of prosperity. The emergence of more inclusive authorities in a "time of war" (*pachakuti* in Aymara [Bertonio 1612:242]) would directly effect this reversal of conditions (*kuti* means revolution, complete reversal [Bouysse-Cassagne 1975:200–203]), endowing emergent leaders with a practical connection to fertiliy and life.

Acknowledgments

I would like to express my gratitude to all the participants in the advanced seminar on "Warfare in Cultural Context" for an extremely enriching intellectual experience and to John Ware for his hospitality at the Amerind Foundation.

Wars, Rumors of Wars, and the Production of Violence

Timothy R. Pauketat

Until recently, most archaeologists thought that warfare in ancient soci-eties was to be isolated and investigated as if one might study kinship, capital, or cancer. Bellicose behaviors, some argued, evolved along-side certain types of political organizations (e.g., Haas 2001; Otterbein 1970). Warfare, others concluded, had real economic causes—imbal-ances of land, labor, or other resources (Keeley 1996; LeBlanc 2003). Biology, still others asserted, explains human aggression and, by exten-sion, organized conflict (e.g., Maschner 1996; but contrast Neves, this volume; Thorpe 2003).

However, since the mid-1990s, theoretical advancements and global developments have altered our views on ancient violence (see Gilchrist 2003; Nielsen and Walker, this volume). That's not to say that the old "positivist" views on warfare—behavioral, economic, or biological—were all wrong (see Nielsen and Walker, this volume). They weren't. But they were wrong to the extent that they stripped peoples of their his-tory and their causal powers in ways that "pacified" the past (see Keeley 1996; Vandkilde 2003).

The pacification of the past involved conceptually removing violence from everyday life and social history in three ways—two theoretical and one methodological. First, earlier studies ignored the transforma-tive dimension of particular historical moments and the commemora-tions of those moments—episodes of confrontation, fortification, or arms production. It was as if warring didn't cause change, general evo-lutionary forces did (Vandkilde 2003). Second, earlier studies decontex-tualized and essentialized "organized violence," making it a thing unto itself, as if comprised of tactical practices and spaces disconnected or far removed from "ordinary" social experience (Thorpe 2001). Third, to be able to speak of violence as a "factor" in cultural change (in accordance with the two preceding theoretical positions), one had to recover "direct" evidence of the extraordinary practices or places. That, in turn, usually meant finding osteological remains showing weapon-induced trauma in situ (e.g., Lambert 2002; Milner 1999). Short of smoking-gun evidence, such as bodies in a ditch at Crow Creek (South Dakota) or on house floors at Mohenjo-Daro (Pakistan) or Casas Grandes (Chihuahua), archaeologists equivocated over the meanings of walls or weapons and, hence, significantly understated the significance of warring in the past (but see Keeley 1996; LeBlanc 1999, 2003).

While understanding warfare is problematic, the pacification of the past also involves a much larger problem: a mislocation of the causes and effects of human violence. So far, most archaeological studies have not adequately contextualized warfare within historically contingent social fields of the ancient past. Nick Thorpe (2001:134) reminds us that "simply accumulating evidence to demonstrate that people have always been in conflict is no longer sufficient. Instead, we have to move to a more nuanced approach which recognizes that early war had its ritual and real aspects, and that it was not something set aside from the rest of social action, but an integral part of it."

To their credit, the contributors to this volume do initiate a more "nuanced" socio-historical approach to the past, the touchstone for their studies being "practice theory" (Nielsen and Walker, this volume). Here, I seek to highlight the successes of the contributors to this book and to extend a few key insights to what Nielsen (this volume) rightly notes as a most "serious" dimension of human history. In order to do this, I'll review the theory and then argue, perhaps counterintuitively, that we need to de-emphasize "warfare" in order to understand ancient violence.

That violence, including any fuller understanding of organized conflicts, can only be understood if we begin analyzing the host of practices involved in the cultural production—which is to say the materiality and spatiality—of all sorts and scales of violence. Important aspects of that historical process include fortification, enclosure, and the segmentation of entire social fields. These are to be understood by comparing genealogies of things, places, and peoples and tacking between the evidences of warring, peacemaking, and daily engagements of people, places, and life.

Theoretical Battleground

Practice theory is a place to begin, a series of precepts and bodies of research that relate objective reality to the "subjectivities" we all walk around with every day (Nielsen and Walker, this volume). The theory rests on the notion that cultural practices (which are not the same thing as behaviors) are the embodiments and enactments of those subjectivities, which is to say cultural dispositions, traditions, or structures (Pauketat 2001). Following Pierre Bourdieu (1977), as do several contributors to this volume, practice theory might lead us to seek the disparate and conflicting motivations for violent acts and warfare among the cultural dispositions (i.e., "habitus") of ancient people.

This sense of practice—and the emphasis on cultural motivations—leads some to imply that our explanations of warfare are contingent on understanding the "generalized notions," "cultural codes," "shared ritual and cosmological narratives," "repeated practices," or "ideologies" of warfare (see the chapters by Arkush, Cobb and Giles, Inomata and Triadan, and Topic and Topic). It even leads to the implication, found in a couple of the chapters, that codes or cultural structures were themselves the agents of change, a position like that taken by Marshall Sahlins (1985).

But that is going too far and, as others have pointed out, this Bourdieu-inspired and Sahlins-esque sense of practice actually perpetuates the (Cartesian) "structure-agency" or "culture-action" dichotomy that Bourdieu sought to overcome (cf. Farnell 1999; R. Joyce 2005; Meskell 1999; Strathern 1988). It seems that, without archaeology's access to genealogies of people, places, and things, nonarchaeological theorizing slips easily back into a Cartesian formula where cultures are shared, learned in group contexts, and held in the minds of people until "expressed" or "materialized" after the fact (as criticized by Nielsen, this volume). And while such a mentalist insistence (and the concomitant Sahlins-esque privileging of text over other lines of evidence) is still common to cultural anthropology, a growing number of archaeologists are recognizing that thinking as a human or, more accurately, being a human simultaneously occurs in material and spatial dimensions—at the interface of the body and the outside world—and not in the head (e.g., Walker, this volume; also Dobres 2000; R. Joyce 2000, 2005; Pauketat 2001).

This is a "phenomenological" position, where culture is not a thing but an experience (see Merleau-Ponty 1962). It is always culture-in-the-making and exists (and is always being remade) out in the open during the moments and in spaces where people experience, reference, or remember who they are and what they've done in the past. Sensibilities and memories are always momentary and subject to reinterpretation, imperfect recollection, and negotiation today and in the past in all social contexts. And those contexts are parts of larger social fields, arenas, or cultural landscapes that have material and spatial dimensions.

In other words, warfare is not best explained by claiming that shared meanings, identities, or ideologies first coalesced to later motivate violent acts. Rather, the "sharing" of perceptions, sensibilities, identities, etc., is best considered as an active "project" (Nielsen, this volume) or series of practices or culture-making moments that resulted in collective overlapping memories. Certain subjectivities thought to motivate social action—for instance, warrior identities, cults, or codes—are more appropriately considered "official" if not also masculine narratives. They are the imagined histories and propaganda of a few men (cf. Cobb and Giles, Inomata and Triadan, Topic and Topic, Wiessner, this volume). Helle Vandkilde (2003) even argues that the priority we give such official ancient narratives is, in part, a function of our own modern-day notions that conflate masculinity, aggression, and national identity.

Certainly, violent encounters, battles, ambuscades, assaults, etc., are only a few of the many culture-making moments or ideological projects of some particular history, albeit the most extreme or intense moments or projects. Likewise, warriors are just one constructed category of persons. The warrior category does not necessarily encapsulate the totality of violent practices that matter in any larger understanding of human history. There were, and are, other "practices of war" constituted "away from battlefields" through public performances and other, less public "daily acts" (Inomata and Triadan, this volume). Such practices were not necessarily aimed at motivating warriors to go to war. In fact, it is as inappropriate to identify the "role" of some belief or identity as it is difficult to identify any pure "practices of war" that didn't also construct cultural identities, genders, or cosmologies (see the chapters by Arkush, Nielsen, Topic and Topic, and Weissner).

Wars

As a result, the authors in this book sometimes seem to talk very little about what Lawrence Keeley (1996) or Steven LeBlanc (2003) would call "real war." In the present case, this lack of discussion, I believe, is mostly a function of the authors' desires to contextualize (not ignore) the "practical" dimensions of warring. Clearly, all the authors appreciate the seriousness of combat and killing, particularly the dramatic social and demographic effects of, say, the Inkan conquest (Arkush, Nielsen, Topic and Topic) or even more recent histories of endemic conflict (Neves, Wiessner). So too did the people of the past. For instance, the Enga's perceptions of the perils and destabilizing potentialities of warring partially explain, according to Polly Weissner (this volume), their ritualized warring and ceremonial exchange. Unchecked, Enga wars transformed lives through a host of social ills—theft, homicide, rape—and demographic reconfigurations.

So, future practice-based and phenomenological approaches to violence need to consider the reasons (or motivations) for warring in terms of the larger historical contexts associated with real war, particularly in terms of demography, migration, and identity (e.g., Pauketat 2003). An important step in this direction is to better integrate within our theories what Abraham Lincoln recognized as war's "awful arithmetic." If there is one thing that will always change history and culture-making trajectories, it is the removal or elimination of the people doing the changing or making. Real wars, we could say, are about killing for demographic and hence cultural effect. Twentieth-century genocides and ethnic cleansings are all-too-obvious examples of that sinister principle.

But the principle is equally applicable at a smaller scale to the study of the ancient past. For instance, among the various regional complexes that comprise the Mississippian subject matter of Charles Cobb and Brett Giles (this volume), there is one called the "Powers phase" in modern-day southeast Missouri. At about AD 1350, the central town—"Powers Fort"—and outliers of the Powers-phase people were attacked. Many thatched-roof houses and temples at the various sites were incinerated (O'Brien 2001; Price and Griffin 1977). Since there were no corpses on floors, it is unclear whether the targets of the attacks were people or the Powers-phase places themselves (see below). Regardless, we may presume that the Powers-phase folks fled and, perhaps, regrouped elsewhere. Whether they relocated en masse to another place or they splintered in many directions of flight is unknown. However, from a practice-theoretic vantage point, we cannot assume that the refugees and émigrés merely transplanted intact to new locations some antecedent set of social relations or cultural practices.

No, displacement means that things were necessarily mixed up, old practices and cultural referents were abandoned, reformulated, and replaced in new landscapes. Migrations might even have "caused" certain warfare practices, in some sense (à *la* Neves, this volume). And survivors—even if they sought to recreate the places or practices of memory—necessarily would have reconfigured their daily social lives in new or altered social environments. Indeed, what followed the Powers-phase war in the fourteenth century was the near-total abandonment of the entire central Mississippi valley (Cobb and Butler 2002; Williams 1990). South of this huge transregional "Vacant Quarter" (below present-day Memphis, Tennessee), subsequent Mississippian towns were also changed. From 1350 on, they were entrenched within their defensive walls. People didn't bother, and probably feared, leaving the settlement to discard human refuse and wastes (see Pauketat 2004).

Perhaps the proximate reasons for such pervasive alterations to midcontinental social fields may be found among the bodies in a ditch at the Crow Creek site in present-day South Dakota. In the late fourteenth or early fifteenth century, warring had become an everyday affair across the Midwest, Mid-South, and eastern Great Plains (see Bamforth 1994; Lambert 2002; Milner 1999; Willey 1990). This was particularly problematic on the Plains, where it appears that migrations of diverse eastern and southern agriculturalists (some of whom were descendants of Mississippians) produced an uneasy ethnicized landscape along the middle Missouri River (see Krause 1998). At the Crow Creek site, likely a proto-Arikara village of fifty lodges and several hundred people, an enemy attacked while workmen were building a new defensive ditch and palisade (Willey 1990; Willey and Emerson 1993).

> In an overwhelming strike, attackers burned the village and massacred its inhabitants. Archaeologists found the remains of at least 486 bodies heaped in a segment of the incomplete fortification ditch. More burned body parts were found inside the burned lodges. Men, women, and children had been arrowshot, clubbed, hacked, and burned to death.

Many of the victims were scalped and their bodies mutilated: tongues cut out, teeth broken out, heads cut off, and bodies dismembered (Willey and Emerson 1993:259). Apparently, many young women had been captured or had escaped, as there were fewer than expected among the dead. A high incidence of carnivore gnawing marks on the human bones indicated that the bodies of the victims had laid dead on the ground for some time before escapees returned or sympathizers arrived to heap the corpses and scattered arms, legs, and heads into the still-open ditch (Willey and Emerson 1993, summarized by Pauketat 2005).

Was this simply an "expression" of tribal warfare, perhaps to be explained by human biology or a struggle over resources? Or do we need to consider the midcontinental history of regional abandonments, migrations, ethnogenesis, and transregional violence to explain the Crow Creek massacre? Moreover, do we need to factor the Crow Creek massacre into our explanations of later Plains history, as practice-based approaches would suggest?

For answers to such questions, we would do well to consider Steven LeBlanc's (1999) or Stephen Lekson's (2002) recent reconsiderations of southwestern warfare. LeBlanc (1999:184–186) and Lekson (2002) have both observed that there was remarkably little violence in the greater San Juan basin during Chaco's heyday (ca. AD 900–1140). At the same time, there were higher incidences of "ritual" killings or mistreatments of the dead within greater Chaco, perhaps indicative of what both researchers suspect may have been officially sanctioned violence. Such practices had historical consequences: Lekson (2002) believes that by controlling violence and redefining who or what constituted a threat, Chacoans altered the practices of many subsequent Puebloan peoples. Post-Chaco warring, he thinks, was based in people's suspicions of each other, suspicions that had been inculcated during the earlier period of Chacoan hegemony. Those suspicions and the collapse of Chaco meant endemic small-scale fighting and feuding.

Rumors of Wars

William Walker (this volume) picks up on Lekson's and LeBlanc's considerations of post-Chaco violence, but with divergent conclusions. The post-Chaco evidence—incinerations of kivas, mutilations of likely presumed witches, and fort-like masonry-room blocks and towers—indicate to Walker that the later warfare was more complex than conveyed by allusions to ideologies of suspicion or altered practices of defense. For Walker, Puebloan conflicts were battles in this world and the spirit world, involving living warriors, ancestors, hero figures, and invisible agents. Hence, weapons and fortifications were in some sense intended for these multi-dimensional fields of conflict. Supposed defensive towers were pieces of larger landscapes of violence that pervaded all aspects of life.

The same depth of field is recognized by Cobb and Giles (for the Precolumbian Southeast) and Nielsen (for the southern Andes). Cobb and Giles posit that Mississippian warring was, in effect, more mythical than real. That is, stories and metaphors of conflict and warrior prowess were the means whereby people (presumably mostly men) understood the world. Similarly, Nielsen contends that various cultural objects, hilltop *pukaras*, *chullpa* shrines, and the imagery of disembodied heads and fertility conflated (through their production, use, and display) everyday practices, agriculture, ancestor veneration, and warring. And whereas Cobb and Giles pin their explanation more on a generic warrior ideology whose dominant symbols betray a pervasive anxiety over bodily mutilation (see also Wiessner, this volume), Nielsen relies more on the cultural history of the Regional Developments Period (AD 1000–1450) to situate his explanation. The importance of both, for my purposes, lies in their recognition of the cultural production of violence (see below).

Such production has often been recognized by others as "ritual warfare," although that unfortunate designation (which loses its meaning outside functionalist approaches to the past) belies both the "reality" and historical ramifications of the violence in question. Of course, all violence and warfare was and is—to variable extent—"ritualized" (from *tinku* in the Andes to the modern-day "conventional" wars with their rules and protocols [see Bell 1997:154–155]). Likewise, all "ritual warfare" known from world history had "real" effects. For instance, the oft-cited "Flower Wars" of the Aztec—where foes met on a field of battle, engaged in limited conflict, and then withdrew—had as a long-term consequence the complete absorption of enemy lands by the Aztec (Hassig 1988).

Thus, small-scale gladiatorial contests or the tinku battles, discussed by Theresa Topic and John Topic (this volume), were not ritual wars from a practice-theory vantage point. Tinku battles pitted against one another enemies whose statuses were defined in larger cosmological and historical terms. Supposed defensive constructions in portions of Peru were less than optimal at defense. Hence, Topic and Topic conclude, battlegrounds and hilltop fortresses—as at Chankillo in the Casma Valley—were stages for theatrical organizing and performance of violence. We should not necessarily expect these pukara fortifications—including those with indefensible open gates—to duplicate the realities of modern-day battlements and weaponry, even as we understand such theatrical spaces to have been "defensive" enclosures as well (Arkush, Topic and Topic, this volume). The same goes for Eduardo Neves' (this volume) embankments and palisades, as at Lago Grande, which are not unambiguously defensive according to modern, western standards.

Instead, and as also discussed by Nielsen and Cobb and Giles, we should remember that certain war-making practices were part and parcel of the sociality of the times, recreating or constructing gender and kinship in the Peruvian case discussed by Topic and Topic, among other things. Such cultural construction was even more obviously the cause and consequence of Maya, Enga, and Amazonian conflicts, as presented here by Inomata and Triadan, Wiessner, and Neves. As in other Puebloan, Mississippian, or Andean cases, battles were fought over life and death, both in figurative and literal senses.

That is, for the Maya, warring recapitulated the cosmological struggles between superhuman characters and the supernatural forces of this world and the next, according to Inomata and Triadan. War was grand spectacle, a matter of shock and awe, and that spectacle was "incorporated"—here again figuratively and literally—into the daily lives and bodies of people: those commoners who were sacrificed, those elites who engaged in auto-sacrifice, and also those nonwarrior spectators who might otherwise have never experienced battle themselves. The ballgame and public sacrificial rites held within Maya cities projected theatrical interpretations of persons living and dead and battles in this life and the next. But so too did the Enga and Tapinamba cases, where warring merged with ceremonial exchange and ceremonious sacrifice or cannibalism, respectively (discussed by Wiessner and Neves, this volume).

Constructing Violence

The extent to which violence was incorporated into societies through lived experience is the extent to which we must rethink the older considerations of real wars and ritual wars using practice-based and phenomenological theories. Organized violence, as implied in the foregoing discussion, cannot be adequately explained without embedding it within the larger fields of cultural experience. And once we begin to do this, recognizing that experience and cultural practices necessarily occupy space and possess a materiality, then it becomes increasingly obvious that violence—organized or not—is a pervasive feature of social life that cannot be segregated into behavioral categories, each explained by invoking some separate external cause. Rather, violence broadly conceived must be analyzed genealogically as a recursive cultural quality of entangled causes and consequences.

Materialities of Violence

That is, the motivations to war are very nearly inseparable from the materiality of social life. As Nielsen illustrates, this inseparability happens through the production and use of cultural objects: slings were weapons of war and pastoralism; trumpets were blown in battle and in order to "fight" the elements that could threaten crops or herds; and metal plaques symbolized the sun and warriors just as pukara fortifications and chullpas did ancestors. In this case, or in Cobb and Giles' case, embodied warriors and warrior imagery created cultural associations, identities, and practices. For instance, internecine violence appears in the fourteenth-century Midwest alongside the widespread imagery of warriors on ordinary farmers' cooking pots. Among the Maya, fired clay figurines—mostly of men (and many of those warriors)—appear in everyday contexts, according to Inomata and Triadan.

Warring and warrior imagery was once thought to be the "result" of warfare, but the logic of practice-based theorizing holds the opposite. The materialities of domestic production itself bring about subjectivities and construct cultures of warring. Parallel arguments have been made elsewhere for the production of armaments in the United States during the nineteenth century's Civil War and the twentieth century's world wars. The manufacture of bullets, canons, bombs, ships, and planes, etc., transformed gender, class, and national identity at least as much as the actual battles of those wars altered world history.

In different ways, three or four contributions to this volume seem at the cusp of significant new understandings of the materiality of violence. Elizabeth Arkush's consideration of community and ethnic identities in terms of "building . . . new communities geared for defense" very nearly suggests the inseparability of community, defense, and cultural practices (see also Topic and Topic, this volume). Arkush notes that the locations of pukaras, one every few kilometers or so, correlates with Colla ethnic territories and "defensive" coalitions. And she sees this reflected in pottery styles, leading me to wonder if Andean potters, like the midwestern farmers noted above, were not actively manufacturing the sentiments of warring and inculcating in their children violent proclivities.

The same might be true of Neves's recognition of Amazonian monumentality and centralized pottery production as an indication of higher-order communal formations. Although he points out that neither big mounds nor nice pots necessarily reflect the "control of labor" by elites, it is yet possible that the construction of monuments, the production of feast foods and cookpots, and especially the building of moats and walls during the sixth through the twelfth centuries effectively allowed a new kind or greater degree of war making. Older theories of warfare and normative views on culture have misled us into assumptions that fortifications were inert "defensive" structures largely devoid of causal power. They were more than this, both as moments of construction and as the materiality of subsequent war-making practices. Taking a lead from Walker's discussion of the agency of things, we could say that fortifications possessed a kind of agency.

Spatialities of Violence

After all, like Neves's earthen and wooden fortifications, most ancient walls and ditches were "communal" projects. Building a wall or digging a ditch synchronized labor practices during brief, intense (liminal) moments of culture making that, in the process and thereafter, had at least as much potential as battles to memorialize some new sense of community. As such, we need to consider fortification also as a process of enclosure, like that described by Mathew Johnson (1996) in late

Medieval Europe. Phenomenologically speaking, walls and ditches constrained bodily movements within their enclosed spaces, creating a different spatiality within sites and perhaps generating segmented regional and transregional landscapes of increasingly segregated and compartmentalized space. Places had the power to define identities and might have sometimes assumed the qualities of communities (as imagined identities) or perhaps nonhuman social "persons," as noted by Nielsen (this volume) for pukaras and chullpa towers (see also Fowler 2004).

In thinking more about the agency of places, we should recognize that there was an offensive (i.e., "proactive") dimension to most any so-called "defensive" construction, whether an intentional design feature or not (Hassig 1998). Ultimately, this may be because fortifications allow people to project a larger fighting force while leaving a smaller one behind to defend their home. Hence, walls and ditches—if not whole settlements built into naturally defensible locations—might have been perceived as, if they were not in fact, threats to their neighbors. As threats, they potentially redefined entire social and political landscapes by reconfiguring the spatiality of offensive-defensive possibility.

Fortifications likely tipped the balance, that is, of regional power relations even as they projected "community identities" through construction. Such identity formations (qua fortifications) might have been based on a variety of subjectivities: fear, insecurity, and resentment. And such subjectivities could have been the basis of resistance within or between fortified communities. The materiality of resistance within a community might have taken the form of built-in weaknesses in the fortification, slow construction rates, or failures to maintain the fortification. Resistance from without might have taken the form of pre-emptive military strikes.

Take, for example, the case of the Galisteo basin, southeast of the San Juan and greater Chaco. Beginning in AD 1250, migrants with diverse cultural backgrounds established a series of composite agricultural pueblos, some referencing (if not actually symbolically inaugurating) a prominent nearby landmark called "Petroglyph Hill" (Snead 2004, 2005). Inhabitants of one hilltop site, situated in a side valley of the Galisteo, added a prominent multi-storied square-room block just after AD 1300, a move that James Snead (2005) now interprets as an attempt by residents to make permanent their place in an otherwise

precarious and dangerous social landscape. The move, however, may
have been interpreted as an offensive one by the pueblo's neighbors.
And it might have been these others who, one September morning just
a couple decades later, attacked and burned what we now call "Burnt
Corn Pueblo" (Snead, personal communication, 2005).

The agency of this place did not end there, however. In succeeding
decades, people avoided the locality. Perhaps invisible agents—ghosts—
inhabited the site. Snead concludes:

> The incineration of Burnt Corn Pueblo was not simply a raid or strike
> against economic rivals but an attack on history—either the real history
> of a group with deep cultural ties to the area or the "created" history of
> an immigrant group. In its place the smoking ruins represented a third
> history, one that persisted long after any direct recollection of the events
> that created it had vanished. We can't know what that story was, but in
> Pueblo tradition the destruction of a community was an inevitable cor-
> relate of the moral failure of its inhabitants . . . (Snead 2008:165)

The Production of Peace

Like the earlier examples of the Powers phase and Crow Creek, the
Burnt Corn Pueblo case points out that violence has a spatiality that
directly impinges on the history of entire cultural landscapes, memories, and peoples (see also Saunders 2003). When we begin to analyze the dimensionality of violence and its cultural production at such
scales, however, we quite easily get entangled in complex webs of cause
and effect. Methodologically speaking, the key to disentangling such
all-encompassing historical processes is a two-pronged approach that
Susan Alt and I have discussed elsewhere (Pauketat and Alt 2005). First,
disentangling historical processes entails examinations of the genealogies of cultural production. What were the causes and effects of certain
practices and at what effective scales?

In the present volume, such an approach is nicely exemplified by
Polly Wiessner's study of Enga history. And while her data are not,
strictly speaking, archaeological (nor her theory phenomenological),
the causes and effects of her Enga history do possess an inherent materiality. It is not difficult to imagine archaeologists retrieving data on the
frequency and extent of bodily trauma through time relative to the rate

and centrality of exchange practices, the emergence of more powerful leaders and gendered divisions of labor, and the construction of central sites for Great Ceremonial Wars.[1]

Second, explanations of those historical processes will derive from tacking between alternate lines of evidence at variable effective scales, just as Wiessner does between violence and exchange. As extreme examples, we might consider such a tacking procedure to move from the scale of a single human body—say that of an arrowshot Neolithic Ice Man or a battle-scarred Paleo-Indian—to that of communal fortifications, regional militarizations, continent-wide population movements, and international arms races. For Wiessner, the tacking procedure takes us from the social ills of unchecked violence (attacks on individuals) to the emergence of the great intertribal exchanges and peacemaking ceremonies in the nineteenth century. In the end, it is clear to Wiessner that you can't understand one without knowing the other. Just as exchange and warfare are opposite sides of her tribal coin, so peace and violence are opposite sides of most all historical coins.

Consider that communal sports and "gaming" were often associated with or described as warring (Bell 1997:154). In Maya, Mississippian, and Andean contexts, ballgames, chunkey playing, and boys' *ch'ajwa* games, respectively, were virtually synonymous with warring (Inomata and Triadan; Cobb and Giles; and Topic and Topic; this volume). Hero figures there, and in the American Southwest (see also Walker, this volume), played games with otherworldly combatants, the outcomes of which might include their execution—a practice embodied by players in this world as well—or the significant redistribution of gambled wealth and power (e.g., Pauketat 2004:63).

Elsewhere, I have argued that the production of a transregional peace in the early Mississippian world—the Pax Cahokiana—stemmed from the spread of a political-religious movement centered at the proto-urban center of Cahokia (Pauketat 2004, 2005; following Hall 1991). Lekson (1999) has argued something very similar for ancient Chaco in the Southwest. In the Cahokia case, the materiality of such a peacemaking movement includes the distribution of likely Cahokia-made chunkey gaming pieces, among other "pieces of Cahokia," to far-off peoples. At and around greater Cahokia, the peace was marked by a lack of fortifications.

So, why consider the lack of conflict when analyzing war? In his summary of the New World's colonial period, Murdo Macleod (1998:130) notes that imperial regimes "do not reduce" warfare. Rather, they "displace" it. In part, Macleod (1998) continues, these regimes "send it to more distant locations"—to the margins of their domains. In other words, like all practices, peacemaking was not the realization of some natural state but was someone's project, and as such must be understood genealogically in different dimensions and at different scales of analysis. As I've already noted, the Pax Chaco both followed and preceded periods of pervasive violence (Lekson 2002). Like Chaco, Cahokia's peace was also marked by human sacrifices and, only in far-off lands, village incinerations (Pauketat 2004). Elsewhere, violence was on the rise in the Mississippianizing world (Milner 1999) and, with the evaporation of Pax Cahokiana after AD 1200, became an internecine struggle over places and identities.

Hence, Pax Cahokiana and Pax Chaco—if not also Middle-Horizon Tiwanaku and Pucará in the Andes, or Formative-period Monte Albán, in Oaxaca, Mexico—were products of violence as *displaced phenomena evident only at sub-continental scales* (Arkush, this volume; Janusek 2004; A. Joyce 2005; LeBlanc 1999; Lekson 1999; Pauketat 2004). Beyond this, there is another significant aspect to understanding such seemingly peaceful domination: there was more violence to hegemonies than there was organized warfare. There are cultural-hegemonic qualities built into any sizeable political formation, be it a fortified village, an ancient state, or some larger imperial or global formation (see Janusek 2004; Pauketat 1994; Smith 2003). In such contexts, says Macleod (1998:130), "violence is not always overt." It is hidden and suppressed (see Scott 1990). And from time to time, it may surface in unsanctioned ways.

Speaking of the colonial period Pax Hispanica, Macleod (1998:130–132) notes that hegemony engendered resistance and brought about other forms of violence, including infanticide, murder, mass suicide, and armed rebellion. Based on the case material reviewed in this book, we might add to such a list (of the violence that accompanies or results from hegemonies): massacres, bodily mutilation, torture, emotional anxiety, domestic violence, and various sorts of immolation and auto-sacrifice (see also Meskell and Joyce 2003:147–157). Little of this violence is recognized by most archaeologies of war, given the latter's usual basis in the official narratives of warriors and combat. But they are as much a part of a practice-based approach to violence as are those rare instances of actual combat.

Indeed, state-sponsored killings, such as the sacrificing of the losers of Maya ball games or the killing of captives in honor of Aztec gods, could be seen as "alternatives" to open conflict. In a sense, the "real" difference between such executions, ceremonious sacrifices, and warring might be an "official" narrative that legitimates the violence. Even the distinctions between legitimate armies and terrorists, or political violence and criminal violence, "are ones made by the governmental forces that created the legal system" (Macleod 1998:131). Our archaeologies of warfare, thus, need to merge with archaeologies of death, hegemony, identity, and social life in general.

Conclusions

Simply put, I must conclude that the archaeology of war or warfare should be folded into a broad-based archaeology of violence. War as an event can be isolated today or in the past, but it cannot be explained in its own terms. Likewise, warfare as a set of practices cannot be explained apart from its historical context using generic biological or behavioral terms. It cannot be disembedded from the larger social fields and cultural histories within which it derives. It is not separable from the social life and social ills of all people, even if defensive walls, weapons, warriors, warrior cults, and battlefields seem distinct and isolatable from the domesticity and rituality of everyday social life.

Yes, violence begets violence, or so it has been said, but it is the historical implications of that begetting process—in terms of the materiality and spatiality of the production of violence—that archaeology needs to take into account. There were and are recursive relationships between social life and human violence. Thus, archaeologists must situate their explanations of warring within larger social fields and cultural histories if we hope to explain the "roots" of war, "real" and imagined. To do this, we must ask new questions: How, where, and at what scale was violence organized, performed, or ritualized and, hence, subsumed within the lived landscapes and daily practices of people? Where and at what scale did wars transform demography and culture making? Did warriors target people or places? How and at what scale did memories of wars, the inculcations of aggression, or the productions of arms alter the identities, sensibilities, and practices of peoples for generations to come? How did hegemony and peacemaking alter or displace violence? How did the building of walls, ditches, and watchtowers also build or divide communities vis-à-vis the "real" objective constraints—in terms of labor and firepower—of any such construction? Did the imagined communities that resulted from the physicality of construction memorialize kin ties, homelands, or some greater order (e.g., the Great Wall of China and empire)? How important to such historical processes were Pax Chaco, Pax Cahokiana, Pax Hispanica, and Pax Tiwanaku, to name a few? Were imperial expansions enabled by such state-sponsored periods of peace or by regional provincializations of landscapes through the enclosure of hilltop sites and the segmentation of community identities?

At one time, not so long ago, such questions were unthinkable. Some archaeologists, lacking direct evidence of traumata, figured that they couldn't even broach the issue of warfare, thereby pacifying the past. Others explained their sometimes-meager evidence of tribal fighting, like that documented by Wiessner, with reference to human behavior or biology. Likewise, explanations of centralized political formations, from chiefdoms to empires, rested on abstract beliefs that certain types of governments were inherently expansionistic, perhaps driven by aggrandizers and politicians whose tactical strategies included organized violence (see Blanton et al. 1996, 1999; Haas 2001). In such studies, few to no linkages were made to the other forms of sanctioned killings, be they sacrificial immolations, the executions of criminals and witches, or the killings of family members by patriarchs. There was little thought to the organizing or displacement of violence within society or to the daily practice and inculcation of aggressive sensibilities. Warring was not situated within larger social fields of war making, and war making was somehow separated from the social histories wherein people lived and produced violence of all sorts.

However, from our present-day vantage points, resting on variations of practice theory and phenomenology, we must now seek the genealogies and experiences of arms production, fortification, theatrical performance, enclosure, militarization, segmentation, community formation, and peacemaking among the places, people, and things that archaeologists regularly recover. We measure such things *through things*, which is to say through the materiality of production and the genealogies of objects. We understand such experiences *as experiences* of bodies in spaces and at places, in turn comprised of people, objects, and unseen forces arrayed in ways that affected the subjectivities of all involved. And, based on our new theoretical starting points, we ask different questions regardless of whether or not we have osteological evidence of violent conflict.

Understanding how war and violence was recursively constructed via practice is about much more than finding fortifications or observing the frequency of arrow wounds and parry fractures. The production of violence against people, places, and things was, and is, entangled in social fields linking everyday practices with extraordinary events and transregional phenomena. Just how this happened, and happens, is the subject for a historically oriented, practice-based or phenomenological archaeology of violence. That it happened, and yet happens, makes comparative-historical studies such as this book of the utmost importance both to archaeology and to a greater understanding of our own human condition.

Acknowledgments

To Axel Nielsen and William Walker, the editors of this volume and the organizers of the earlier symposia at the Society for American Archaeology meeting and the Amerind Foundation, I would like to express my most sincere gratitude for jobs very well done. I likewise thank John Ware, at Amerind, for his patience and oversight. For his latest thinking on and guided tour of Burnt Corn Pueblo, I am grateful to James Snead. Ditto to Ross Hassig for his comparative insights into ancient warfare generally, to Ken Sassaman for thinking out loud about violence in the eastern woodlands, and to Tom Emerson for his thoughts on Crow Creek. He and Susan Alt provided helpful comments on an earlier version of this chapter, and the contributors to this volume, in one form or another, also provided feedback. Of course, the usual disclaimers for the mistakes or misstatements in this chapter apply. They are mine. However, the responsibility to fully understand the deeper historical significance and contemporary relevance of the issues broached herein is all of ours.

Notes

Introduction

1. In all fairness, Ferguson's detailed study of Yanomami warfare (1995) is very sensitive to historical complexities and takes far more seriously the nuances of practice than his "synthetic materialist model" would prescribe.

Chapter 2

1. The number refers to the catalog number of the Maya Vase Data Base created by Justin Kerr, available at http://research.famsi.org/kerrmaya.html.

Chapter 6

1. *Akali taiyoko ongo kunao napenge.*
2. The spelling of "Te" is in fact "Tee," according to current Enga orthography.
3. It is difficult to determine the causes of Enga wars, whether past or present, because "causes" of wars given by the Enga are the incidents that trigger violence after tension has built up over years. Moreover, motivations change as wars progress, adding ever new reasons to fight.
4. Only Meggitt's (1977:16–21) section on warfare between phratries (tribes), what we call the Great Ceremonial Wars, is incomplete in that it describes the wars, but not their history or raison d'etre or the exchanges that followed.
5. Enga life involves much verbal and physical violence; however, the vast majority of conflicts are solved through mediation (Talyaga 1982). Why a few incidents escalate into armed conflict depends on historical relations, tensions of the time, the degree of success of immediate attempts at mediation, and the ulterior motivations of those involved.

6. These figures are compiled from many sources. Data for 1885–1915 come from our historical records and include only fights that had significant enough outcomes to be remembered in historical traditions (Wiessner and Tumu 1998). Figures for 1961–1970 come from the studies of Meggitt (1977), and those for 1971–1980 come from both the studies of Meggitt (1977) and from provincial records as reported by Bryant Allen and Rick Giddings (1982). The figures for 1981–1990 come from studies by William Wormsley and Michael Toke (personal communication) and Village Courts records; those for 1990 on come from Village Courts records. I am grateful to Bernard Letakali and Anton Yangupin for making recent Village Court records available to us.

7. In contrast, the ceremonial wars of the Dani appear to have put order to conventional wars (Larson 1987).

8. Occasions when the Enga called on sons of Great War leaders at public events to step forward and replace their fathers are recalled in Enga oral historical traditions.

Chapter 7

1. The bola was a weapon made with a cord attached to several stone weights by means of a medial groove or a hole, thrown in both warfare and hunting.

2. Chullpas are clearly associated with multiple burials (e.g., Rydén 1947). While most sites contain much smaller numbers, Oscar Ayca Gallegos (1995:135) reports the cleaning of two chullpas at Ayrampuni/Cacse, on Lago Umayo, which contained thirty-four and thirty-two mummified individuals respectively, including both sexes and children. Cave burials were always multiple and could also house large numbers: the Molino-Chilacachi cave burial contained at least 133 men, women, and children, interred in more than one episode (de la Vega et al 2002). Slab-cist graves were much smaller but also usually contained the remains of several individuals (Bustina Menéndez 1960; Revilla and Uriarte 1985; M. Tschopik 1946).

3. Pukaras in this map were identified mainly through examining air photos and through ground sighting, but they also include all pukaras mentioned in the archaeological literature for the area.

4. See Elizabeth Arkush (2005). This time frame is admittedly a long one for discussing alliance relationships that may have shifted as pukaras

came in and out of use. However, the broad outlines of the spatial patterns discussed here appear to be robust because they are supported by other evidence. Unfortunately, the imprecision of carbon dates and the long-lived ceramic styles of the LIP may hinder the construction of any finer-grained chronology of pukara use. Here, pukaras are treated as essentially contemporaneous.

5. In this discussion, "alliance relationships" are not necessarily meant to denote politically equal status. A pukara may have dominated its allies politically, exacting tribute or service. An allied cluster may in fact have been one center with a number of outposts. In addition, violent conflict may have sometimes occurred within an alliance group; however, it is expected that conflict between alliance groups was more frequent or intense than within.

6. Grab-bag surface collections of ceramics were made, with the conscious attempt to collect a large number of diagnostic ceramics that seemed representative of the assemblage at the site. At sites with particularly dense ceramics, one or a few collection blocks were judgmentally placed and all diagnostics within them were collected; at sites with very sparse ceramics, all diagnostics seen were collected. In a test of the methodology, systematic collections of about 10 percent of the surface at Llongo (S4), one medium-sized pukara with very dense surface artifacts, corresponded well in ceramic type frequency to grab-bag collections made earlier at the same site. While systematic surface collections at all pukaras would have been preferable, they were not possible due to time and labor constraints, and this test indicates that the collections from the project can be considered an acceptable basis for the broad spatial patterns identified here.

7. Catharine Julien (1983, 1993), Thérèse Bouysse-Cassagne (1978, 1986), and scholars following them (Saignes 1986; Spurling 1992; Torero 1987) have used one principal source to reconstruct the spatial extent of the Collas and neighboring groups during the late Inka and the early Colonial periods: the list of *capitanías de mita* given by Luis Capoche ([1585] 1959), spatial parcels used for labor recruitment for the Potosí mines. This source lists individual encomiendas with their ethnic affiliation (Colla, Lupaca, Cana) and further subdivides the basin into Urcosuyu and Umasuyu sides.

8. For simplicity, this section and the accompanying maps lump together the tribute assessments of Aymaras and Urus.

Chapter 8

1. We will refer to these two periods collectively as "late prehispanic history" (ca. AD 1000–1535). I use this category sometimes because it is difficult to give objects and contexts a more precise chronology.

2. The only other drawing where Guamán Poma (1980:226) represents men—pectorals are always associated with males—carrying these objects is in his depiction of the Coia Raimi or "solemn feast of the queen" in September, when men paraded "armed as if they were going to fight a war" (fig. 8.5).

3. Curiously, González does not pursue this association between mirrors and war in his general interpretation of metal plaques, apparently because "their small size gave them no practical utility as defensive elements" (1992:215).

4. These multiple uses of the same architectural form define a semantic field that is very similar to the one just outlined for mountains, features with which chullpas seem to be related—in Atacama at least—through the systematic orientation of their openings toward prominent peaks or Mallkus (Berenguer et al. 1984).

5. The fact that these later versions of this icon are referred to in the literature as "the sacrificer" reflects the reluctance of many Andean archaeologists to accept the possibility of warfare in the precolumbian past.

6. For other references to the metamorphosis of warriors into animals, see Guamán Poma (1980:122, 132–133).

7. In the case of San Pedro, there was a direct connection with the Tiwanaku core.

8. As has been repeatedly pointed out, osteological traumas provide the best evidence of physical violence. Bioarchaeological studies for this period are still scant in the area, but they already show traces of violence (e.g., Mendonça et al. 1992; Torres-Rouff et al. 2005).

9. Platt makes clear that his account represents "a *model* of social dynamics, not a sequence of historical events" (1987:95). Archaeological evidence from the Circumpuna area, however, suggests that his model is probably a good synthetic description of actual "historical events."

10. In historically and ethnographically documented cases, the tinku confronted members of the two moieties ("Alasaya-Majasaya" in

Aymara, "Anansaya-Urinsaya" in Quechua) that characterize the dual structure of many Andean polities, placing institutional limits on the expansive ambitions of individual segments (Platt 1987:83).

11. For examples of the persistence of violent conflicts over territorial boundaries that were never resolved by segmentary dynamics, see Xavier Izko (1992).

Chapter 9

1. Indeed, Kenneth Sassaman's forthcoming re-analysis of the Middle-Late Archaic period of the American Midsouth draws on precisely these sorts of data using diachronic and transregional evidence of mound construction, mortuary practices, exchange, territoriality, and traumatic injury (personal communication, 2005).

REFERENCES CITED

Abreu, Maria Emília V. 2001. *Estudo dos Padrões de uso do espaço do sítio arqueológico Osvaldo (AM-IR-09)*. Unpublished report submitted to the Foundation for the Support of Research of the State of São Paulo.

Abu-Lughod, Lila. 1991. Writing Against Culture. In *Recapturing Anthropology: Working in the Present*, edited by Richard G. Fox, pp. 137–162. School of American Research Press, Santa Fe.

Adair, James. [1775] 1930. *The History of the American Indians*. Edited by Samuel Cole Williams. Watauga Press, Johnson City, Tenn.

Adams, E. Charles. 1991. *The Origin and Development of the Pueblo Katsina Cult*. University of Arizona Press, Tucson.

Adams, Robert McCormick. 1966. *The Evolution of Urban Society*. Aldine, Chicago.

Akrich, Madeleine. 1992. The De-Scription of Technical Objects. In *Shaping Technology/Building Society: Studies in Sociotechnical Change*, edited by Weibe E. Bijker and John Law, pp. 205–224. MIT Press, Cambridge.

Albert, Bruce. 1989. Yanomami "Violence": Inclusive Fitness or Ethnographer's Representation? *Current Anthropology* 30(5): 637–640.

———. 1990. On Yanomami Warfare: Rejoinder. *Current Anthropology* 31(5): 538–563.

Aldunate, Carlos, and Victoria Castro. 1981. *Las chullpa de Toconce y su relación con el poblamiento altiplánico en el Loa Superior, Período Tardío*. Ediciones Kultrun, Santiago.

Allen, Bryant, and Rick Giddings. 1982. Land Disputes and Violence in Enga. In *Enga: Foundations for Development*, vol. 3, edited by Bruce Carrad, David Lea, and Kundapen Talyaga, pp. 179–197. Department of Geography, University of New England, Armidale.

Allen, Catherine J. 2002. *The Hold Life Has: Coca and Cultural Identity in an Andean Community*, 2nd ed. Smithsonian Institution Press, Washington.

Allen, Tim, and John Eade. 1996. Anthropological Approaches to Ethnicity and Conflict in Europe and Beyond. *International Journal on Minority and Group Rights* 4(3–4): 217–246.

Alva, Walter, and Christopher Donnan. 1994. *Royal Tombs of Sipan*. University of California, Los Angeles.

Ambrose, Stephen E. 1997. *Citizen Soldiers: The U.S. Army from the Normandy Beaches to the Bulge to the Surrender of Germany, June 7, 1944–May 7, 1945*. Touchstone, New York.

Ambrosetti, Juan B. 1906. Arqueología de la Puna de Atacama. *Revista del Museo de La Plata* 7:3–34.

Ames, Roger T. 1993. Introduction. *Sun-Tzu, The Art of Warfare: The First English Translation Incorporating the Recently Discovered Yin-Ch'üeh-Shan Texts.* Translated by Roger T. Ames. Ballantine Books, New York.

Amorim, Antonio Brandão de. 1926. Lendas em Nheengatú e Português. *Revista do Instituto Histórico e Geográfico Brasileiro* 154(100): 9–475.

Anderson, David G. 1994. *The Savannah River Chiefdoms: Political Change in the Late Prehistoric Southeast.* University of Alabama Press, Tuscaloosa.

———. 1999. Examining Chiefdoms in the Southeast: An Application of a Multiscalar Analysis. In *Great Towns and Regional Polities in the Prehistoric American Southwest and Southeast*, edited by Jill E. Neitzel, pp. 95–107. University of New Mexico Press, Albuquerque.

Aoyama, Kazuo. 1999. Ancient Maya State, Urbanism, Exchange, and Craft Specialization: Chipped Stone Evidence from the Copán Valley and the La Entrada Region, Honduras. *Memoirs in Latin American Archaeology*, no. 12. University of Pittsburgh, Pittsburgh.

Appadurai, Arjun. 1996. *Modernity at Large: Cultural Dimension of Globalization.* University of Minnesota Press, Minneapolis.

Archer, Margaret. 2000. *Being Human: The Problem of Agency.* Cambridge University Press, Cambridge.

Archivo Nacional Quito, Indigenas. 22XI [1566] 1680. Yndios de telimbela. Don Joseph Pilco y consortes casiques de Sant Lorenso sobre las tierras de Telinvela.

Arellano López, Jorge. 2002. Reconocimiento Arqueológico em la cuenca del rio Orthon, Amazonia Boliviana. Museu Jacinto Jijón, Quito y Caamaño / Taraxacum, Washington.

Arkush, Elizabeth. 2005. Colla Fortified Sites: Warfare and Regional Power in the Late Prehispanic Titicaca Basin, Peru. PhD dissertation, University of California, Los Angeles.

———. 2006. Collapse, Conflict, Conquest: Warfare and Divergent Outcomes in the Late Prehispanic Andean Highlands. In *The Archaeology of Warfare: Prehistories of Raiding and Conquest*, edited by Elizabeth Arkush and Mark W. Allen, pp. 286–335. University of Florida Press, Gainesville.

———. 2008. War, Causality, and Chronology in the Titicaca Basin. *Latin American Antiquity* 19(4): 339–373.

Arkush, Elizabeth, and Charles Stanish. 2005. Interpreting Conflict in the Ancient Andes. *Current Anthropology* 46:3–28.

Arsenault, Daniel. 1994. Symbolisme, rapports sociaux et pouvoir dans les contextes sacrificiels de la société mochica (Pérou précolombien): Une étude archéologique et iconographique. PhD dissertation, anthropolody department, University of Montreal.

Aschero, Carlos A. 1979. Aportes al estudio del arte rupestre de Inka Cueva 1 (Departamento de Humahuaca, Jujuy). Actas de las Jornadas de Arqueología del Noroeste Argentino. *Antiquitas* 2:419–459. Buenos Aires.

———. 2000a. Figuras humanas, camélidos y espacio en la interacción circumpuneña. In *Arte en las rocas: Arte rupestre, menhires y piedras de colores en Argentina*, edited by María Mercedes Podestá and María de Hoyos, pp. 15–44. Sociedad Argentina de Antropología y Asociación Amigos del INAPL, Buenos Aires.

———. 2000b. El poblamiento del territorio. In *Nueva historia Argentina*, vol. 1, edited by Myriam Noemí Tarragó, pp. 17–59. Sudamericana, Buenos Aires.

Ashmore, Malcolm, Robin Wooffitt, and Stella Harding. 1994. Humans and Others, Agents and Things. In *Humans and Others: The Concept of "Agency" and its Attribution*, edited by Malcom Ashmore, Robin Wooffitt, and Stella Harding. *American Behavioral Scientist* 37:733–740.

Aunger, Robert. 1999. Culture as Consensus: Against Idealism/Contra Consensus. *Current Anthropology* 40:S93–S101.

Ayca Gallegos, Oscar. 1995. *Sillustani*. Instituto de Arqueologia del Sur, Tacna, Peru.

Bamforth, Douglas B. 1994. Indigenous People, Indigenous Violence: Precontact Warfare on the North American Great Plains. *Man* 29:95–115.

Bandelier, Adolph. 1890. Final Report of Investigations among the Indians of the Southwestern United States, Part I. *Papers of the Archaeological Institute of America, American Series*, no. 4, Cambridge.

———. 1910. *The Islands of Titicaca and Koati*. The Hispanic Society of America, New York.

Barnett, Catherine. 1990. Of Masks and Marauders: An Underground Market and the Supernatural World Collide in an Ancient Hopi Village. *Art & Antiques* 7(7): 99–109, 140, 142, 144–148.

Barreda Murillo, Luis. 1958. *Sitios archeológicos Kollao en Nuñoa (Melgar-Puno)*. Bachelor's thesis, Universidad Nacional de Cuzco, Cusco, Peru.

Barrera Vásquez, Alfredo. 1965. *El libro de los cantares de Dzitbalché*. Instituto Nacional de Antropología e Historia, Serie investigaciones, 9. Instituto Nacional de Antropología e Historia, México, D.F.

Barth, Fredrik. 1969. Introduction. In *Ethnic Groups and Boundaries: The Social Organization of Culture Difference*, edited by Fredrik Barth, pp. 1–38. Little, Brown and Company, Boston.

———. 1994. A Personal View of Present Tasks and Priorities in Cultural Social Anthropology. In *Assessing Cultural Anthropology*, edited by Robert Brofsky, pp. 349–360. McGraw-Hill, New York.

Beckerman, Stephen. 1979. The Abundance of Protein in Amazonia: A Reply to Gross. *American Anthropologist* 81:533–560.

Bell, Catherine. 1997. *Ritual: Perspectives and Dimensions*. Oxford University Press, Oxford.

Bennett, Wendell C. 1933. Archaeological Hikes in the Andes. *Natural History* 33(2): 163–174.

———. 1948. A Reappraisal of Peruvian Archaeology. *Memoirs of the Society for American Archaeology*, no. 4, Menasha.

———. 1950. Cultural Unity and Disunity in the Titicaca Basin. *American Antiquity* 16(2): 89–98.

Bentley, G. Carter. 1987. Ethnicity and Practice. *Comparative Studies in Society and History* 29:24–55.

Berdan, Frances, and Patricia R. Anawalt (editors). 1992. *Codex Mendoza*. University of California Press, Berkeley.

Berenguer, José. 1998. La iconografía del poder en Tiwanaku y su rol en la integración de zonas de frontera. *Boletín del Museo Chileno de Arte Precolombino* 7:19–37. Santiago.

———. 1999. El evanescente lenguaje del arte rupestre en los Andes atacameños. In *El Arte Rupestre en los Andes de Capricornio*, edited by José Berenguer and Francisco Gallardo, p. 56. Museo Chileno de Arte Precolombino, Santiago.

———. 2004. *Caravanas, interacción y cambio en el desierto de Atacama*. Sirawi Ediciones, Santiago.

Berenguer, José, Carlos Aldunate, and Victoria Castro. 1998. Orientación orográfica de las chulpas en Likan: La importancia de los cerros en la Fase Toconce. In *Simposio Culturas Atacameñas*, edited by Grete Mostny, pp. 174–220. Museo Arqueológico Gustavo Le Paige, San Pedro de Atacama.

Bertonio, Padre Ludovico. [1612] 1984. *Vocabulario de la lengua Aymara*. Museo Nacional de Etnografía y Folklore Instituto Francés de Estudios Andinos, Cochabamba.

Betanzos, Juan de. [1551–7] 1996. *Narrative of the Inkas*. University of Texas Press, Austin.

Bey, George J., III, Graig A. Hanson, and William M. Ringle. 1997. Classic to Postclassic at Ek Balam, Yukatan: Architectural and Ceramic Evidence for Defining the Transition. *Latin American Antiquity* 8:237–254.

Binford, Michael, Alan Kolata, Mark Brenner, John Janusek, Matthew Seddon, Mark Abbott, and Jason Curtis. 1997. Climate Variation and the Rise and Fall of an Andean Civilization. *Quaternary Research* 47:235–248.

Blanton, Richard E., Gary M. Feinman, Stephan A. Kowalewski, and Peter N. Peregrine. 1996. A Dual-Processual Theory for the Evolution of Mesoamerican Civilization. *Current Anthropology* 37:1–31.

Blanton, Richard E., Gary M. Feinman, Stephen A. Kowalewski, and Linda M. Nicholas. 1999. *Ancient Oaxaca*. Cambridge University Press, Cambridge.

Blitz, Jon H. 1988. Adoption of the Bow in Prehistoric North America. *North American Archaeologist* 9:123–145.

Bloch, Maurice. 1986. *From Blessing to Violence: History and Ideology in the Circumcision Ritual of the Merina of Madagascar.* Cambridge University Press, Cambridge.

Blom, Deborah E., John W. Janusek, and Jane E. Buikstra.2003. A Re-Evaluation of Human Remains from Tiwanaku. In *Tiwanaku and its Hinterland: Archaeology and Paleoecology of an Andean Civilization*, Urban and Rural Archaeology, vol. 2, edited by Alan L. Kolata, pp. 435–446. Smithsonian Institution Press, Washington, D.C.

Boast, Robin. 1997. A Small Company of Actors: A Critique of Style. *Journal of Material Culture* 2:173–198.

Boehm, Christopher. 1999. *Hierarchy in the Forest: The Evolution of Egalitarian Behavior.* Harvard University Press, Cambridge.

Bonavia, Duccio. 1985. *Mural Painting in Ancient Peru.* Translated by Patricia J. Lyon. Indiana University Press, Bloomington.

Boone, Elizabeth H. (editor). 1984. *Ritual Human Sacrifice in Mesoamerica.* Dumbarton Oaks Research Library and Collection, Washington, D.C.

Bordo, Susan. 1993. *Unbearable Weight: Feminism, Western Culture and the Body.* University of California Press, Berkeley.

———. 1997. "Material Girl": The Effacements of Postmodern Culture. In *The Gender/Sexuality Reader: Culture, History, Political Economy*, edited by Roger N. Lancaster and Micaela di Leonardo, pp. 335–358. Routledge, New York.

Bourdieu, Pierre. 1977. *Outline of a Theory of Practice.* Cambridge University Press, Cambridge.

———. 1990. *The Logic of Practice.* Stanford University Press, Stanford.

Bourget, Steve. 2001. Rituals of Sacrifice: Its Practice at Huaca de la Luna and Its Representation in Moche Iconography. In *Moche Art and Archaeology in Ancient Peru*, edited by Joanne Pillsbury, pp. 88–109. National Gallery of Art, Washington, D.C.

———. 2005. Who Were the Priests, the Warriors, and the Prisoners? A Peculiar Problem of Identity in Moche Culture and Iconography, North Coast of Peru. In *Us and Them: Archaeology and Ethnicity in the Andes*, edited by Richard M. Reycraft, pp. 73–85. Costen Institute of Archaeology, University of California, Los Angeles.

Bouysse-Cassagne, Thérèse. 1975. *La identidad aymara: Aproximación histórica (siglo XV, siglo XVI).* Hisbol, La Paz.

———. 1978. L'éspace aymara: Urco et uma. *Annales: Economies, Sociétés, Civilisations* 5–6:1057–1080. Paris.

———. 1986. Urco and Uma: Aymara Concepts of Space. In *Anthropological History of Andean Politics*, edited by J. V. Murra, Nathan Wachtel, and Jacques Revel, pp. 201–227. Cambridge University Press, Cambridge.

Bouysse-Cassagne, Thérèse, and Olivia Harris. 1987. Pacha: En torno al pensamiento aymara. In *Tres reflexiones sobre el pensamiento aymara*, pp. 11–59. Hisbol, La Paz.

Bovon, Anne. 1963. La representation des guerriers perses et la notion de Barbare dans la 1re moitie du Ve siecle. *Bulletin de Correspondance Hellenique* 87(2): 579–602.

Bradley, Bruce A. 1992. Pitchers to Mugs: Chacoan Revival at Sand Canyon Pueblo. *Kiva* 61:241–255.

Bradley, Richard. 2000. *An Archaeology of Natural Places*. Routledge, New York.

Brain, Jeffrey P., and Philip Phillips. 1996. *Shell Gorgets: Styles of the Late Prehistoric and Protohistoric Southeast*. Peabody Museum Press, Harvard University, Cambridge.

Braswell, Geoffrey E. 2003. Introduction: Reinterpreting Early Classic Interaction. In *The Maya and Teotihuacan*, edited by Geoffrey E. Braswell, pp. 1–44. University of Texas Press, Austin.

Bridges, Patricia S. 1996. Warfare and Mortality at Koger's Island, Alabama. *International Journal of Osteoarchaeology* 6:66–75.

Bridges, Patricia S., Keith P. Jacobi, and Mary Lucas Powell. 2000. Warfare-Related Trauma in the Late Prehistory of Alabama. In *Bioarchaeological Studies of Life in the Age of Agriculture: A View from the Southeast,* edited by Patricia M. Lambert, pp. 35–63. University of Alabama Press, Tuscaloosa.

Brose, David S. 1989. From the Southeastern Ceremonial Complex to the Southern Cult: "You Can't Tell the Players Without a Program." In *The Southeastern Ceremonial Complex: Artifacts and Analysis*, edited by Patricia Galloway, pp. 27–37. University of Nebraska Press, Lincoln.

Brown, James A. 1976. The Southern Cult Reconsidered. *Midcontinental Journal of Archaeology* 2:115–135.

———. 1996. *The Spiro Ceremonial Center: The Archaeology of Arkansas Valley Caddoan Culture in Eastern Oklahoma*. Museum of Anthropology, Memoirs, vol. 2, no. 29. University of Michigan, Ann Arbor.

———. 2004. The Cahokian Expression: Creating Court and Cult. In *Hero, Hawk, and Open Hand*, edited by Robert V. Sharp, pp. 105–123. Yale University Press, New Haven.

Brownrigg, Leslie Ann. 1972. El papel de los ritos de pasaje en la integración social de los Cañaris Quichuas del Austral Ecuatoriano. In *Actas y memorias del XXXIX congreso internacional de americanistas*, vol. 6, edited by Rosalía Avalos de Matos and Rogger Ravines, pp. 92–99. Lima.

Brubaker, Rogers, and David D. Laitin. 1998. Ethnic and Nationalist Violence. *Annual Review of Sociology* 24:423–452.

Brumann, Christoph. 1992. Writing for Culture: Why a Successful Concept Should Not Be Discarded. *Current Anthropology* 40:S1–S27.

Bullock, Peter Y. (editor). 1998. *Deciphering Anasazi Violence*. HRM Books, Santa Fe.

Bunzel, Ruth. 1932. Introduction to Zuni Ceremonialism. In the Forty-seventh
Annual Report of the Bureau of American Ethnology, pp. 467–544.
Smithsonian Institution, Washington, D.C.

Burger, Richard L. 1992. *Chavín and the Origins of Andean Civilization*. Thames
and Hudson, New York.

Bustina Menéndez, David. 1959. Estudios arqueológicos en la provincia de Ayaviri, Departamento de Puno. In *Antiguo Peru: Espacio y tiempo*. Works
presented at Peruvian Archaeology Week, pp. 349–350. Editorial Juan
Mejia Baca, Lima.

Butler, Judith. 1993. *Bodies that Matter: On the Discursive Limits of "Sex."* Routledge, New York.

Capoche, Luis. [1585] 1959. Relación general de la villa imperial de Potosí, un
capítulo inédito en la historia del nuevo mundo. *Biblioteca de Autores
Españoles* 112:9–221. Madrid.

Carneiro, Robert L. 1970. A Theory of the Origin of the State. *Science*
169:733–738.

———. 1981. The Chiefdom: Precursor of the State. In *The Transition to Statehood
in the New World*, edited by Grant D. Jones and Robert R. Kautz, pp.
37–79. Cambridge University Press, Cambridge.

———. 1987. Further Reflections on Resource Concentration and its Role in
the Rise of the State. In *Studies in the Neolithic and Urban Revolutions
(The V. Gordon Childe Colloquium, 1986)*, edited by Linda Manzanilla,
pp. 245–260. British Archaeological Reports International Series 349,
Oxford.

———. 1990. Chiefdom-Level Warfare as Exemplified in Fiji and the Cauca
Valley. In *The Anthropology of War*, edited by Jonathan Haas, pp. 190–
211. Cambridge University Press, Cambridge.

———. 1995. The History of Ecological Interpretations of Amazonia: Does
Roosevelt Have it Right? In *Indigenous Peoples and the Future of Amazonia: An Ecological Anthropology of and Endangered World*, edited by Leslie Sponsel, pp. 45–70. University of Arizona Press, Tucson.

Carr, Christopher, and Jill E. Nietzel (editors). 1995. *Style, Society, and Person:
Archaeological and Ethnological Perspectives*. Plenum, New York.

Carr, Pat. 1979. Mimbres Mythology. *Southwestern Studies Monograph*, no. 56. University of Texas at El Paso.

Cassin, Elena. 1981. The Death of the Gods. In *Mortality and Immortality: The
Anthropology and Archaeology of Death*, edited by S. C. Humphreys and
Helen King, pp. 317–325. Academic Press, London.

Castillo, Luis Jaime. 2001. The Last of the Mochicas: A View from the Jequetepeque
Valley. In *Moche Art and Archaeology in Ancient Peru*, edited by Joanne
Pillsbury, pp. 307–332. National Gallery of Art, Washington, D.C.

Chagnon, Napoleon. 1968. *Yanomamo: The Fierce People*. Holt, Rinehart, and Winston, New York.

————. 1968. Yanomamö Social Organization and Warfare. In *War: the Anthropology of Armed Conflict and Aggression*, edited by Morton Fried, Marvin Harris, and Robert Murphy, pp. 85–91. Natural History Press, Garden City, New York.

————. 1983. *Yanomamo, The Fierce People*, 3rd ed. Holt, Rinehart, and Winston, New York.

————. 1988. Life Histories, Blood Revenge, Warfare in a Tribal Population, *Science* 239:985–992.

Chase, Diane Z., and Arlen F. Chase. 1998. The Architectural Context of Caches, Burials, and Other Ritual Activities for the Classic Period Maya (as reflected at Caracol Belize). In *Function and Meaning in Classic Maya Architecture*, edited by Stephen D. Houston, pp. 299–332. Dumbarton Oaks Research Library and Collection, Washington, D.C.

Chase, Diane Z., and Arlen F. Chase. 2002. Classic Maya Warfare and Settlement Archaeology at Caracol, Belize. *Estudios de Cultura Maya* 22:33–51.

Chávez, Sergio J. 1992. The Conventionalized Rules in Pucara Pottery Technology and Iconography: Implications for Socio-Political Developments in the Northern Lake Titicaca Basin. PhD dissertation, Michigan State University.

Chávez, Sergio J., and Karen L. Mohr Chávez. 1975. A Carved Stela from Taraco, Puno, Peru, and the Definition of an Early Style of Stone Sculpture from the Altiplano of Peru and Bolivia. *Ñawpa Pacha* 13:45–83.

Chernela, Janet. 1983. *Hierarchy and Economy of the Uanano (Kotiria) Speaking People of the Middle Uaupés Basin*. PhD dissertation, Columbia University, New York.

Chervin, Arthur. 1913. Aymaras and Quichuas: A Study of Bolivian Anthropology. *Proceedings of the International Congress of Americanists* 18(1): 63–74.

Child, Mark B. n.d. The Dynamics of Classic Maya Warfare. Manuscript on file.

Chowning, Ann. 1979. Leadership in Melanesia. *Journal of Pacific History* 14:66–84.

Cicourel, Aaron V. 1993. Aspects of Structural and Processual Theories of Knowledge. In *Bourdieu: Critical Perspectives*, edited by Craig Calhoun, Edward LiPuma, and Moishe Postone, pp. 89–115. University of Chicago Press, Chicago.

Cieza de Leon, Pedro. [1550] 1984. *La crónica del Perú*. Historia 16, Madrid.

————. [1550] 1985. *El señorío de los Inkas*. Madrid, Historia 16, Madrid.

————. [1553] 1985. *La crónica del Perú: Segunda parte*. Pontificia Universidad Católica del Perú y Academia Nacional de la Historia, Lima.

Cigliano, Eduardo M. 1973. *Tastil, una ciudad preincaica argentina*. Ediciones Cabargon, Buenos Aires.

Clastres, Pierre. 1974. *A sociedade contra o estado: Pesquisas de antropologia política*. Francisco Alves, Rio de Janeiro.

Clayton, Lawrence A., Vernon James Knight, Jr., and Edward C. Moore (editors). 1993. *The De Soto Chronicles, The Expedition of Hernando de*

Soto to North America in 1539–1543. University of Alabama Press, Tuscaloosa.

Clifford, James. 1986. Introduction: Partial Truths. In *Writing Culture: The Poetics and Politics of Ethnography*, edited by James Clifford and George Marcus, pp. 1–26. University of California Press, Berkeley.

Cobb, Charles R., and Brian M. Butler. 2002. The Vacant Quarter Revisited: Late Mississippian Abandonment of the Lower Ohio Valley. *American Antiquity* 67:625–641.

Cobo, Bernabé. [1653] 1979. *History of the Inka Empire*. University of Texas Press, Austin.

Coe, Michael D. 1975. *Classic Maya Pottery at Dumbarton Oaks*. Dumbarton Oaks Research Library and Collection, Washington, D.C.

Coe, William R. 1990. *Tikal Report no. 14: Excavations in the Great Plaza, North Terrace and North Acropolis of Tikal*. University Museum, University of Pennsylvania, Philadelphia.

Coggins, Clemency C., and Shane, O. C. 1984. *Cenote of Sacrifice: Maya Treasures from the Sacred Well at Chichen Itzá*. University of Texas Press, Austin.

Cohen, Amanda. 2003. *Formative Period Domestic and Ritual Architecture of the Pucara Valley, Peru*. Paper presented at the Institute of Andean Studies meeting, Berkeley.

Cohen, Ronal. 1984. Warfare and State Formation: Wars Make States and States Make War. In *Warfare, Culture, and Environment*, edited by R. Brian Ferguson, pp. 329–358. Academic Press, Orlando.

Colas, Pierre R., and Alexander Voss. 2000. A Game of Life and Death: The Maya Ballgame. In *Maya: Divine Kings of the Rain Forest*, edited By N. Grube, pp. 186–191. Könemann, Cologne.

Collingwood, Robin George. 1994. *The Idea of History: Revised Edition with Lectures, 1926–1928*, edited by Jan van der Dussen. Oxford University Press, Oxford.

Connell, Robert W. 1995. *Masculinities*. University of California Press, Berkeley.

Connerton, Paul. 1989. *How Societies Remember*. Cambridge University Press, Cambridge.

Cordy-Collins, Alan. 1983. The Cerro Sechín Massacre: Did it Happen? *Ethnic Technology Notes*, no. 18. San Diego Museum of Man, San Diego, California.

———. 1992. Archaism or Tradition? The Decapitation Theme in Cupisnique and Moche Iconography. *Latin American Antiquity* 3(3): 206–220.

Covey, R. Alan, and Donato Amado Gonzalez. 2008. *Imperial Transformations in Sixteenth-Century Yucay*. Memoirs of the Museum of Anthropology, University of Michigan, vol. 44, Ann Arbor.

Cowgill, George. 1976. Teotihuacan, Internal Militaristic Competition, and the Fall of the Classic Maya. In *Maya Archaeology and Ethnohistory*, edited by Norman Hammond and Gordon R. Willey, pp. 51–62. University of Texas Press, Austin.

————. 2000. Rationality and Contexts in Agency Theory. In *Agency in Archae-ology*, edited by Marcia-Anne Dobres and John E. Robb, pp. 51–60. Routledge, London.

Creel, Darrell, and Roger Anyon. 2003. New Interpretations of Mimbres Public Architecture and Space: Implications for Cultural Change. *American Antiquity* 68:67–92.

Crown, Patricia. 1994. *Ceramics and Ideology: Salado Polychrome Pottery*. University of New Mexico Press, Albuquerque.

Crown, Patricia L., and W. H. Wills. 2003. Modifying Pottery and Kivas at Chaco: Pentimento, Restoration, or Renewal? *American Antiquity* 68:511–532.

Csordas, Thomas J. 2000. The Body's Career in Anthropology. In *Anthropologi-cal Theory Today*, 2nd ed., edited by H. L. Moore, pp. 172–205. Polity Press, Cambridge.

Cushing, Frank Hamilton. 1923. Origin Myth from Oraibi. *Journal of American Folklore* 36(140): 163–170.

Dahlin, Bruce. 2000. The Barricade and Abandonment of Chunchucmil: Impli-cations for Northern Maya Warfare. *Latin American Antiquity* 11:293–298.

Darling, Andrew. 1998. Mass Inhumation and the Execution of Witches in the American Southwest. *American Anthropologist* 100:732–752.

DeBoer, Warren, Keith Kintigh, and Arthur Rostoker. 1996. Ceramic Seriation and Site Reoccupation in Lowland South America. *Latin American Antiq-uity*, 7(3): 263–278.

DeCerteau, Michel. 1984. *The Practice of Everyday Life*. University of California Press, Berkeley.

Deetz, James. 1990. Landscapes as Cultural Statements. In *Earth Patterns: Essays in Landscape Archaeology*. Edited by William M. Kelso and Rachel Most, pp. 1–4. University Press of Virginia, Charlottesville.

DeLeonardis, Lisa. 2000. The Body Context: Interpreting Early Nasca Decapitated Burials. *Latin American Antiquity* 11:363–386.

Demarest, Arthur A. 1978. Interregional Conflict and "Situations Ethics" in Classic Maya Warfare. In *Codex Wauchope: A Tribute Roll*, edited by M. Giar-dina, B. Edmonson, and W. Creamer, pp. 101–111. Human Mosaic, vol. 12, Tulane University, New Orleans.

————. 1984. Overview: Mesoamerican Human Sacrifice in Evolutionary Perspec-tive. In *Ritual Human Sacrifice in Mesoamerica*, edited by Elizabeth H. Boone, pp. 227–247. Dumbarton Oaks Research Library and Collec-tion, Washington, D.C.

————. 1997. The Vanderbilt Petexbatun Regional Archaeological Project 1989–1994: Overview, History, and Major Results of a Multidisciplinary Study of the Classic Maya Collapse. *Ancient Mesoamerica* 8:209–228.

Demarest, Arthur A., and Antonia E. Foias. 1993. Mesoamerican Horizons and the Cultural Transformations of Maya Civilization. In *Latin Ameri-*

can Horizons, edited by Don S. Rice, pp. 147–191. Dumbarton Oaks Research Library and Collection, Washington, D.C.

Demarest, Arthur A., Matt O'Mansky, Claudia Wolley, Dirk Van Tuerenhout, Takeshi Inomata, Joel Palka, and Héctor Escobedo. 1997. Classic Maya Defense Systems and Warfare in the Petexbatun Region: Archaeological Evidence and Interpretation. *Ancient Mesoamerica* 8(2): 229–254.

DeMarrais, Elizabeth, Luis J. Castillo, and Timothy Earle. 1996. Ideology, Materialization, and Power Strategies. *Current Anthropology* 37:15–31.

Denevan, William. 1992. Native American Population in 1492: Recent Research and Revised Hemispheric Estimate. In *The Native Population of the Americas in 1492*, 2nd ed., edited by William Denevan, pp. xvii–xxxviii. University of Wisconsin Press, Madison.

———. 2001. *Cultivated Landscapes of Native Amazonia and the Andes.* Oxford University Press, New York.

Dennen, Johan M. G. van der. 1995. *The Origin of War*, vols. 1 and 2. Origin Press, Groningen.

DePratter, Chester B. 1991. *Late Prehistoric and Early Historic Chiefdoms in the Southeastern United States.* Garland, New York.

Diaz-Granados, Carol, and James R. Duncan. 2000. *The Petroglyphs and Pictographs of Missouri.* University of Alabama Press, Tuscaloosa.

Diaz-Granados, Carol, Marvin W. Rowe, Marian Hyman, James R. Duncan, and John R. Southon. 2001. AMS Radiocarbon Dates for Charcoal from Three Missouri Pictographs and their Associated Iconography. *American Antiquity* 66:481–492.

Dickson, D. Bruce. 1981. The Yanomamo of the Mississippi Valley? Some Reflections on Larson (1972), Gibson (1974) and Mississippian Warfare in the Southeastern United States. *American Antiquity* 46:909–916.

Dietler, Michael, and Ingrid Herbich. 1998. Habitus, Techniques, Style: An Integrated Approach to the Social Understanding of Material Culture and Boundaries. In *The Archaeology of Social Boundaries*, edited by M. T. Stark, pp. 232–263. Smithsonian Institution Press, Washington, D.C.

Dillehay, Tom D. 2001. Town and Country in Late Moche Times: A View from Two Northern Valleys. In *Moche Art and Archaeology in Ancient Peru*, edited by Joanne Pillsbury, pp. 259–283. National Gallery of Art, Washington, D.C.

Dirks, Nicholas B., Geoff Eley, and Sherry Ortner. 1994. Introduction. In *Culture/Power/History: A Reader in Contemporary Social Theory*, edited by Nicholas B. Dirks, Geoff Eley, and Sherry Ortner, pp. 3–45. Princeton University Press, Princeton.

Divale, William, Frosine Chameris, and Deborah Gangloff. 1976. War, Peace and Marital Residence in Pre-Industrial Societies. *Journal of Conflict Resolution* 20:57–78.

Dobres, Marcia-Anne. 2000. *Technology and Social Agency.* Blackwell, Oxford.

Dobres, Marcia-Anne, and John Robb. 2000. Agency in Archaeology: Paradigm or Platitude? In *Agency in Archaeology*, edited by Marcia-Anne Dobres and John E. Robb, pp. 3–17. Routledge, New York.

Donatti, Patrícia B. 2003. *A Arqueologia da Margem Norte do Lago Grande, Amazonas.* Dissertação de Mestrado, Programa de Pós-Graduação em Arqueologia, Museu de Arqueologia e Etnologia, Universidade de São Paulo.

Donnan, Christopher B. 1978. *Moche Art of Peru.* University of California, Museum of Cultural History, Los Angeles.

———(editor). 1985. *Early Ceremonial Architecture in the Andes.* Dumbarton Oaks Research Library and Collection, Washington, D.C.

———. 2001. Moche Ceramic Portraits. In *Moche Art and Archaeology in Ancient Peru*, edited by Joanne Pillsbury, pp. 69–87. National Gallery of Art, Washington, D.C.

———. 2004. *Moche Portraits from Ancient Peru.* University of Texas Press, Austin.

Dornan, Jennifer L. 2002. Agency in Archaeology: Past, Present, and Future Directions. *Journal of Archaeological Method and Theory* 9:303–329.

Dreyfus, Simone. 1983–84. Historical and Political Anthropological Interconnections: The Multi-Linguistic Indigenous Polity of the "Carib" Islands and the Mainland Coast from the 16th to the 18th Century. *Antropológica*, 59–62:39–55.

Dumarest, Noël. 1919. Notes on Cochiti, New Mexico. *American Anthropological Association Memoirs*, vol. 6, no. 3. Lancaster, Pennsylvania.

Duviols, Pierre. 1979. Un symbolisme de l'occupation, de l'amenagement et de l'exploitation de l'espace: Le monolithe "huanca" et sa fonction dans les Andes préhispaniques. *L'Homme* 19(2): 7–31.

Dye, David H. 1995. Feasting with the Enemy: Mississippian Warfare and Prestige Goods Circulation. In *Native American Interactions*, edited by Kenneth E. Sassaman and Michael. S. Nassaney, pp. 289–316. University of Tennessee Press, Knoxville.

———. 2002. Warfare in the Protohistoric Southeast 1500–1700. In *Between Contacts and Colonies*, edited by Cameron Wesson and Mark Rees, pp. 126–141. University of Alabama Press, Tuscaloosa.

———. 2004. Art, Ritual, and Chiefly Warfare in the Mississippian World. In *Hero, Hawk, and Open Hand*, edited by Robert V. Sharp, pp. 191–206. Yale University Press, New Haven.

Earle, Timothy K. 1997. *How Chiefs Come to Power: The Political Economy in Prehistory.* Stanford University Press, Stanford.

Echo-Hawk, Roger C. 2000. Ancient History in the New World: Integrating Oral Traditions and the Archaeological Record of Deep Time. *American Antiquity* 65(2): 267–290.

Eggan, Fred. 1967. From History to Myth: A Hopi Example. In *Studies in Southwestern Ethnolinguistics*, edited by Dell H. Hymes and William E. Bittle, pp. 33–53. Mouton, Paris.

Ehrenreich, Barbara. 1997. *Blood Rites: Origins and History of the Passions of War.* Henry Holt, New York.

Ehrenreich, Robert M., Carole L. Crumley, and Janet E. Levy (editors). 1995. *Heterarchy and the Analysis of Complex Societies.* American Anthropological Association, Washington, D.C.

Eller, Jack David. 1999. *From Culture to Ethnicity to Conflict: An Anthropological Perspective on International Ethnic Conflict.* University of Michigan Press, Ann Arbor.

Ellis, Florence H. 1951. Patterns of Aggression and the War Cult in Southwestern Pueblos. *Southwestern Journal of Anthropology* 7:177–201.

Ember, Carol R., and Melvin Ember. 1992. Resource Unpredictability, Mistrust, and War. *Journal of Conflict Resolution* 36:242–262.

———. 1994. Cross-Cultural Studies of War and Peace: Recent Achievements and Future Possibilities. In *Studying War: Anthropological Perspectives*, edited by Stephen P. Reyna and R. E. Downs, pp. 185–208. Gordon and Breach, Amsterdam.

Emberling, Geoff. 1997. Ethnicity in Complex Societies: Archaeological Perspectives. *Journal of Archaeological Research* 5(4): 295–344.

Erickson, Clark. 2000. An Artificial Landscape-Scale Fishery in the Bolivian Amazon. *Nature* 408:190–193.

Fahsen, Federico. 2002. *Rescuing the Origins of Dos Pilas Dynasty: A Savage of Hieroglyphic Stairway #2, Structure L5–49.* Report presented to the Foundation for the Advancement of Mesoamerican Studies, Inc. http://www.famsi.org/reports/01098.

Farage, Nádia. 1991. *As Muralhas dos Sertões: Os povos indígenas no rio Branco e a colonização.* Paz e Terra, Rio de Janeiro.

Farnell, Brenda. 1999. Moving Bodies, Acting Selves. *Annual Review of Anthropology* 28:341–373.

Faulkner, Charles R. 1997. Four Thousand Years of Native American Cave Art in the Southern Appalachians. *Journal of Cave and Karst Studies* 59(3): 148–153.

Fausto, Carlos. 1992. Fragmentos de história e cultura Tupinambá: Da etnologia como instrumento crítico de conhecimento etno-histórico. In *História dos índios no Brasil*, edited by M. Carneiro da Cunha, pp. 381–396. Cia. das Letras / FAPESP / SMC, São Paulo.

———. 1999. Of Enemies and Pets: Warfare and Shamanism in Amazonia. *American Ethnologist* 26(4): 933–966.

———. 2001. *Inimigos Fiéis: História, guerra e xamanismo na Amazônia*, Editora da Universidade de São Paulo, São Paulo.

Feil, Daryl K. 1984. *Ways of Exchange: The Enga Tee of Papua New Guinea.* University of Queensland Press, St. Lucia.

———. 1987. *The Evolution of Highland Papua New Guinea Societies.* Cambridge University Press, Cambridge.

Fenton, Steve. 2003. *Ethnicity.* Blackwell, Cambridge.

Ferguson, R. Brian. 1990. Explaining War. In *The Anthropology of Warfare*, edited by J. Haas, pp. 26–55. Cambridge University Press, Cambridge.

———. 1994. The General Consequences of War: An Amazonian Perspective. In *Studying War*, edited by Stephen P. Reyna and R. Downs, pp. 85–111. Gordon and Breach, New York.

———. 1995. *Yanomami Warfare: A Political History*. School of American Research Press, Santa Fe.

———. 1997. Violence and War in Prehistory. In *Troubled Times: Violence and Warfare in the Past*, edited by Debra L. Martin and David W. Frayer, pp. 321–355. Gordon and Breach, Amsterdam.

———. 2000. The Causes and Origins of "Primitive Warfare." *Anthropological Quarterly* 73(3): 159–164.

———. 2001. Materialist, Cultural and Biological Theories on Why Yanomami Make War. *Anthropological Theory* 1:99–116.

Ferguson, R. Brian (editor). 1984. *Warfare, Culture, and Environment*. Academic Press, Orlando.

Ferguson, R. Brian, and Neil Whitehead. 1992. The Violent Edge of Empire. In *War in the Tribal Zone: Expanding States and Indigenous Warfare*, edited by R. Brian Ferguson and Neil Whitehead, pp. 1–30. School for American Research Press, Santa Fe.

Ferguson, William M., and Arthur H. Rohn. 1987. *Anasazi Ruins of the Southwest in Color*. University of New Mexico Press, Albuquerque.

Fewkes, Jesse Walter. 1911. Preliminary Report on a Visit to Navajo National Monument Arizona. Bureau of American Ethnology Bulletin 50. Washington, D.C.

———. 1893. Awatobi: An Archaeological Verification of a Tusayan Legend. *American Anthropologist* 6:363–371.

Flanagan, James. 1989. Hierarchy in Simple "Egalitarian" Societies. *Annual Review of Anthropology* 18:245–266.

Flannery, Kent, and Joyce Marcus. 1999. Cognitive Archaeology. In *Contemporary Archaeology in Theory: A Reader*, edited by Robert Preucel and Ian Hodder, pp. 350–363. Blackwell, Oxford.

———. 2000. Formative Mexican Chiefdoms and the Myth of the "Mother Culture." *Journal of Anthropological Archaeology* 19:1–37.

———. 2004. The Origin of War: New ^{14}C Dates from Ancient Mexico, *Proceedings of the National Academy of Science*, 100(20): 11801–11805.

Flores Ochoa, Jorge A. 1979. *Pastoralists of the Andes*. Translated by Ralph Bolton. Institute for the Study of Human Issues, Philadelphia.

Fordred-Green, Lesley, David Green, and Eduardo G. Neves. 2003. Indigenous Knowledge and Archaeological Science: The Challenges of Public Archaeology in the Reserva Uaçá. *Journal Of Social Archaeology* 3(3): 366–398.

Foucault, Michel. [1970] 1999. *El Orden del Discurso*. Tusquets Editores, Barcelona.

Fowler, Chris. 2004. *The Archaeology of Personhood: An Anthropological Approach.* Routledge, London.

Frazer, James G. 1980. *The Golden Bough: A Study in Magic and Religion.* Macmillan, London.

Fried, Morton. 1960. On the Evolution of Social Stratification and the State. In *Culture and History: Essays in Honor of Paul Radin,* edited by Stanley Diamond, pp. 713–731. Columbia University Press, New York.

Freidel, David A. 1986. Maya Warfare: An Example of Peer Polity Interaction. In *Peer-Polity Interaction and the Development of Sociopolitical Complexity,* edited by Colin Renfrew and J. F. Cherry, pp. 93–108. Cambridge University Press, Cambridge.

Freidel, David A., and Linda Schele. 1989. Dead Kings and Living Temples: Dedication and Termination Rituals Among the Ancient Maya. In *Word and Image in Maya Culture,* edited by William F. Hanks and Don S. Rice, pp. 233–243. University of Utah Press, Salt Lake City.

Friedman, Jonathan. 1994. *Cultural Identity and Global Process.* Sage, London.

Frye, Kirk Lawrence. 1997. Political Centralization in the Altiplano Period in the Southwestern Titicaca Basin (Appendix 2). In *Archaeological Survey in the Juli-Desaguadero Region of Lake Titicaca Basin, Southern Peru,* edited by Charles Stanish, pp. 129–141. Field Museum of Natural History, Chicago.

Futrell, Alison. 1997. *Blood in the Arena: The Spectacle of Roman Power.* University of Texas Press, Austin.

Galloway, Patricia (editor). 1989. *The Southeastern Ceremonial Complex: Artifacts and Analysis.* University of Nebraska Press, Lincoln.

García Azcárate, Jorgelina. 1996. Monolitos-Huancas: Un intento de explicación de las piedras de Tafí (Rep. Argentina). *Chungara* 28:159–174.

Geertz, Armin W. 1994. *The Invention of Prophecy: Continuity and Meaning in Hopi Indian Religion.* University of California Press, Berkeley.

Geertz, C. 1963. The Integrative Revolution: Primordial Sentiments and Civil Politics in the New States. In *Old Societies and New States,* edited by Clifford Geertz. New York, the Free Press.

Gell, Alfred. 1998. *Art and Agency: An Anthropological Theory.* Clarendon Press, Oxford.

Ghezzi, Ivan. 2006. Religious Warfare at Chankillo. In *Andean Archaeology III: North and South,* edited by Wiliam H. Isbell and Helaine Silverman, pp. 67–84. Springer Science, New York.

Gibson, Jon. 1974. Aboriginal Warfare in the Protohistoric Southeast: An Alternative Perspective. *American Antiquity* 39:130–133.

Giddens, Anthony. 1979. *Central Problems in Social Theory: Action, Structure, and Contradiction in Social Analysis.* University of California Press, Berkeley.

———. 1984. *The Constitution of Society: Outline of a Theory of Structuration.* University of California Press, Berkeley.

Giesso, Martin. 2003. Stone Tool Production in the Tiwanaku Heartland. In *Tiwanaku and its Hinterland: Archaeology and Paleoecology of an Andean Civilization*, vol. 2, Urban and Rural Archaeology, edited by Alan L. Kolata, pp. 363–383. Smithsonian Institution Press, Washington, D.C.

Gill, Richardson B. 2000. *The Great Maya Droughts: Water, Life, and Death.* University of New Mexico Press, Albuquerque.

Gillespie, Susan D. 1991. Ballgames and Boundaries. In *The Mesoamerican Ball Game*, edited by Vernon L. Scarborough and David R. Wilcox, pp. 317–345. University of Arizona Press, Tucson.

Gil-White, Francisco J. 1997. How Thick Is Blood? The Plot Thickens . . . If Ethnic Actors Are Primordialists, What Remains of the Circumstantialist/Primordialist Controversy? *Ethnic and Racial Studies* 22(5): 789–820.

Glazer, Nathan, and Daniel P. Moynihan. 1975. Introduction. In *Ethnicity: Theory and Experience*, edited by Nathan Glazer and Daniel P. Moynihan, pp. 1–26. Harvard University Press, Cambridge.

Goldfrank, Esther S. 1948. The Impact of Situation and Personality on Four Hopi Emergence Myths. *Southwestern Journal of Anthropology* 4:241–262.

Goldman, Irving. 1963. *The Cubeo: Indians of the Northwest Amazon.* University of Illinois Press, Urbana.

Gomes, Denise M. C. 2001. Santarém: Symbolism and Power in the Tropical Forest. In *Unknown Amazon: Nature in Culture in Ancient Brazil*, edited by Colin McEwan, Cristina Barreto, and Eduardo Neves, pp. 134–155. British Museum Press, London.

Gómez Otero, Julieta, and Silvia Dahinten. 1999. Evidencias de contactos interétnicos en el siglo XVI en Patagonia: Informe preliminar sobre el sitio enterratorio Rawson (Chubut). In *Actas del XII Congreso Nacional de Arqueología Argentina*, vol. 3, pp. 44–53. La Plata.

González, Alberto R. 1979. Pre-Columbian Metallurgy of Northwest Argentina: Historical Development and Cultural Process. In *Precolumbian Metallurgy of South America*, edited by Elizabeth Benson, pp. 133–202. Dumbarton Oaks Research Library and Collection, Washington, D.C.

———. 1992. *Las placas metálicas de los Andes del sur: Contribución al estudio de las religiones precolombinas.* Ava-Materialien 46. Philipp Von Zabern, Mainz.

———. 1998. *Cultura la Aguada: Arqueología y diseños.* Filmediciones Valero, Buenos Aires.

González Holguín, Diego. [1608] 1952. *Vocabulario de la lengua general de todo el Perú llamada lengua Quichua o del Inka.* Universidad Nacional Mayor de San Marcos, Lima.

Gorbak, Celina, Mirtha Lischetti, and Carmen Paula Muñoz. 1962. Batallas rituales del Chiaraje y del Tocto de la provincia de Kanas (Cuzco-Perú). *Revista del Museo Nacional*, vol. 21, Lima.

Gordon, Robert, and Mervyn Meggitt. 1985. *Law and Order in the New Guinea Highlands*. University Press of New England, Hanover.

Gose, Peter. 1994. *Deathly Waters and Hungry Mountains: Agrarian Ritual and Class Formation in an Andean Town*. University of Toronto Press, Toronto.

Gottdiener, Mark. 1995. *Postmodern Semiotics: Material Culture and the Forms of Post-Modern Life*. Blackwell, Oxford.

Graham, Ian. 1967. *Archaeological Explorations in El Peten, Guatemala*. Middle American Research Institute Publication 33. Tulane University, New Orleans.

———. 1975. *Corpus of Maya Hieroglyphic Inscriptions*. Peabody Museum of Archaeology and Ethnology, Harvard University, Cambridge.

Graham, John A. 1990. Monumental Sculpture and Hieroglyphic Inscriptions. *Excavations at Seibal, Department of Peten, Guatemala*. Peabody Museum of Archaeology and Ethnology Memoirs 17, vol. 17, no. 1. Cambridge University Press, Cambridge.

Grube, Nikolai, and Simon Martin. 2000. The Dynastic History of the Maya. In *Maya: Divine Kings of the Rain Forest*, edited by Nikolai Grube, pp. 149–171. Könemann, Cologne.

Gruszczynska-Ziótkowska, Anna. 1995. *El poder del sonido: El papel de las crónicas españolas en la etnomusicología andina*. Biblioteca Abya-Yala 24. Cayambe, Ecuador.

Guaman Poma de Ayala, Felip. 1980a. *El primer nueva corónica y buen gobierno*, critical edition edited by John V. Murra and Rolena Adorno. Siglo Veintiuno Editores, México, D.F.

———. [1615] 1980b. *Nueva crónica y buen gobierno*. Siglo XXI, Mexico, D.F.

Gudemos, Mónica L. 1998. *Antiguos sonidos: El material arqueológico musical del museo Dr. Eduardo Casanova, Tilcara, Jujuy (Rep. Argentina)*. Instituto Interdisciplinario Tilcara, Tilcara.

Guilaine, Jean, and Jean Zammit. 2005. *The Origins of War: Violence in Prehistory*. Blackwell, Oxford.

Gutierrez, Mary Ellen. 1990. The Maya Ballgame as a Metaphor for Warfare. *Mexicon* 12(6): 105–108.

Haas, Jonathan. 1990. Warfare and the Evolution of Tribal Polities in the Prehistoric Southwest. In *The Anthropology of War*, edited by Jonathan Haas, pp. 171–189. Cambridge University Press, Cambridge.

———. 2001. Warfare and the Evolution of Culture. In *Archaeology at the Millennium: A Sourcebook*, edited by Gary Feinman and T. Douglas Price, pp. 329–350. Kluwer / Plenum, New York.

Haas, Jonathan, and Winifred Creamer. 1993. *Stress and Warfare among the Kayenta Anasazi of the 13th Century A.D.* Fieldiana, Anthropology New Series, no. 21. Field Museum of Natural History, Chicago.

———. 1997. Warfare among the Pueblos: Myth, History, and Ethnography. *Ethnohistory* 44:235–261.

Hall, Jonathan M. 1997. *Ethnic Identity in Greek Antiquity*. Cambridge University Press, Cambridge.

Hall, Robert L. 1989. The Cultural Background of Mississippian Symbolism. In *The Southeastern Ceremonial Complex: Artifacts and Analysis*, edited by Patricia Galloway, pp. 239–278. University of Nebraska Press, Lincoln.

———. 1991. Cahokia Identity and Interaction Models of Cahokia Mississippian. In *Cahokia and the Hinterlands: Middle Mississippian Cultures of the Midwest*, edited by Thomas E. Emerson and R. Barry Lewis, pp. 3–34. University of Illinois Press, Urbana.

Hamilakis, Yannis. 1999. Food Technologies/Technologies of the Body: The Social Context of Wine and Oil Production and Consumption in Bronze Age Crete. *World Archaeology* 38(1): 38–54.

Hamilakis, Yannis, Mark Pluciennik, and Sarah Tarlow (editors). 2002. *Thinking Through the Body: Archaeologies of Corporality*. Kluwer Academic/Plenum, New York.

Harrison, Simon. 1989. The Symbolic Construction of Aggression and War in a Sepik River Society. *Man*, n.s., 24:583–599.

Hartmann, Roswith. 1972. Otros datos sobre las llamadas "batallas rituales." In *Actas y memorias del XXXIX Congreso Internacional de Americanistas*, vol. 6, edited by Rosalía Avalos de Matos and Rogger Ravines. *Congreso Internacional de Americanistas*, Lima.

———. 1978. Más noticias sobre el "Juego del Pucara." In *Amerikanistische studien/Estudios americanistas II, collectanea instituti anthropos 21*, edited by Roswith Hartmann and Udo Oberem, pp. 202–218. Ibero-Amerikanische Institut, Bonn.

Harvey, Graham. 2006. *Animism: Respecting the Living World*. Columbia University Press, New York.

Hassig, Ross. 1988. *Aztec Warfare: Imperial Expansion and Political Control*. University of Oklahoma Press, Norman.

Hastorf, Christine A. 1993. *Agriculture and the Onset of Political Inequality Before the Inka*. Cambridge University Press, Cambridge.

Hastorf, Christine (editor). 1999. *Early Settlement in Chiripa, Bolivia*. University of California Press, Berkeley.

Hathcock, Roy. 1976. *Ancient Indian Pottery of the Mississippi River Valley*. Hurley Press, Camden, Arizona.

———. 1983. *The Quapaw and Their Pottery*. Hurley Press, Camden, Arizona.

Heckenberger, Michael, James B. Petersen, and Eduardo G. Neves. 1999. Village Size and Permanence in Amazonia: Two Archaeological Examples from Brazil. *Latin American Antiquity* 10(4): 533–576.

Heckenberger, Michael, Afukaka Kuikuro, Urissapa T. Kuikuro, J. Christian Russell, Morgan Schmidt, Carlos Fausto, and Bruna Franchetto. 2003. Amazonia 1492: Pristine Forest or Cultural Parkland. *Science* 301:1710–1714.

Hemming, John. 1970. *The Conquest of the Inkas*. Harcourt Brace & Co, New York.

Hill, J. D. 1995. Ritual and Rubbish in the Iron Age of Wessex: A Study on the Formation of a Specific Archaeological Record. *BAR International Series* 242, British Archaeological Reports, Oxford.

Hocquenghem, Anne Marie. 1987. *Iconografía Mochica*. Pontificia Universidad de Católica del Perú, Lima.

Hodder, Ian. 1991. *Reading the Past: Current Approaches to Interpretation in Archaeology*, 2nd ed. Cambridge University Press, Cambridge.

———. 1999. *The Archaeological Process*. Blackwell, Oxford.

Hopkins, D. 1982. Juego de Enemigos. *Allpanchis* 20:167–187. Cuzco.

Horton, Robin. 1993. *Patterns of Thought in Africa and the West*. Cambridge University Press, Cambridge.

Houston, Stephen D. 1993. *Hieroglyphs and History at Dos Pilas: Dynastic Politics of the Classic Maya*. University of Texas Press, Austin.

———. 2002. Impersonation, Dance, and the Problem of Spectacle Among the Classic Maya. Paper presented at the annual conference of the Society for American Archaeology, Denver.

Houston, Stephen, Héctor Escobedo, Perry Hardin, Richard Terry, David Webster, Mark Child, Charles Golden, Kitty Emery, and David Stuart. 1999. Between Mountain and Sea: Investigations at Piedras Negras, Guatemala. *Mexicon* 21(1): 10–17.

Howard, Calvin D. 1974. The Atlatl: Function and Performance. *American Antiquity* 39:102–104.

Hudson, Charles. 1976. *The Southeastern Indians*. University of Tennessee Press, Knoxville.

Hugh-Jones, Stephen. 1979. *The Palm and the Pleiades: Initiation and Cosmology in Northwest Amazonia*. Cambridge University Press, Cambridge.

———. 1995. Inside-Out and Back-to-Front: The Androgynous House in Northwest Amazon. In *About the House: Lévi-Strauss and Beyond*, edited by Janet Carsten and Stephen Hugh-Jones, pp. 226–252. Cambridge University Press, Cambridge.

Hyslop, John. 1976. An Archaeological Investigation of the Lupaqa Kingdom and its Origins. PhD dissertation, Columbia University, New York.

———. 1990. *Inka Settlement and Planning*. University of Texas Press, Austin.

Inomata, Takeshi. 1995. Archaeological Investigations at the Fortified Center of Aguateca, El Petén, Guatemala: Implications for the Study of the Classic Maya Collapse. PhD dissertation, Vanderbilt University.

———. 1997. The Last Day of a Fortified Classic Maya Center: Archaeological Investigations at Aguateca, Guatemala. *Ancient Mesoamerica* 8(2): 337–352.

———. 2001. The Classic Maya Royal Palace as a Political Theater. In *Reconstruyendo la ciudad maya: El urbanismo en la sociedades antigua*, edited by Andrés Ciudad Ruiz, María Josefa Iglesias Ponce de León, and María del Carmen Martínez Martínez, pp. 341–362. Sociedad Española de Estudios Mayas, Madrid.

———. 2003. War, Destruction, and Abandonment: The Fall of the Classic Maya Center of Aguateca, Guatemala. In *The Archaeology of Settlement Abandonment in Middle America*, edited by Takeshi Inomata and Ronald Webb, pp. 43–60. University of Utah Press, Salt Lake City.

———. 2006. Politics and Theatricality in Maya Society. In *Theaters of Power and Community: Archaeology of Performance*, edited by Takeshi Inomata and Lawrence Coben, pp. 187–221. AltaMira Press, Walnut Creek.

———. 2006. Plazas, Performers, and Spectators: Political Theaters of the Classic Maya. *Current Anthropology* 47(5): 805–842.

Inomata, Takeshi, and Lawrence Coben. 2006. Overture: An Invitation to the Archaeological Theater. In *Archaeology of Performance: Theaters of Power, Community, and Politics*, edited by Takeshi Inomata and Lawrence Coben, pp. 11–46. AltaMira Press, Lanham.

Inomata, Takeshi, and Daniela Triadan. 2003. El espectáculo de la muerte en las tierras bajas mayas. In *Antropología de la eternidad: La muerte en la civilización maya*, edited by Andrés Ciudad Ruiz, Mario Humberto Ruz, and Maria Josefa Iglesias Ponce de Leon, pp. 195–207. Sociedad Española de Estudios Mayas and Centro de Estudios Mayas, Universidad Autónoma de México, Madrid.

Inomata, Takeshi, Erick Ponciano, Richard Terry, Estela Pinto, Daniela Triadan, and Harriet F. Beaubien. 2001. In the Palace of the Fallen King: The Excavation of the Royal Residential Complex at the Classic Maya Center of Aguateca, Guatemala. *Journal of Field Archaeology* 28(3–4): 287–306.

Inomata, Takeshi, Erick Ponciano, Oswaldo Chinchilla, Otto Román, Véronique Breuil-Martínez, and Oscar Santos. 2004. An Unfinished Temple at the Classic Maya Center of Aguateca, Guatemala. *Antiquity* 78(302): 798–811.

Isbell, William H. 1997. *Mummies and Mortuary Monuments: A Postprocessual Prehistory of Central Andean Social Organization.* University of Texas Press, Austin.

Isbell, William H., and Helaine Silverman. 2006. Rethinking the Central Andean Co-Tradition. In *Andean Arcaheology III: North and South*, edited by Wiliam H. Isbell and Helaine Silverman, pp. 497–518. Springer Science, New York.

Izko, Xavier. 1992. *La doble frontera: Identidad, política y ritual en el altiplano central.* Hisbol, La Paz.

Janusek, John W. 1999. Craft and Local Power: Embedded Specialization in Tiwanaku Cities. *Latin American Antiquity* 10(2): 107–131.

———. 2004. *Identity and Power in the Ancient Andes: Tiwanaku Cities Through Time.* Routledge, New York.

Jenkins, Richard. 1992. *Pierre Bourdieu.* Routledge, London.

Johnson, Allen, and Timothy Earle. 1987. *The Evolution of Human Societies*. Stanford University Press, Stanford.

Johnson, Matthew. 1996. *An Archaeology of Capitalism*. Blackwell, Oxford.

Jones, Sian. 1997. *The Archaeology of Ethnicity: Constructing Identities in the Past and Present*. Routledge, London.

Joyce, Arthur A. 2004. Sacred Space and Social Relations in the Valley of Oaxaca. In *Mesoamerican Archaeology: Theory and Practice*, edited by Julia A. Hendon and Rosemary Joyce, pp. 192–216. Blackwell, Oxford.

Joyce, Rosemary A. 1998. Performing the Body in Pre-Hispanic Central America. *RES* 43:147–165.

———. 2000a. *Gender and Power in Prehispanic Mesoamerica*. University of Texas Press, Austin.

———. 2000b. Heirlooms and Houses: Materiality and Social Memory. In *Beyond Kinship: Social and Material Reproduction in House Societies*, edited by Rosemary A. Joyce and Susan D. Gillespie, pp. 189–212. University of Pennsylvania Press, Philadelphia.

———. 2005. Archaeology of the Body. *Annual Review of Anthropology* 34:139–158.

Julien, Catherine. 1983. *Hatunqolla: A View of Inka Rule from the Lake Titicaca Region*. University of California Press, Berkeley.

———. 1993. Finding a Fit: Archaeology and Ethnohistory of the Inkas. In *Provincial Inka*, edited by Michael Malpass, pp. 177–233. University of Iowa Press, Iowa City.

Keane, Webb. 2005. Signs Are Not the Garb of Meaning: On the Social Analysis of Material Things. In *Materiality*, edited by Daniel Miller, pp. 182–205. Duke University Press, Durham.

Keegan, John. 1976. *The Face of Battle*. Viking Press, New York.

Keeley, Lawrence H. 1996. *War Before Civilization*. Oxford University Press, New York.

Kelly, Raymond C. 1993. Constructing Inequality: The Fabrication of a Hierarchy of Virtue among the Etoro. University of Michigan Press, Ann Arbor.

———. 2000. *Warless Societies and the Origin of War*. University of Michigan Press, Ann Arbor.

Kidder, Alfred V. 1924. An Introduction to the Study of Southwestern Archaeology. Published for the Department of Archaeology, Phillips Academy. Yale University Press, New Haven.

King, Adam. 2003. *Etowah: The Political History of a Chiefdom Capital*. University of Alabama Press, Tuscaloosa.

Knauft, Bruce. 1991. Violence and Sociality in Human Evolution. *Current Anthropology* 32:391–428.

Knight, Vernon James, Jr. 1986. The Institutional Organization of Mississippian Religion. *American Antiquity* 51(4): 675–687.

Knight, Vernon James, Jr., James A. Brown, and George E. Lankford. 2001. On the Subject Matter of Southeastern Ceremonial Complex Art. *Southeastern Archaeology* 20:129–141.

Kolata, Alan. 1993. *The Tiwanaku: Portrait of an Andean Civilization*. Blackwell, Cambridge.

———. 2003. Tiwanaku Ceremonial Architecture and Urban Organization. In *Tiwanaku and its Hinterland: Archaeology and Paleoecology of an Andean Civilization*, vol. 2, Urban and Rural Archaeology, edited by Alan L. Kolata, pp. 175–201. Smithsonian Institution Press, Washington, D.C.

Kolb, Michael, and Boyd Dixon. 2002. Landscapes of War: Rules and Conventions of Conflict in Ancient Hawai'i (and Elsewhere). *American Antiquity* 67(3): 514–534.

Krause, Richard A. 1998. A History of Great Plains Prehistory. In *Archaeology on the Great Plains*, edited by W. Raymond Wood, pp. 48–86. University of Kansas Press, Lawrence.

Küchler, Susanne. 1988. Malangan: Objects, Sacrifice and the Production of Memory. *American Ethnologist* 15:625–637.

———. 1999. The Place of Memory. In *The Art of Forgetting*, edited by Adrian Forty and Susanne Küchler, pp. 53–72. Berg, Oxford.

Kumu, Umusin P., and Tolaman Kenhíri. 1980. *Antes o mundo não existia*. Cultura, São Paulo.

Kus, Susan. 1992. Towards an Archaeology of Body and Soul. In *Representation in Archaeology*, edited by Jean-Claude Gardin and Christopher Peebles, pp. 168–177. Indiana University Press, Bloomington.

Kyle, Donald. 1998. *Spectacles of Death in Ancient Rome*. Routledge, London.

Lafferty, Robert H., III. 1973. *An Analysis of Prehistoric Southeastern Fortifications*. Master's thesis, Department of Anthropology, Southern Illinois University, Carbondale.

Lakau, Andrew. 1994. *Customary Land Tenure and Alienation of Customary Land Rights among the Kaina, Enga Province, Papua New Guinea*. PhD dissertation, Department of Geographical Studies, University of Queensland.

Lambert, Patricia M. 2002. The Archaeology of War: A North American Perspective. *Journal of Archaeological Research* 10:207–241.

LaMotta, Vincent M., and Michael B. Schiffer. 1999. Formation Process of House Floor Assemblages. In *The Archaeology of Household Activities*, edited by Penelope Allison, pp. 19–29. Routledge, London.

Landa, Diego de. 1938. *Relación de las cosas de Yucatán*, 7th ed. Editorial P. Robredo, México, D.F.

Lankford, George E. 1992. Red and White: Some Reflections on Southeastern Symbolism. *Southern Folklore* 50(1): 53–80.

Larson, Lewis H., Jr. 1972. Functional Considerations of Warfare in the Southeast During the Mississippi Period. *American Antiquity* 37:383–392.

Latcham, Ricardo E. 1938. *Arqueología de la Región Atacameña.* Universidad de Chile, Santiago.

Lathrap, Donald W. 1973. Gifts of the Cayman: Some Thoughts on the Subsistence Basis of Chavin. In *Variation in Anthropology*, edited by Donald W. Lathrap and J. Douglas, pp. 91–105. Illinois Archaeological Survey, Urbana, Illinois.

———. 1982. Jaws: The Control of Power in the Early Nuclear American Ceremonial Center. In *Early Ceremonial Architecture in the Andes*, edited by Christopher B. Donnan, pp. 241–267. Dumbarton Oaks Research Library and Collection, Washington, D.C.

Latour, Bruno. 1992. Where are the Missing Masses? The Sociology of a Few Mundane Artifacts. *In Shaping Technology/Building Society: Studies in Sociotechnical Change*, edited by Weibe E. Bijker and John Law, pp. 225–258. MIT Press, Cambridge.

———. 1993. *We Have Never Been Modern.* Harvester Wheatsheaf, Hemel Hempstead.

———. 1994. Pragmatologies: A Mythical Account of How Humans and Nonhumans Swap Properties. *American Behavioral Scientist* 37: 791–808.

———. 2005. *Reassembling the Social: An Introduction to Actor-Network-Theory.* Oxford University Press, Oxford.

Lau, George F. 2004. Object of Contention: An Examination of Recuay-Moche Combat Imagery. *Cambridge Archaeological Journal* 14(2): 163–184.

LeBlanc, Steven A. 1999. *Prehistoric Warfare in the American Southwest.* University of Utah Press, Salt Lake City.

———. 2003a. *Constant Battles: Why We Fight.* St. Martin's Griffin Press, New York.

———. 2003b. *Constant Battles: The Myth of the Peaceful, Noble Savage.* St. Martin's Press, New York.

LeBlanc, Steven A., and Glen E. Rice (editors). 2001. *Deadly Landscapes: Case Studies in Prehistoric Southwestern warfare.* University of Utah Press, Salt Lake City.

Lee, Richard. 1993. *The Dobe Ju/'hoansi.* Harcourt Brace, New York.

Lekson, Stephen H. 1999. *The Chaco Meridian: Centers of Political Power in the Ancient Southwest.* AltaMira Press, Walnut Canyon, California.

———. 2002. War in the Southwest, War in the World. *American Antiquity* 67:607–624.

Lesure, Richard. 2005. Linking Theory and Evidence in an Archaeology of Human Agency: Iconography, Style, and Theories of Embodiment. *Journal of Archaeological Method and Theory* 12:237–255.

Levi-Strauss, Claude. 1979. *Myth and Meaning.* Schocken Books, New York.

Levy, Jerrold E. 1992. *Orayvi Revisited: Social Stratification in an "Egalitarian" Society.* School of American Research Press, Santa Fe.

Lewis, R. Barry, and Charles Stout (editors). 1998. *Mississippian Towns and Sacred Places*. University of Alabama Press, Tuscaloosa.

Leyenaar, Tedd J. J., and Lee Allen Parsons. 1988. *Ulama: The Ballgame of the Maya and Aztecs*. Spruyt, Van Mangtem and De Does, Leiden, Netherlands.

Lightfoot, Ricky R., Mary C. Etzkorn, and Mark Varien. 1993. Excavations. In *The Duckfoot Site, vol. 1: Descriptive Archaeology*, edited by Ricky C. Lightfoot and Mary C. Etzkorn, pp. 15–129. Ocassional Paper no. 3, Crow Canyon Archaeological Center, Cortez, Colorado.

Lightfoot, Ricky R., and Kristin A. Kuckleman. 2001. A Case of Warfare in the Mesa Verde Region. In *Deadly Landscapes: Case Studies in Prehistoric Southwestern Warfare*, edited by Glen. E. Rice and Steven A. LeBlanc, pp. 51–64. University of Utah Press, Salt Lake City.

Lima, Helena P. 2004. *Cronologia da Amazônia central: O significado da variabilidade dafase Manacapuru*. Unpublished report submitted to the Fundação de Amparoà Pesquisa do Estado de São Paulo, São Paulo.

Lima, Luiz Fernando E. 2003. *Levantamento arqueológico das áreas de interflúvio na área de confluência dos rios Negro e Solimões*. Master's thesis, Archaeology and Ethnology Museum, Sao Paulo University, Sao Paulo.

Linton, Ralph. 1944. Nomadic Raids and Fortified Pueblos. *American Antiquity* 10:28–32.

Lipe, William D. 1995. The Depopulation of the Northern San Juan: Conditions in the Turbulent 1200s. *Journal of Anthropological Archaeology* 14:143–169.

Liu, James H., and Mark W. Allen. 1999. The Evolution of Political Complexity in Maori Hawke's Bay: Archaeological History and its Challenge to Psychological Theory. *Group Dynamics: Theory, Research and Practice* 3:64–80.

Lock, Margaret. 1993. Cultivating the Body: Anthropology and the Epistemologies of Bodily Practice and Knowledge. *Annual Review of Anthropology* 22:133–155.

Lowie, Robert. 1948. The Tropical Forests: An Introduction. In *Handbook of South American Indians*, vol. 3, edited by Julian Steward, pp. 1–56. Bureau of American Ethnology, Bulletin 143, Smithsonian Institution Press, Washington, D.C.

Lozano Machuca, Juan. [1581] 1992. Carta del factor de Potosí . . . al virrey del Perú, en donde se describe la provincia de los Lípez. Potosí, 8 de Noviembre de 1581. *Estudios Atacameños* 11:30–34.

Lucas, Gavin M. 1996. Of Death and Debt: A History of the Body in Neolithic and Early Bronze Age Yorkshire. *Journal of European Archaeology* 4:99–188.

Lucero, Lisa J. 2006. *Water and Ritual the Rise and Fall of Classic Maya Rulers*. University of Texas Press, Austin.

Lutonsky, Anthony F. 1998. *Implements of Close Encounter: Shock Weapon Systems of the Pre-Historic Southwest*. Paper presented at the meeting of the Arizona Archaeological Council in Flagstaff Arizona, October 1998, in a symposium entitled Archaeology and Architecture of Tactical Sites.

Lynn, John A. 2003. *Battle: A History of Combat and Culture*. Westview, Boulder, Colorado.

Machado, Juliana Salles. 2005. *O processo de formação de estruturas artificiais na Amazônia central: Um estudo de caso do sítio Hatahara*. Unpublished master's thesis, Programa de Pós-Graduação em Arqueologia, Museu de Arqueologia e Etnologia, Universidade de São Paulo, São Paulo.

Mackey, James C., and R. C. Green. 1979. Largo-Gallina Towers: An Explanation. *American Antiquity* 44:144–154.

Mackey, James C., and Sally J. Holbrook. 1978. Environmental Reconstruction and the Abandonment in the Largo-Gallina Area, New Mexico. *Journal of Field Archaeology* 5:29–49.

Macleod, Murdo J. 1998. Some Thoughts on the Pax Colonial, Colonial Violence, and Perceptions of Both. In *Native Resistance and the Pax Colonial in New Spain*, edited by Susan Schroeder, pp. 129–142. University of Nebraska Press, Lincoln.

Malinowski, Bronislaw. 1926. *Myth in Primitive Psychology*. W. W. Norton, New York.

Malotki, Ekkehart (editor). 1993. *Hopi Ruin Legends*. University of Nebraska Press, Lincoln.

Mannheim, Bruce. 1991. *The Language of the Inka since the European Invasion*. Texas Linguistics Series, University of Texas Press, Austin.

Manzanilla, Linda, and Eric Woodward. 1990. Restos humanos asociados a la pirámide de Akapana (Tiwanaku, Bolivia). *Latin American Antiquity* 1(2): 133–149.

Marafioti, Roberto. 2004. *Charles S. Peirce: El extasis de los signos*. Editorial Biblos, Buenos Aires.

Marcus, Joyce, and Kent V. Flannery. 2004. The Coevolution of Ritual and Society: New [14]C Dates from Ancient Mexico. *Proceedings of the National Academy of Sciences* 101(52): 18257–18261.

Marcus, Michelle I. 1993. Incorporating the Body: Adornment, Gender, and Social Inequality in Ancient Iraq. *Cambridge Archaeological Journal* 3(2): 157–178.

Marret, Robert R. 1909. *Threshold of Religion*. Methuen, London.

Martin, Debra L., and David W. Frayer (editors). 1997. *Troubled Times: Violence and Warfare in the Past*. Gordon and Breach, Amsterdam.

Martin, Simon. 2000. Warfare among the Classic Maya. In *Maya: Divine Kings of the Rain Forest*, edited by Nikolai Grube, pp. 175–185. Könemann, Cologne.

Martin, Simon, and Nikolai Grube. 2000. *Chronicle of the Maya Kings and Queens*. Thames and Hudson, London.

Martínez, José Luis. 1995. *Autoridades en los Andes, los atributos del señor*. Pontificia Universidad Católica del Perú, Lima.

Maschner, Herbert D. G. (editor). 1996. *Darwinian Archaeologies*. Plenum Press, New York.

Mason, Ronald J. 2000. Archaeology and Native North American Oral Traditions. *American Antiquity* 65(2): 239–266.

Matheny, Ray. 1983. Investigations at Edzná, Campeche, Mexico. Papers of the New World Archaeological Foundation, no. 46. Brigham Young University, Provo.

Mayer, Enrique. 2002. *The Articulated Peasant: Household Economies in the Andes.* Westview Press, Boulder.

McEwan, Colin, Cristiana Barreto, and Eduardo G. Neves (editors). 2001. *Unknown Amazon: Culture in Nature in Ancient Brazil.* British Museum Press, London.

McGregor, John C. 1941. *Southwestern Archaeology.* John Wiley and Sons, New York.

McIntosh, Susan Keech. 1999. Pathways to Complexity: An African Perspective. In *Pathways to Complexity in Africa,* edited by Susan McIntosh, pp. 1–19. Cambridge University Press, Cambridge.

Meggers, Betty J. 1990. Reconstrução do comportamento locacional pré-histórico na Amazônia. *Boletim do Museu Paraense Emilio Goeldi, série Antropologia* 6(2): 183–203.

———. 1993–95. Amazonia on the Eve of European Contact: Ethnohistorical, Ecological, and Anthropological Perspectives. *Revista de Arqueología Americana* 8:91–115.

Meggers, Betty, Ondemar Dias, Eurico Miller, and Celso Perota. 1988. Implications of Archaeological Distributions in Amazonia. In *Proceedings of a Workshop on Neotropical Distribution Patterns,* edited by Paul Vanzolini and W. Ronald Heyer, pp. 275–294. Academia Brasileira de Ciências, Rio de Janeiro.

Meggitt, Mervyn. 1965. *The Lineage System of the Mae-Enga of New Guinea.* Barnes and Noble, New York.

———. 1972. System and Sub-System: The "Te" Exchange Cycle among the Mae Enga. *Human Ecology* 1:111–123.

———. 1974. "Pigs are our Hearts!" The Te exchange Cycle among the Mae Enga of New Guinea. *Oceania* 44:165–203.

———. 1977. *Blood Is Their Argument.* Mayfield, Palo Alto.

Mendonça, Osvaldo, Asunción Bordach, and Silvia G. Valdano. 1992. Reconstrucción del comportamiento biosocial en el pukará de Tilcara (Jujuy): Una propuesta heurística. *Cuadernos* 3:144–154. San Salvador de Jujuy.

Mercado De Peñalosa, Don Pedro. [1586] 1885. Relación de la provincia de los pacajes. In *Relációnes geográficas de Indias: Peru,* vol. 2, edited by Marcos Jiménez de la Espada, pp. 51–64. Ministerio de Fomento, Madrid, Tip. de M. G. Hernández.

Merleau-Ponty, Marcel. 1962. *The Phenomenology of Perception.* Routledge and Kegan Paul, London.

Merwin, Bruce W. 1934. An Aboriginal Village Site in Union County. *Journal of the Illinois State Historical Society* 28:78–91.

Meskell, Lynn M. 1996. The Somatization of Archaeology: Institutions, Discourses and Corporeality. *Norwegian Archaeological Review* 29(1): 1–16.

———. 1999. *Archaeologies of Social Life: Age, Sex, and Class in Ancient Egypt*. Basil Blackwell Press, Oxford.

———. 2000. Writing the Body in Archaeology. In *Reading the Body*, edited by Alison E. Rautman, pp. 13–21. University of Pennsylvania Press, Philadelphia.

———. 2004. *Object Worlds in Ancient Egypt: Material Biographies Past and Present*. Berg, Oxford.

Meskell, Lynn M., and Rosemary A. Joyce. 2003. *Embodied Lives: Figuring Ancient Maya and Egyptian Experience*. Routledge, London.

Meyer, Peter. 1990. Human Nature and the Function of War in Social Evolution: A Critical Review of a Recent Form of the Naturalistic Fallacy. In *Sociobiology and Conflict: Evolutionary Perspectives on Competition, Cooperation, Violence and Warfare*, edited by Johan M. G. van der Dennen and V. S. E. Falger, pp. 227–240. Chapman and Hall, London.

Miller, Mary E. 1985. A Re-Examination of the Mesoamerican Chacmool. *Art Bulletin* 67(1): 7–17.

———. 1986. *The Murals of Bonampak*. Princeton University Press, Princeton.

Milner, George R. 1995. Osteological Evidence for Prehistoric Warfare. In *Regional Approaches to Mortuary Analysis*, edited by Lane Anderson Beck, pp. 221–244. Plenum Press, New York.

———. 1999. Warfare in Prehistoric and Early Historic North America. *Journal of Archaeological Research* 7:105–151.

Milner, George R., Eve Anderson, and Virginia G. Smith. 1991. Warfare in Late Prehistoric West-Central Illinois. *American Antiquity* 56:581–603.

Miranda, Cristóbal de. [1583] 1906. Relación de los corregimientos y otros oficios que se proveen en los reynos e provincios del Piru, en el distrito y gobernación del Vissorrey dellos. In *Juicio de Límites*, vol. 1, edited by Victor M. Maúrtua, pp. 168–280. Imprenta de Henrich, Barcelona.

Mishkin, Bernard. 1940. *Rank and Warfare among the Plains Indians*. University of Washington Press, Seattle.

Mock, Shirley Boteler (editor). 1998. *The Sowing and the Dawning: Termination, Dedication, and Transformation in the Archaeological and Ethnographic Record of Mesoamerica*. University of New Mexico Press, Albuquerque.

Molinié-Fioravanti, Antoinette. 1988. Sanglantes et fertiles frontières. A propos des batailles rituelles andines. *Journal de la Société des Américanistes* 74:49–70.

Montgomery, Barbara. 1993. Ceramic Analysis as a Tool for Discovering Processes of Pueblo Abandonment. In *Abandonment of Sites and Regions*, edited by Catherine M. Cameron and Steven A. Tompka, pp. 157–164. Cambridge University Press, Cambridge.

Montgomery, Ross, Watson Smith, and John O. Brew. 1949. Franciscan Awatovi. *Papers of the Peabody Museum of American Archaeology and Ethnology*, vol. 36. Harvard University, Cambridge.

Montserrat, Dominic. 1998. *Changing Bodies, Changing Meanings*. Routledge, New York.

Moore, Henrietta. 1999. Whatever Happened to Women and Men? Gender and Other Crises in Anthropology. In *Anthropological Theory Today*, edited by Henrietta Moore, pp. 151–171. Polity Press, Cambridge.

Moore, Jerry D. 1996. *Architecture and Power in the Ancient Andes: The Archaeology of Public Buildings*. Cambridge University Press, Cambridge.

Morais, Claide P. 2005. Levantamento arqueológico das áreas de entorno do Lago do Limão, Iranduba, Am. Unpublished report submitted to the Fundação de Amparo à Pesquisa do Estado de São Paulo.

Moreira, Ismael P., and Angelo B. Moreira. 1994. *Mitologia Tariana*. Instituto Brasileiro de Patrimônio Cultural, Manaus.

Morris, Earl H. 1941. Prayer Sticks in Walls of Mummy Cave Tower, Canyon Del Muerto. *American Antiquity* 6:227–230.

Morse, Dan F., and Phyllis A. Morse. 1983. *Archaeology of the Central Mississippi Valley*. Academic Press, New York.

Moseley, Michael E. 1983. Patterns of Settlement and Preservation in the Viru and Moche Valleys. In *Prehistoric Settlement Patterns: Essays in Honor of Gordon R. Willey*, edited by Evon Z. Vogt and Richard M. Leventhal, pp. 423–442. University of New Mexico Press, Albuquerque/Peabody Museum of Archaeology and Ethnology, Cambridge.

———. 2001. *The Inkas and their Ancestors: The Archaeology of Peru*, rev. ed. Thames and Hudson, New York.

Mostny, Grete. 1958. Máscaras, tubos y tabletas para rapé y cabezas trofeo entre los atacameños. In *Miscelánea Paul Rivet*, vol. 2, pp. 379–392. Universidad Nacional Autónoma de México, México, D.F.

Muller, Jon. 1978. The Kincaid Settlement System. In *Mississippian Settlement Patterns*, edited by Bruce D. Smith, pp. 269–292. Academic Press, New York.

———. 1986. Serpents and Dancers: Art of the Mud Glyph Cave. In *The Prehistoric Native American Art of Mud Glyph Cave*, edited by Charles H. Faulkner, pp. 36–80. University of Tennessee Press, Knoxville.

———. 1989. The Southern Cult. In *The Southeastern Ceremonial Complex: Artifacts and Analysis*, edited by Patricia Galloway, pp. 11–26. University of Nebraska Press, Lincoln.

Munita, Casimiro, and Emílio Soares. 2003. O levantamento das fontes de argila e a análise por ativação de amostras de argila e cerâmicas arqueológicas. In *Levantamento arqueológico da área de confluência dos rios Negro e Solimões, estado do Amazonas: Continuidade das escavações, análise da composição química das cerâmicas e montagem de um sistema de informações geográfi-*

cas, edited by Eduardo G. Neves, pp. 110–122. Activities report presented to the Fundação de Amparo à Pesquisa do Estado de São Paulo.

Murra, John V. 1975. *Formaciones económicas y políticas en el mundo andino*. Instituto de Estudios Peruanos, Lima.

———. 1982. The *Mit'a* Obligations of Ethnic Groups to the Inka State. In *The Inka and Aztec States 1400–1800: Anthropology and History*, edited by George A. Collier, Renato I. Rosaldo, and John D. Wirth, pp. 237–262. Academic Press, New York.

Murúa, Fray Martín de. [1611] 1962. *Historia general del Perú, origen y descendencia de los Inkas*. Instituto Gonzalo Fernández de Oviedo, Madrid.

Myers, Thomas. 1992. Agricultural Limitations of the Amazon in Theory and Practice. *World Archaeology* 24(1): 82–97.

Nabokov, Peter. 2002. *A Forest of Time: American Indian Ways of History*. Cambridge University Press, Cambridge.

Naroll, Raoul and William T. Divale. 1976. Natural Selection in Cultural Evolution: Warfare Versus Peaceful Diffusion. *American Ethnologist* 3:97–128.

Nassaney, Michael S., and Kendra Pyle. 1999. The Adoption of the Bow and Arrow in Eastern North America: A View from Central Arkansas. *American Antiquity* 64:243–263.

Neira Avedaño, Máximo. 1967. Informe preliminar de las investigaciones arqueológicos en el Departmento de Puno. *Anales del Instituto de Estudios Socio Economicos 1(1): 107–164*.

Netherly, Patricia. 1977. *Local Level Lords on the North Coast of Peru*. PhD dissertation, Cornell University, Ithaca, New York.

———. 1990. Out of Many, One: The Organization of Rule in the North Coast Polities. In *The Northern Dynasties: Kingship and Statecraft in Chimor*, edited by Michael E. Moseley and Alana Cordy-Collins, pp. 461–487. Dumbarton Oaks Research Library and Collection, Washington, D.C.

Neves, Eduardo G. 1995. Village Fissioning in Amazonia: A Critique of Monocausal Determinism. *Revista do Museu de Arqueologia e Etnologia da Universidade de São Paulo* 5:195–209.

———. 1998. *Paths in Dark Waters: Archaeology as Indigenous History in the Upper Rio Negro Basin, Northwest Amazon*. Unpublished PhD dissertation, Department of Anthropology, Indiana University.

———. 1999. Changing Perspectives in Amazonian Archaeology. In *Archaeology in Latin America*, edited by G. Politis and B. Alberti, pp. 216–243. Routledge, London.

———. 2000. *Levantamento arqueológico da área de confluência dos rios Negro e Solimões, estado do Amazonas, Junho 1999–Agosto 2000*. Activities report presented to the Fundação de Amparo à Pesquisa do Estado de São Paulo.

———. 2001. Indigenous Historical Trajectories in the Upper Rio Negro Basin. In *Unknown Amazon: Nature in Culture in Ancient Brazil*, edited by Colin

McEwan, Cristiana Barreto, and Eduardo Neves, pp. 266–286. British Museum Press, London.

———(editor). 2003. *Levantamento arqueológico da área de confluência dos rios Negro e Solimões, estado do Amazonas: Continuidade das escavações, análise da composição química das cerâmicas e montagem de um sistema de informações geográficas.* Activities report presented to the Fundação de Amparo à Pesquisa do Estado de São Paulo.

———. 2006. *Arqueologia da Amazônia.* Jorge Zahar, Rio de Janeiro.

Neves, Eduardo G., James B. Petersen, Robert N. Bartone, and Carlos Augusto da Silva. 2003. Historical and Socio-Cultural Origins of Amazonian Dark Earths. In *Amazonian Dark Earths: Origin, Properties, Management*, edited by Johannes Lehmann, Dirse C. Kern, Bruno Glaser, and William I. Woods, pp. 29–50. Kluwer, Netherlands.

Neves, Eduardo G., and Adriana Schmidt Dias. 2003. A identidade na mudança: Transformações e reprodução na arte pré-colombiana das Terras Altas e Terras Baixas da América do Sul. In *4ª Bienal de Artes Visuais do Mercosul*, edited by N. Aguilar, pp. 47–69. Fundação Bienal de Artes Visuais do Mercosul, Porto Alegre.

Neves, Eduardo G., James B. Petersen, Robert N. Bartone, and Michael Heckenberger. 2004. The Timing of Terra Preta Formation in the Central Amazon: Archaeological Data from Three Sites. In *Amazonian Dark Earths: Explorations in Space and Time*, edited by W. Glaser and W. Woods, pp. 125–134. Springer Verlag, Berlin.

Neves, Eduardo G., and James B. Petersen. 2006. The Political Economy of Pre-Columbian Amerindians: Landscape Transformations in Central Amazonia. In *Time and Complexity in Historical Ecology: Studies in the Neotropical Lowlands*, edited by William Balée and Clark Erickson. Columbia University Press, New York.

Nielsen, Axel E. 1995. Architectural Performance and the Reproduction of Social Power. In *Expanding Archaeology*, edited by James M. Skibo, William H. Walker, and Axel E. Nielsen, pp. 47–66. University of Utah Press, Salt Lake City.

———. 2001. Evolución social en Quebrada de Humahuaca (AD 700–1536). In *Historia Argentina Prehispánica*, vol. 1, edited by E. Berberián and A. Nielsen, pp. 171–264. Editorial Brujas, Córdoba.

———. 2002. Asentamientos, conflicto y cambio social en el Altiplano de Lípez (Potosí, Bolivia). *Revista Española de Antropología Americana* 32:179–205. Madrid.

———. 2006. Plazas para los antepasados: Descentralización y poder corporativo en las formaciones políticas preincaicas de los Andes circumpuneños. *Estudios Atacameños* 31:63–89. San Pedro de Atacama, Chile.

Nielsen, Axel. E. and William H. Walker. 1999. Comquista ritual y dominación politica en el Tawantinsuyu: El caso de los amarillos (Jujuy, Argentina).

In *Sed non satiata: La teoría social en la arqueología latinoamericana contemporánea*, edited by A. Zarankin and F. Acuto, pp. 153–169. Ediciones del Tridente, Buenos Aires.

North, Douglas. 1990. *Institutions, Institutional Change and Economic Performance.* Cambridge University Press, Cambridge.

Nöth, Winfried. 1990. *Handbook of Semiotics.* Indiana University Press, Bloomington.

Núñez, Lautaro. 1987. El tráfico de metales en el area centro-Sur Andina: Factos y expectativas. *Cuadernos del Instituto Nacional de Antropología* 12:73–107.

———. 1991. *Cultura y conflicto en los oasis de San Pedro de Atacama.* Santiago, Editorial Universitaria.

Núñez, Lautaro, and Tom Dillehay. 1979. *Movilidad giratoria, armonía social y desarrollo en los Andes Meridionales: Patrones de tráfico e interacción económica.* Universidad Católica del Norte, Antofagasta.

Nuñez del Prado, Juan Víctor. 1974. The Supernatural World of the Quechua of Southern Peru as Seen from the Community of Qotobamba. In *Native South Americans: Ethnology of the Least Known Continent*, edited by Patricia J. Lyon, pp. 238–250. Little, Brown and Company, Boston.

Núñez Regueiro, Víctor. 1974. Conceptos instrumentales y marco teórico en relación al análisis del desarrollo cultural del noroeste Argentino. *Revista del Instituto de Antropología* 5:169–190. Córdoba.

Nuttall, Zelia. 1903. *The Book of the Life of the Ancient Mexican Containing an Account of Their Rites and Superstitions.* University of California Press, Berkeley.

Oberschall, Anthony. 2000. The Manipulation of Ethnicity: From Ethnic Cooperation to Violence and War in Yugoslavia. *Ethnic and Racial Studies* 23(6): 982–1001.

O'Brien, Michael J. (editor). 2001. Mississippian Community Organization: The Powers Phase in Southeastern Missouri. Kluwer Academic/Plenum Publishers, New York.

Ogburn, Dennis. 2004. *Human Trophies in the Late Prehispanic Andes: Display, Propaganda and Reinforcement of Power among the Inkas and Other Societies.* Paper presented at the sixty-ninth annual meeting of the Society for American Archaeology, Montreal.

Olsen, Bjørnar. 2003. Material Culture After Text: Re-Membering Things. *Norwegian Archaeological Review* 36(2): 87–104.

Ortner, Sherry. 1984. Theory in Anthropology Since the Sixties. *Society for Comparative Study of Society and History* 26:126–166.

———. 2001. Practice, Power and the Past. *Journal of Social Archaeology* 1:271–278.

Otterbein, Keith F. 1968. Internal War: A Cross-Cultural Study. *American Anthropologist* 70:277–289.

———. 1970. *The Evolution of War: A Cross-Cultural Study.* HRAF Press, New Haven.

———. 1994. *Feuding and Warfare: Selected Works of Keith F. Otterbein*. Gordon and Breach, Langhorne, Pennsylvania.

———. 1999. A History of Research on Warfare in Anthropology. *American Anthropologist* 101:794–805.

Pachacuti Yamqui Salcamaygua, and Joan de Santa Cruz. 1993. *Relación de antigüedades deste reyno del Piru*. Edited by P. Duviols and C. Itier. Instituto Francés de Estudios Andinos-Centro de Estudios Regionales Andinos "Bartolomé de Las Casas," Cusco.

Parmentier, Richard J. 1997. The Pragmatic Semiotics of Culture. *Semiotica* 116:1–113.

Parsons, Elsie Clews. 1922. Oraibi in 1920: Contributions to Hopi History. *American Anthropologist* 24:253–298.

———. 1924. The Scalp Ceremonial of Zuni. *Memoirs of the American Anthropological Association* 31. Menasha.

———. [1939] 1996. *Pueblo Indian Religion*, 2 vols. Bison Books, University of Nebraska Press, Lincoln.

Parsons, Jeffrey R., and Charles M. Hastings. 1988. The Late Intermediate Period. In *Peruvian Prehistory*, edited by Richard W. Keatinge, pp. 190–229. Cambridge University Press, Cambridge.

Pärssinnen, Martti, Ari Siiriäinen, and Antti Korpisaari. 2003a. Fortifications Related to the Inka Expansion. In *Western Amazonia–Amazonia Ocidental: Multidisciplinary Studies on Ancient Expansionistic Movements, Fortifications and Sedentary Life*, edited by Martti Pärssinnen and Antti Korpisaari, pp. 29–72. Renvall Institute Publications 14, University of Helsinki, Helsinki.

Pärssinnen, Martti, Alceu Ranzi, Sanna Saunaluoma, and Ari Siiriäinen. 2003b. Geometrically Patterned Ancient Earthworks in the Rio Branco Region of Acre, Brazil: New Evidence of Ancient Chiefdom Formations in Amazonian Interfluvial *Terra Firme* Environment. In *Western Amazonia–Amazonia Ocidental: Multidisciplinary Studies on Ancient Expansionistic Movements, Fortifications and Sedentary Life*, edited by Martti Pärssinnen and Antti Korpisaari, pp. 97–133. Renvall Institute Publications 14, University of Helsinki, Helsinki.

Pauketat, Timothy R. 1994. *The Ascent of Chiefs: Cahokia and Mississippian Politics in Native North America*. University of Alabama Press, Tuscaloosa.

———. 1999. America's Ancient Warriors. *MHQ: The Quarterly Journal of Military History* 11(4): 50–55.

———. 2001. Practice and History in Archaeology: An Emerging Paradigm. *Anthropological Theory* 1:73–98.

———. 2003. Resettled Farmers and the Making of a Mississippian Polity. *American Antiquity* 68:39–66.

———. 2004. Ancient Cahokia and the Mississippians. Cambridge University Press, Cambridge.

————. 2005. The Forgotten History of the Mississippians. In *North American Archaeology*, edited by Timothy R. Pauketat and Diana D. Loren, pp. 187–212. Blackwell, Oxford.

Pauketat, Timothy R., and Susan M. Alt. 2005. Agency in a Postmold? Physicality and the Archaeology of Culture-Making. *Journal of Archaeological Method and Theory* 12:213–236.

Peirce, Charles S. 1931–1958. *Collected Papers*, 8 vols. Edited by Charles Hartshorne, Paul Weiss, and Arthur W. Burks. Harvard University Press, Cambridge.

Pérez de Arce, José. 1995. *Música en la piedra: Música prehispánica y sus ecos en Chile actual*. Museo Chileno de Arte Precolombino, Santiago.

Pérez Gollán, José A. 2000. El jaguar en llamas (La religión en el antiguo Noroeste argentino). *Nueva Historia Argentina*, vol. 1, edited by Myriam Tarragó, pp. 229–256. Sudamericana, Buenos Aires.

Petersen, James B., Eduardo G. Neves, and Michael J. Heckenberger. 2001. Gift from the Past: Terra Preta and Prehistoric Indigenous Occupation in Amazonia. In *Unknown Amazon: Nature in Culture in Ancient Brazil*, edited by C. McEwan, Cristiana Barreto, and Eduardo Neves, pp. 86–105. British Museum Press, London.

Petersen, James B., Eduardo G. Neves, Robert N. Bartone, Manuel A. Arroyo-Kalin, and Fernando W. Costa. 2004. *An Overview of Indigenous Cultural Chronology in the Central Amazon*. Paper presented at the annual meeting of the Society for American Archaeology, Montreal.

Phillips, Phillip, and James A. Brown. 1978. *Pre-Columbian Shell Engravings at Spiro*. Peabody Museum, Harvard University, Cambridge.

Pizarro, Pedro. [1571] 1965. *Relación del descubrimiento y conquista de los reinos del Peru*. Biblioteca de Autores Españoles, Madrid.

————. [1571] 1978. *Relación del descubrimiento y conquista de los reinos del Peru*. Pontificia Universidad Católica del Peru, Lima.

Platt, Tristan. 1986. Mirrors and Maize: The Concept of *Yanantin* Among the Macha of Bolivia. In *Anthropological History of Andean Polities*, edited by John V. Murra, Nathan Wachtel, and Jacques Revel, pp. 228–259. Cambridge University Press, Cambridge.

————. 1987a. The Andean Soldiers of Christ: Confraternity Organization, the Mass of the Sun and Regenerative Warfare in Rural Potosi (18th–20th Centuries). *Journal de la Société des Américanistes* 73:139–191. Musée de L'Homme, Paris.

————. 1987b. Entre *Ch'axwa* y *Muxsa*: Para una historia del pensamiento político aymara. In *Tres reflexiones sobre el pensamiento andino*, pp. 61–132. Hisbol, La Paz.

Plourde, Aimée M., and Charles Stanish. 2001. *Formative Period Settlement Patterning in the Huancané-Putina River Valley, Northeastern Titicaca Basin*. Paper presented at the annual meeting of the Society for American Archaeologists, New Orleans.

Pollard, Joshua. 2001. The Aesthetics of Depositional Practice. *World Archaeology* 33(2): 315–333.

Pollock, Susan, and Reinhard Bernbeck. 2000. And They Said, Let Us Make Gods in Our Image: Gendered Ideologies in Ancient Mesopotamia. In *Reading the Body*, edited by Alison E. Rautman, pp. 150–164. University of Pennsylvania Press, Philadelphia.

Porro, Antonio. 1985. Mercadorias e rotas de comércio intertribal na Amazônia. *Revista do Museu Paulista*, n.s., 30:7–12.

Preucel, Robert W. 2006. *Archaeological Semiotics*. Blackwell, Malden.

Price, James E. 1978. The Settlement Pattern of the Powers Phase. In *Mississippian Settlement Patterns*, edited by Bruce D. Smith, pp. 201–231. Academic Press, New York.

Price, James E., and James B. Griffin. 1979. *The Snodgrass Site of the Powers Phase of Southeast Missouri*. Museum of Anthropology Anthropological Papers 66. University of Michigan, Ann Arbor.

Puleston, Denise. 1983. *The Settlement Survey of Tikal.* Tikal Reports no. 13, University Museum Publications, University of Pennsylvania, Philadelphia.

Quilter, Jeffrey. 1990. The Moche Revolt of the Objects. *Latin American Antiquity* 1(1): 42–65.

Rautman, Alison E. (editor). 2000. *Reading the Body*. University of Pennsylvania Press, Philadelphia.

Redfield, James. 1994. *Nature and Culture in the Iliad*. Duke University Press, Durham, North Carolina.

Redmond, Elsa M. 1994. *Tribal and Chiefly Warfare in South America*. Memoirs of the Museum of Anthropology, no. 28. University of Michigan, Ann Arbor.

Reichert, Raphael Xavier. 1989. A Moche Battle and the Question of Identity. In *Cultures in Conflict: Current Archaeology Perspective*, edited by Diana C. Tkaczuk and David C. Vivian, pp. 86–89. Proceedings of the twentieth annual Chacmool Conference, University of Calgary, Calgary.

Revilla Becerra, Rosanna Liliana, and Mauro Alberto Uriarte Paniagua. 1985. *Investigación arqueológica en la zona de Sillustani-sector Wakakancha-Puno*. Bachelor's thesis, Universidad Católica Santa Maria.

Reyna, Stephen P. 1994a. Preface: Studying War, an Unfinished Project of the Enlightenment. In *Studying War: Anthropological Perspectives*, edited by Stephen P. Reyna and R. E. Downs, pp. 29–65. Gordon and Breach, Amsterdam.

————. 1994b. A Mode of Domination Approach to Organized Violence. In *Studying War: Anthropological Perspectives*, edited by Stephen P. Reyna and R. E. Downs, pp. 29–65. Gordon and Breach, Amsterdam.

Ribeiro, Darcy. 1957. Convívio e contaminação. *Sociologia* 18(1): 3–50.

Rice, Don S. 1986. The Peten Postclassic: A Settlement Perspective. In *Late Lowland Maya Civilization: Classic to Postclassic*, edited by Jeremy Sabloff and W. Andrews V, pp. 301–345. University of New Mexico Press, Albuquerque.

Rice, Don S., and Prudence M. Rice. 1981. Muralla de Leon: A Lowland Maya Fortification. *Journal of Field Archaeology* 8:272–288.

Richards, Colin, and Julian Thomas. 1984. Ritual Activity and Structured Deposition in Later Neolithic Wessex. In *Neolithic Studies: A Review of Some Current Research*, edited by Richard Bradley and Julie Gardiner, pp. 189–218. British Archaeological Reports series 133, Oxford.

Ringle, William M., and George J. Bey, III. 2001. Post-Classic and Terminal Classic Courts of the Northern Maya Lowlands. In *Royal Courts of the Ancient Maya, Volume 2: Data and Case Study*, edited by Takeshi Inomata and Stephen D. Houston, pp. 266–307. Westview Press, Boulder.

Robarchek, Clayton, and Carole Robarchek. 1998. *Waorani: The Contexts of Violence and War*. Harcourt Brace, Orlando.

Robb, John. 1997. Violence and Gender in Early Italy. In *Troubled Times: Violence and Warfare in the Past*, edited by Debra L. Martin and David W. Frayer, pp. 111–144. Gordon and Breach, Amsterdam.

Robbins, Joel. 1994. Equality as Value: Ideology in Dumont, Melanesia and the West. *Social Analysis* 36:21–70.

Rodrigues Ferreira, Alexandre. 1983. *Viagem filosófica ao Rio Negro*. Museu Paraense Emilio Goeldi / CNPq, Belém.

Romans, Bernard. [1775] 1999. *A Concise Natural History of East and West Florida*. Edited by Kathryn E. H. Braund. University of Alabama Press, Tuscaloosa.

Roosevelt, Anna. 1987. Chiefdoms in the Amazon and Orinoco. In *Chiefdoms in the Americas*, edited by Robert Drennan and Carlos Uribe, pp. 153–185. University Presses of America, Lanham.

———. 1989. Resource Management in Amazonia Before the Conquest. *Advances in Economic Botany* 7:30–62.

———. 1991. *Moundbuilders of the Amazon: Geophysical Archaeology on Marajó Island, Brazil*. Academic Press, San Diego.

———. 1993. The Rise and Fall of the Amazonian Chiefdoms. *L'Homme* 33(2–4): 255–282.

———. 1995. Early Pottery in the Amazon: Twenty Years of Scholarly Obscurity. In *The Emergence of Pottery. Technology and Innovation in Ancient Societies*, edited by William K. Barnett and John Hoopes, pp. 115–131. Smithsonian Institution Press, Washington D.C.

———. 1999. The Development of Prehistoric Complex Societies: Amazonia, A Tropical Forest. In *Complex Polities in the Ancient Tropical World*, vol. 9, edited by Elisabeth Bacus and Lisa Lucero, pp. 13–33. Archaeological Papers of the American Anthropological Association, Arlington, Virginia.

Roscoe, Paul. 2000. "New Guinea Leadership as Ethnographic Analogy." *Journal of Archaeological Method and Theory* 7(2): 79–126.

Rose, Dan. 1990. *Living the Ethnographic Life*. Sage, Newbury Park.

Rostworowski de Diez Canseco, María. 1970. El repartamiento de doña Beatriz Coya, en el valle de Yucay. *Historia y Cultura* 4:153–268.

———. 1988. *Historia del Tahuantinsuyu*. Instituto de Estudios Peruanos, Lima.

Rowe, John H. 1946. Inka Culture at the Time of the Spanish Conquest. In *Handbook of South American Indians*, vol. 2, edited by Julian Steward, pp. 183–330. Smithsonian Institution Press, Washington, D.C.

———. 1985. Probanza de los Inkas Nietos de Conquistadores. *Historica* 9(2): 193–245.

Runciman, W. G. 1998. Greek Hoplites, Warrior Culture, and Indirect Bias. *Journal of the Royal Anthropological Institute* 4(4): 731–745.

Rydén, Stig. 1944. *Contributions to the Archaeology of the Rio Loa Region*. Elanders Boktryckeri Actiebolag, Göteborg.

———. 1947. *Archaeological Researches in the Highlands of Bolivia*. Erlanders Boktryckeri Aktiebolog, Gothenburg, Sweden.

Sackett, J. R. 1990a. Isocrestism and Style: A Clarification. *Journal of Anthropological Archaeology* 5:266–277.

———. 1990b. Style and Ethnicity in Archaeology: The Case for Isochrestism. Cambridge University Press, Cambridge.

Sackschewsky, Marvin, D. Gruenhagen, and J. Ingebritson. 1970. The Clan Meeting in Enga Society. In *Exploring Enga Culture: Studies in Missionary Anthropology*, edited by Paul Brennan, pp. 51–101. Kristen Press, Wapenamanda, Papua New Guinea.

Sahlins, Marshall. 1961. The Segmentary Lineage: An Organization of Predatory Expansion. *American Anthropologist* 63:322–345.

———. 1976. *Culture and Practical Reason*. University of Chicago Press, Chicago.

———. 1981. *Historical Metaphors and Mythical Realities: Structure in the Early History of the Sandwich Islands Kingdom*. University of Michigan Press, Ann Arbor.

———. 1985. *Islands of History*. University of Chicago Press, Chicago.

———. 1999. Two or Three Things that I Know about Culture. *Journal of the Royal Anthropological Institute*, n.s., 5:399–421.

Saignes, Thierry. 1986. The Ethnic Groups in the Valleys of Larecaja: From Descent to Residence. In *Anthropological History of Andean Polities*, edited by John V. Murra, Nathan Wachtel and Jacques Revel. Cambridge University Press, Cambridge.

Salazar-Burger, Lucy, and Richard L. Burger. 1982. La araña en la iconografía del horizonte temprano en la costa norte del Perú. In *Beiträge zur Allegemeinen und Vergleichenden Archäologie*, pp. 213–253. vol. 4. German Archaeological Institute Commission for General and Comparative Archaeology.

Sancho de Hoz, Pedro. [1534] 2004. *Relación de la conquista del Peru*. Asociación Amigos de la Historia de Calahorra, Calahorra.

San Pedro, Fray Juan de. [1560] 1992. *La persecución del demonio: Crónica de los primeros agustinos en el norte del Perú (1560)*. Colección Nuestra América, vol. 1. Editorial Algazara S. L., Málaga.

Santa Cruz Pachacuti Yamqui, Joan. [1615] 1928. *Historia de los Inkas y relación de su gobierno*. Sanmarti, Lima.

Santo Tomás, Fray Domingo de. [1560] 1951. *Lexicon o vocabulario de la lengua general del Perú*. Universidad Nacional de San Marcos, Lima.

Sarmiento de Gamboa, Pedro. [1572] 1988. *Historia de los Inkas*. Miraguano/Polifemo, Madrid.

Saunders, Nicholas J. 2003. Crucifix, Calvary, and Cross: Materiality and Spirituality in Great War Landscapes. *World Archaeology* 35(1): 7–21.

Sawyer, Allan. 1961. Paracas and Nasca Iconography. In *Essays in Precolumbian Art and Archaeology*, edited by S. K. Lothrop, pp. 269–298. Harvard University Press, Cambridge.

Scarborough, Vernon L. 1983. A Preclassic Maya Water System. *American Antiquity* 48:720–744.

Schaafsma, Polly. 1994. The Prehistoric Kachina Cult and Its Origins as Suggested by Southwestern Rock Art. In *Kachinas in the Pueblo World*, edited by Polly Schaafsma, pp. 63–79. University of New Mexico Press, Albuquerque.

———. 2000. *Warrior, Shield, and Star: Imagery and Ideology of Pueblo Warfare*. Western Edge Press, Santa Fe, New Mexico.

Schaan, Denise P. 2001. Into the Labyrinths of Marajoara Pottery: Status and Cultural Identity in Prehistorica Amazonia. In *Unknown Amazon: Nature in Culture in Ancient Brazil*, edited by C. McEwan, C. Barreto, and Eduardo Neves, pp. 108–133. British Museum Press, London.

———. 2004. *The Camutins Chiefdom*. Unpublished PhD dissertation, Department of Anthropology, University of Pittsburgh.

Schele, Linda, and David Freidel. 1990. *A Forest of Kings: The Untold Story of the Ancient Maya*. William Morrow, New York.

Schele, Linda, and Mary E. Miller. 1986. *The Blood of Kings: Dynasty and Ritual in Maya Art*. George Brazillier, New York.

Schiappacasse, Virgilio, Victoria Castro, and Hans Niemeyer. 1989. Los desarrollos regionales en el norte grande (1000–1400 D.C.). In *Culturas de Chile: Prehistoria*, edited by Jorge Hidalgo, Virgilio Schiappacasse, and Hans Niemeyer, pp. 181–220. Andrés Bello, Santiago.

Schiffer, Michael B. 2003. Drawing the Lightning Down. *Benjamin Franklin and Electrical Technology in the Age of Enlightenment*. University of California Press, Berkeley.

———. 1987. *Formation Processes of the Archaeological Record*. Albuquerque: University of New Mexico Press.

Schiffer, Michael B., and Andrea Miller. 1999. *The Material Life of Human Beings: Artifacts, Behavior, and Communication*. Routledge, New York.

Schiffer, Michael B., Tamara C. Butts, and Kimberly Grimm. 1994. *Taking Charge: The Electric Automobile in America.* Smithsonian Institution Press, Washington, D.C.

Schroeder, Sissel. 2003. Mississippian Political Organization: Economy Versus Ideology. *The Review of Archaeology* 23(2): 6–13.

Schulman, Albert. 1950. Pre-Columbian Towers in the Southwest. *American Antiquity* 4:288–297.

Scott, James C. 1990. *Domination and the Arts of Resistance: Hidden Transcripts.* Yale University Press, New Haven, Connecticut.

Seed, Patricia. 1991. "Failing to Marvel": Atahualpa's Encounter with the Word. *Latin American Research Review* 26(1): 7–32.

Shanks, Michael. 1995. Art and an Archaeology of Embodiment: Some Aspects of Archaic Greece. *Cambridge Archaeological Journal* 5:207–244.

Shennan, Stephen (editor). 1989. *Archaeological Approaches to Cultural Identity.* Unwin-Hyman, London.

Shils, Edward A. 1957. Center and Periphery: Essays in Macrosociology. In *Selected Papers of Edward Shils*, vol. 2, Cambridge University Press, Cambridge.

Sievert, April K. 2003. Spiro Painted Maces and Shell Cups: The Scientific Use of Artifacts without Context. In *Theory, Method, and Practice in Modern Archaeology*, edited by Robert J. Jeske and Douglas K. Charles, pp. 182–194. Praeger, Westport, Connecticut.

Sillitoe, Paul. 1978. Big Men and War in New Guinea. *Man* 13:352–372.

Silverman, Helaine. 2004. Introduction: Space and Time in the Central Andes. In *Andean Archaeology*, edited by Helaine Silverman, pp. 1–15. Blackwell Publishing, Malden.

Simek, Jan F., Alan Cressler, Charles H. Faulkner, Todd M. Ahlman, Brad Creswell, and Jay D. Franklin. 2001. The Context of Late Prehistoric Cave Art: The Art and Archaeology of 11th Unnamed Cave, Tennessee. *Southeastern Archaeology* 20:142–153.

Skeat, Walter W. 1912. Snakestones and Stone Thunderbolts as Subjects for Systematic Investigation. *Folklore* 23:45–80.

Smith, Adam T. 2003. *The Political Landscape: Constellations of Authority in Early Complex Polities.* University of California Press, Berkeley.

Smith, Maria Ostendorf. 2003. Beyond Palisades: The Nature and Frequency of Late Prehistoric Deliberate Trauma in the Chickamauga Reservoir of East Tennessee. *American Journal of Physical Anthropology* 121:303–318.

Snead, James E. 2004. Ancestral Pueblo Settlement Dynamics: Landscape, Scale, and Context in the Burnt Corn Community. *Kiva* 69:243–269.

————. 2005. History, Place, and Social Power in the Galisteo Basin, AD 1250–1325. Paper presented at the annual meeting of the Society for American Archaeology. Salt Lake City, Utah.

Solometo, Julie. 2006. The Dimensions of War: Conflict and Culture Change in Central Arizona. In *The Archaeology of Warfare: Prehistories of Raiding and Conquest*, edited by E. N. Arkush and M. W. Allen, pp. 24–65. University Press of Florida, Gainesville.

Sperber, Dan. 1992. Culture and Matter. In *Representations in Archaeology*, edited by Jean-Claude Gardin and Christopher S. Peebles, pp. 56–65. Indiana University Press, Bloomington.

Spurling, Geoffrey E. 1992. The Organization of Craft Production in the Inka State: The Potters and Weavers of Milliraya. PhD dissertation, Cornell University.

Stahl, Peter W. 2002. Paradigms in Paradise: Revising standard Amazonian Prehistory. *The Review of Archaeology* 23(2): 39–51.

Standen, Vivian, and Calogero M. Santoro. 1994. Patapatane-1: Temprana evidencia funeraria en los Andes de Arica (norte de Chile) y sus correlaciones. *Chungara* 26:165–183.

Stanish, Charles. 2001. The Origin of State Societies in South America. *Annual Review of Anthropology* 30:41–64.

———. 2003. *Ancient Titicaca: The Evolution of Complex Society in Southern Peru and Northern Bolivia*. University of California Press, Berkeley.

Stanish, Charles, Edmundo de la Vega, Lee Steadman, Cecilia Chávez Justo, Kirk Frye, Luperio Onofre Mamani, Matthew Seddon, and Percy Calisaya Chuquimia. 1997. *Archaeological Survey in the Juli-Desaguadero Region of the Lake Titicaca Basin, Southern Peru*. Field Museum of Natural History, Chicago.

Stark, Miriam T. (editor). 1998. *The Archaeology of Social Boundaries*. Smithsonian Institution Press, Washington, D.C.

Steinen, Karl T. 1992. Ambushes, Raids, and Palisades: Mississippian Warfare in the Interior Southeast. *Southeastern Archaeology* 11:132–139.

Stephen, Alexander M. 1936. Hopi Journal of Alexander M. Stephen, edited by Elsie Clews Parsons. *Columbia University Contributions to Anthropology* 23. Columbia University Press, New York.

Steward, Julian. 1948. Culture Areas of the Tropical Forests. In *Handbook of South American Indians*, vol. 3, edited by Julian Steward, pp. 883–903. Bureau of American Ethnology, Bulletin 143, Smithsonian Institution Press, Washington, D.C.

Strathern, Marilyn. 1988. *The Gender of the Gift: Problems with Women and Problems with Society in Melanesia*. University of California Press, Berkeley.

Strong, John A. 1989. The Mississippian Bird Man Theme in Cross-Cultural Perspective. In *The Southeastern Ceremonial Complex: Artifacts and Analysis*, edited by Patricia Galloway, pp. 211–238. University of Nebraska Press, Lincoln.

Stuart, David. 2000. "The Arrival of Strangers": Teotihuacan and Tollan in Classic Maya History. In *Mesoamerica's Classic Heritage: Teotihuacan to the Aztecs*, edited by David Carrasco, Lindsay Jones, and Scott Sessions, pp. 465–513. University of Colorado Press, Boulder.

Suhler, Charles, and David Freidel. 2000. Rituales de terminación: Implicaciones de la guerra maya. In *La guerra entre los antiguos Mayas: Memoria de la Primera Mesa Redonda de Palenque*, edited by Silvia Trejo, pp. 73–104. Instituto Nacional de Antropología e Historia, México, D.F.

Swanton, John R. 1928. Social Organization and the Social Usages of the Indians of the Creek Confederacy. In *Forty-Second Annual Report of the Bureau of American Ethnology, 1924–1925*, pp. 279–325. Washington, D.C.

———. 1946. *The Indians of the Southeastern United States*. Bulletin 137. Bureau of American Ethnology, Washington, D.C.

Tallman, Sean D. 2004. The Osteological Evidence for Violence and Health in a Middle Mississippian Settlement from the Central Illinois River Valley. Master's thesis, Department of Anthropology, State University of New York, Binghamton.

Talyaga, Kundapen. 1982. The Enga Yesterday and Today: A Personal Account. In *Enga: Foundations for Development*, vol. 3, edited by Bruce Carrad, David Lea, and Kundapen Talyaga, pp. 59–75. University of New England, Armidale.

Tapia Pineda, Félix. 1978a. *Contribuciónes al estudio de la cultura precolombina en elaltiplano Peruano*. Instituto Nacional de Arqueología, La Paz, Bolivia.

———. 1978b. Investigaciones arqueológicas en Kacsili. *Pumapunku* 12:7–37.

———. 1985. Contribución a la investigación arqueológica en los valles de Sandia y Carabaya, en el departamento de Puno–Peru. Grupo de Arte Utaraya, Puno, Peru.

Tarlow, Sarah. 2000. Emotion in Archaeology. *Current Anthropology* 41:713–746.

Taube, Karl. 1988. A Study of Classic Maya Scaffold Sacrifice. In *Maya Iconography*, edited by Elizabeth Benson and Gillett Griffin, pp. 331–351. Princeton University Press, Princeton.

———. 2000. The Turquoise Hearth: Fire, Self Sacrifice, and the Central Mexican Cult of War. In *Mesoamerica's Classic Heritage: From Teotihuacan to the Aztecs*, edited by Davíd Carrasco, Lindsay Jones, and Scott Sessions, pp. 269–340. University Press of Colorado, Boulder.

Tedlock, Dennis. 1985. *Popol Vuh: The Mayan Book on the Dawn of Life*. Simon and Schuster, New York.

Tello, Julio C. 1956. *Arqueología del valle de Casma: Culturas Chavín, Santa, o Huaylas, Yunga y Sub-Chimu*. Publicaciones Antropológicas del Archivo Tello, vol. 1. Universidad Nacional Mayor de San Marcos, Lima.

Thomas, Cyrus. 1894. Report on the Mound Explorations of the Bureau of Ethnology. *Twelfth Annual Report of the Bureau of American Ethnology, 1890–1891*. Washington, D.C.

Thompson, L. G., E. Moseley-Thompson, J. F. Bolzan, and B. R. Koci. 1985. A 1500-Year Record of Tropical Precipitation in Ice Cores from the Quelccaya Ice Cap, Peru. *Science* 229:971–973.

Thorpe, Nick. 2001. A War of Words. *Cambridge Archaeological Journal* 11:132–134.

————. 2003. Anthropology, Archaeology, and the Origin of Warfare. *World Archaeology* 35:145–165.

Tierney, Patrick. 2000. *Darkness in El Dorado: How Scientists and Journalists Devastated the Amazon*, W. W. Norton, New York.

Tilley, Christopher Y. 1994. *A Phenomenology of Landscape: Places, Paths, and Monument.* Oxford, Berg.

Toledo, Francicso de. [1570] 1940. Información hecha por orden de Don Francicso de Toledo en suvisita de las Provincias del Perú. In *Don Francisco de Toledo, Supremo Organizador del Peru, su Vida, Su Obra [1515–1582]*, vol. 2, edited by R. Levillier, pp. 14–37. Espasa-Calpe, Buenos Aires.

Topic, John R. 1989. The Ostra Site: The Earliest Fortified Site in the New World? (with an appendix by Theresa Topic). In *Cultures in Conflict: Current Archaeological Perspectives,* edited by Diana Claire Tkaczuk and Brian C. Vivian Chacmool, pp. 215–228. Archaeological Association of the University of Calgary, Calgary.

————. 1990. Craft Production in the Kingdom of Chimore. In *The Northern Dynasties: Kingship and Statecraft in Chimor,* edited by Michael E. Moseley and Alana Cordy-Collins, pp. 145–176. Dumbarton Oaks Research Library and Collection, Washington, D.C.

————. 1992. Las Huacas de Huamachuco: Precisiones en torno a una imagen indígena de un paisaje andino. In *La Persecución del Demonio: Crónica de lo Primeros Agustinos en el Norte del Peru (1560)*, by Fray Juan de San Pedro, pp. 41–99. Editorial Algazara y Centro Andino y Mesoamericano de Estudios, Malaga and México.

————. 2003. From Stewards to Bureaucrats: Architecture and Information Flow at Chan Chan, Peru. *Latin American Antiquity* 14(3): 243–274.

————. 2008. El santuario de Catequil: Estructura y agencia; Hacia el entendimiento de los oráculos andinos. In *Oráculos y divinación en los Andes,* edited by Marco Curatola and Mariusz S. Ziolkowski, pp. 71–95. Pontificia Universidad Católica del Perú, Lima.

Topic, John, and Theresa Topic. 1987. The Archaeological Investigation of Andean Militarism: Some Cautionary Observations. In *The Evolution of the Andean* State, edited by Jonathan Haas, Sheila Pozorski, and Thomas Pozorski, pp. 47–55. Cambridge University Press, Cambridge.

————. 1997a. La guerra Mochica. *Revista Arqueològico SIAN* 4:10–12. Universidad Nacional Mayor de Trujillo.

————. 1997b. Hacia una comprensión conceptual de la guerra andina. In *Arqueología, Antropología e Historia en los Andes: Homenaje a María Rostworowski,* edited by Rafael Varón Gabai and Javier Flores Espinoza, pp. 567–590. Instituto de Estudio Peruanos, Lima.

————. 2001. Death and the Regeneration of Life: Warfare and Ancestor Worship in the Andes. Paper presented at the annual meeting of the Society for American Archaeology, New Orleans.

Topic, Theresa Lange. 1990. Territorial Expansion and the Kingdom of Chimor. In *The Northern Dynasties: Kingship and Statecraft in Chimor*, edited by Michael E. Moseley and Alana Cordy-Collins, pp. 177–194. Dumbarton Oaks Research Library and Collection, Washington, D.C.

———. 1991. The Middle Horizon in Northern Peru. In *Huari Administrative Structure: Prehistoric Monumental Architecture and State Government*, edited by William H. Isbell and Gordon F. McEwan, pp. 233–246. Dumbarton Oaks Research Library and Collection, Washington, D.C.

Torero, Alfredo. 1987. Lenguas y pueblos altiplanicos en torno al Siglo XVI. *Revista Andina* 5(2): 329–405

Torres, Constantino M. 1987. *The Iconography of South American Snuff Trays and Related Paraphernalia*. Etnologiska Studier 37, Gothenburg, Sweden.

Torres-Rouff, Christina, María A. Costa-Junqueira, and Agustín Llagostera. 2005. Violence in Times of Change: The Late Intermediate Period in San Pedro de Atacama. *Chungará* 37:75–83.

Treherne, Paul. 1995. The Warrior's Beauty: The Masculine Body and Self-Identity in Bronze-Age Europe. *Journal of European Archaeology* 3(1): 105–144.

Triadan, Daniela. 2007. Warriors, Nobles, Commoners and Beasts: The Figurines of Aguateca. *Latin American Antiquity* 18(3): 269–294.

Trigger, Bruce G. 1989. *A History of Archaeological Thought*. Cambridge University Press, Cambridge.

———. 1990. Maintaining Economic Equality in Opposition to Complexity: An Iroquoian Case Study. In *The Evolution of Political Systems: Socio-Politics in Small-Scale Sedentary Societies*, edited by Steadman Upham, pp. 1–17. Cambridge University Press, Cambridge.

Trik, Helen, and Michael E. Kampen. 1983. *Tikal Report No. 31: The Graffiti of Tikal*. University Museum Monograph 57. University Museum, University of Pennsylvania, Philadelphia.

Trubitt, Mary Beth D. 2003. Warfare and Palisade Construction at Cahokia. In *Theory, Method, and Practice in Modern Archaeology*, edited by Robert J. Jeske and D. K. Charles, pp. 149–181. Praeger, Westport, Connecticut.

Tschopik, Harry. 1946. The Aymara. In *Handbook of South American Indians*, vol. 2, edited by Julian Steward, pp. 501–574. Smithsonian Institution Press, Washington, D.C.

Tschopik, Marion H. 1946. Some Notes on the Archaeology of the Department of Puno, Peru, vol. 27, No. 3. Papers of the Peabody Museum of American Archaeology and Ethnology. Harvard University, Cambridge.

Turner, Bryan S. 1996. *The Body and Society: Explorations in Social Theory*, 2nd ed. Sage, London.

———. 2003. Foreword. In *Embodied Lives: Figuring Ancient Maya and Egyptian Experience*, by Lynn M. Meskell and Rosemary A. Joyce, pp. xiii-xx. Routledge, London.

Turner, Christy G., II, and Jacqueline Turner. 1999. *Man Corn: Cannibalism and Violence in the Prehistoric Southwest*. University of Utah Press, Salt Lake City.

Turner, Edith. 1992. *Experiencing Ritual: A New Interpretation of African Healing*. University of Pennsylvania Press, Philadelphia.

Turney-High, Harry H. 1949. *Primitive War: Its Practice and Concepts*. University of South Carolina Press, Columbia.

Turton, David. 1997. War and Ethnicity: Global Connections and Local Violence in Northeast Africa and Former Yugoslavia. *Oxford Development Studies* 25(1): 77–94.

Tuzin, Donald F. 2001. *Social Complexity in the Making: A Case Study among the Arapesh of New Guinea*. Routledge, New York.

Tylor, Edward B. 1875. *Primitive Culture*. Murray, London.

Uchendu, Victor. 1965. *The Igbo of Southeast Nigeria*. Harcourt Brace Jovanovich Publishers, Fort Worth.

Urton, Gary. 1993. Moieties and Ceremonialism in the Andes: The Ritual Battles of the Carnival Season in Southern Peru. In *El Mundo Ceremonial Andino*, Senri Ethnological Studies no. 37, edited by Luis Millones and Yoshio Onuki, pp. 117–142. Nakanishi Printing, Japan.

———. 1994. Actividad ceremonial y división de mitades en el mundo andino: Las batallas rituales en los carnavales del Sur del Perú. In *El Mundo Ceremonial Andino*, edited by Luis Millones and Yoshio Onuki, pp. 117–142. Editorial Horizonte, Lima.

Vaca de Castro, Cristobal. [1543] 1908. Ordenanzas de tambos de vaca de Castro. *Revista Historica* 3:427–492, Lima.

Vandkilde, Helle. 2003. Commemorative Tales: Archaeological Responses to Modern Myth, Politics, and War. *World Archaeology* 35(1): 126–144.

Van Dyke, Ruth M. 2004. Formalizing the Chacoan Landscape: McElmo Style Architecture in Context. Unpublished manuscript.

Van Horne, Wayne William. 1993. *The Warclub: Weapon and Symbol in Southeastern Indian Societies*. Unpublished PhD dissertation, Department of Anthropology, University of Georgia, Athens.

Vansina, Jan. 1985. *Oral Tradition as History*. University of Wisconsin Press, Madison.

Varien, Mark D., William D. Lipe, Michael A. Adler, Ian M. Thompson, and Bruce A. Bradley. 1996. Southwestern Colorado and Southeastern Utah Settlement Patterns: A.D. 1100–1300. In *The Prehistoric Pueblo World, A.D. 1150–1350*, edited by Michael A. Adler, pp. 86–113. University of Arizona Press, Tucson.

Vayda, Andrew. 1961. Expansion and Warfare among Swidden Agriculturalists. *American Anthropologist* 63:346–358.

Vega, Edmundo M. de la. 1990. Estudio arqueologico de Pucaras o poblados amuralladas de cumbre en territorio Lupaqa: El caso de Pucara-Juli. Bachelor's thesis, Universidad Católica Santa Maria.

Vega, Edmundo M. de la, Kirk L. Frye, and Cecilia Chávez J. 2002. La cueva funeraria de Molino-Chilacachi (Acora), Puno. *Gaceta Arqueológica Andina* 26:121–137.

Vega, Garcilaso de la. [1609] 1966. *Royal Commentaries of the Inkas and General History of Peru*. Austin, University of Texas Press.

Verano, John W. 1986. A Mass Burial of Mutilated Individuals at Pacatnamu. In *The Pacatnamu Papers*, vol. 1, edited by Guillermo A. Cock and Christopher B. Donnan. Museum of Cultural History, Los Angeles.

———. 1995. Where do They Rest? The Treatment of Human Offerings and Trophies in Ancient Peru. In *Tombs for the Living: Andean Mortuary Practices*, edited by Tom Dillehay, pp. 189–227. Dumbarton Oaks Research Library and Collection, Washington, D.C.

———. 2001. War and Death in the Moche World: Osteological Evidence and Visual Discourse. In *Moche Art and Archaeology in Ancient Peru*, edited by Joanne Pillsbury. National Gallery of Art, Washington, D.C.

Vignati, Milciades A. 1930. Los cráneos trofeo de las sepulturas indígenas de la quebrada de Humahuaca (provincia de Jujuy). *Archivos del Museo Etnográfico* 1. Buenos Aires.

Vivante, Armando. 1973. El craneoutilitario de Tastil. In *Tastil, una ciudad preincaica Argentina*, edited by Eduardo M. Cigliano, pp. 623–633. Ediciones Cabargon, Buenos Aires.

Viveiros de Castro, Eduardo. 1992. *From the Enemy's Point of View: Humanity and Divinity in an Amazonian Society*. University of Chicago Press, Chicago.

———. 1996. Images of Nature and Society in Amazonian Ethnology. *Annual Review of Anthropology* 25:179–200.

———. 2002a. Perspectivismo e multinaturalismo na América indígena. In *A inconstância da alma selvagem e outros ensaios de antropologia*, edited by Viveiros de Castro, pp. 347–399. Cosac e Naify, São Paulo.

———. 2002b. Xamanismo e sacrifício. In *A inconstância da alma selvagem e outros ensaios de antropologia*, edited by Viveiros de Castro, pp. 457–472. Cosac e Naify, São Paulo.

Vivian, R. Gwinn. 1990. *The Chaco Prehistory of the San Juan Basin*. Academic Press, San Diego.

Waddell, Eric. 1972. *The Mound Builders: Agricultural Practices, Environment and Society in the Central Highlands of New Guinea*. University of Washington Press, Seattle.

Wade, Peter. 1999. Working Culture: Making Cultural Identities in Cali, Colombia. *Current Anthropology* 40:449–472.

Walker, William H. 2001. Ritual Technology in an Extranatural World. In *Anthropological Perspectives on Technology*, edited by Michael B. Schiffer, pp. 87–106. University of New Mexico Press, Albuquerque.

———. 1999. Ritual, Life Histories, and the Afterlives of People and Things. *Journal of the Southwest* 41(3): 383–405.

————. 1998. Where Are the Witches of Prehistory? *Journal of Archaeological Method and Theory* 5:245–308.

————. 2002. Stratigraphy and Practical Reason. *American Anthropologist* 104(1): 159–177.

Walker, William H., and Lisa J. Lucero. 1995. Ceremonial Trash? In *Expanding Archaeology*, edited by James M. Skibo, William H. Walker, and Axel E. Nielsen, pp. 67–79. University of Utah Press, Salt Lake City.

————. 2000. The Depostional History of Ritual and Power. In *Agency in Archaeology*, edited by Marcia-Anne Dobres and John Robb, pp. 130–147. Routledge, London.

Walker, William H., James M. Skibo, and Axel E. Nielsen. 1995. Introduction. In *Expanding Archaeology*, edited by James M. Skibo, William H. Walker, and Axel E. Nielsen, pp. 1–12. University of Utah Press, Salt Lake City.

Waring, A. J., Jr., and Preston Holder. [1945] 1968. A Prehistoric Ceremonial Complex in the Southeastern United States. In *The Waring Papers: The Collected Works of Antonio J. Waring, Jr.*, edited by Stephen Williams, pp. 9–29. Peabody Museum, Harvard University, Cambridge.

Watson, Patty Jo, Steven A. LeBlanc, and Charles Redman. 1980. Aspects of Zuni Prehistory: Preliminary Report on Excavations and Survey in the El Morro Valley of New Mexico. *Journal of Field Archaeology* 7:201–218.

Weber, Max. 1976. *The Protestant Ethic and the Spirit of Capitalism*, translated by Talcott Parsons. Charles Scribner's Sons, New York.

Webster, David L. 1976. *Defensive Earthworks at Becan, Campeche, Mexico*. Publication 41. Middle American Research Institute, Tulane University, New Orleans.

————. 1977. Warfare and the Evolution of Maya Civilization. In *The Origins of Maya Civilization*, edited by Richard E. W. Adams, pp. 335–372. University of New Mexico Press, Albuquerque.

————. 1979. *Cuca, Chacchob, Dzonot Ake: Three Walled Northern Maya Centers*. Department of Anthropology, Pennsylvania State University, University Park.

Webster, David, Jay Silverstein, Timothy Murtha, Horacio Martinez, and Kirk Straight. 2004. *The Tikal Earthworks Revisited*. Occasional Paper 28. Department of Anthropology, Pennsylvania State University, University Park.

Weiner, Annette B. 1992. *Inalienable Possessions: The Paradox of Keeping-While-Giving*. University of California Press, Berkeley.

White Deer, Gary. 1997. Return of the Sacred: Spirituality and the Scientific Imperative. In *Native Americans and Archaeologists: Stepping Stones to Common Ground*, edited by Nina Swindler, Kurt E. Dongoske, Roger Anyon, and Alan S. Downer, pp. 37–43. AltaMira Press, Walnut Creek, California.

White, Leslie A. 1935. The Pueblo of Santo Domingo, New Mexico. *American Anthropological Association Memoirs*, vol. 63. American Anthropological Association, Lancaster, Pennsylvania.

———. 1942. The Pueblo of Santa Ana, New Mexico. *American Anthropological Association Memoirs*, vol. 60. American Anthropological Association, Menasha, Wisconsin.

Whitehead, Neil. 1990a. The Snake Warriors—Sons of the Tiger's Teeth: A Descriptive Analysis of Carib Warfare, ca. 1500–1820. In *The Anthropology of War*, edited by Jonathan Haas, pp. 146–170. Cambridge University Press, Cambridge.

———. 1990b. Carib Ethnic Soldiering in Venezuela, the Guianas, and the Antilles. *Ethnohistory* 37(4): 357–385.

Whiteley, Peter M. 1988. *Deliberate Acts: Changing Hopi Culture through the Oraibi Split*. University of Arizona Press, Tucson.

———. 2002. Archaeology and Oral Tradition: The Scientific Importance of Dialogue. *American Antiquity* 67(3): 405–415.

Wiessner, Polly. 1983. Style and Social Information in Kalahari San Projectile Points. *American Antiquity* 49(2): 253–76.

———. 1990. Is There a Unity to Style? In *The Uses of Style in Archaeology*, edited by Margaret Conkey and Christine Hastorf, pp. 105–112. Cambridge University Press, Cambridge.

———. 1996. Leveling the Hunter: Constraints on the Status Quest in Foraging Societies. In *Food and the Status Quest*, edited by Polly Wiessner and W. Schiefenhövel, pp. 171–191. Berghahn Books, Oxford.

———. 2002. The Vines of Complexity: Egalitarian Structures and the Institutionalization of Inequality among the Enga. *Current Anthropology* 43(2): 233–269.

———. 2006. From Spears to M-16s: Testing the Imbalance of Power Hypothesis among the Enga. *Journal of Anthropological Research* 62:165–191.

Wiessner, Polly, and Akii Tumu. 1998. *Historical Vines: Enga Networks of Exchange, Ritual, and Warfare in Papua New Guinea*. Smithsonian Institution Press, Washington, D.C.

———. 1999. A Collage of Cults. *Canberra Anthropology* 22(1): 34–65.

Wilcox, David R. 1979. The Warfare Implications of Dry-Laid Masonry Walls on Tumamoc Hill. *The Kiva* 45(1–2): 15–38.

Wilcox, David, and Jonathan Haas. 1994. The Scream of the Butterfly: Competition and Conflict in the Prehistoric Southwest. In *Themes in Southwestern Prehistory*, edited by George J. Gumerman, pp. 211–238. School of American Research Press, Santa Fe.

Wilcox, David R., Gerald Robertson Jr., and J. Scott Wood. 2001. Antecedents to Perry Mesa: Early Pueblo III Defensive Refuge Systems in West-Central Arizona. In *Deadly Landscapes: Case Studies in Prehistoric Southwestern Warfare*, edited by Glen E. Rice and Steven A. LeBlanc, pp. 109–140. University of Utah Press, Salt Lake City.

Willey, Gordon R. 1953. Prehistoric Settlement Patterns in the Virú Valley, Perú. *Bureau of American Ethnology*, Bulletin 155, Smithsonian Institution Press, Washington, D.C.

Willey, Patrick. 1990. *Warfare on the Great Plains*. Garland, New York.

Willey, Patrick, and Thomas E. Emerson. 1993. The Osteology and Archaeology of the Crow Creek Massacre. *Plains Anthropologist* 38:227–269.

Williams, Raymond. 1977. *Marxism and Literature*. Oxford University Press, Oxford.

Williams, Stephen. 1990. The Vacant Quarter and Other Late Events in the Lower Valley. In *Towns and Temples Along the Mississippi*, edited by David H. Dye, pp. 170–180. University of Alabama Press, Tuscaloosa.

Wilshusen, Richard H. 1986. The Relationship Between Abandonment Mode and Ritual Use in Pueblo I Anasazi Protokivas. *Journal of Field Archaeology* 13:245–254.

Wilshusen, Richard H., and Scott G. Ortman. 1999. Rethinking the Pueblo I Period in the San Juan Drainage: Aggregation, Migration, and Cultural Diversity. *The Kiva* 64:369–399.

Wilson, David J. 1988. *Settlement Patterns in the Lower Santa Valley, Peru: A Regional Perspective on the Origins and Development of Complex North Coast Society*. Smithsonian Institution Press, Washington, D.C.

———. 1995. Prehispanic Settlement Patterns in the Casma Valley, North Coast of Peru: Preliminary Results to Date. *Journal of the Steward Anthropological Society* 23(1–2): 189–227.

Wilson, Margo, and Martin Daly. 1985. Competitiveness, Risk Taking, and Violence: The Young Male Syndrome. *Ethology and Sociobiology* 6:59–73.

Wolf, Eric R. 1982. *Europe and the People Without History*. University of California Press, Berkeley.

Woodbury, Richard A. 1959. A Reconstruction of Pueblo Warfare in the Southwestern United States. *Proceedings of the International Congress of Americanists* 33:400–401.

Woods, William I., and John M. McCann. 1999. The Anthropogenic Origin and Persistence of Amazonian Dark Earths. *Yearbook of the Conference of Latin American Geographers* 25:7–14.

Wormsley, William, and Michael Toke. 1985. *The Enga Law and Order Project*. Report to the Enga Provincial Government, Wabag, Papua New Guinea.

Wrangham, Richard W., and Dale Peterson, 1996. *Demonic Males: Apes and the Origin of Human Violence*. Houghton Mifflin, Boston.

Wright, Henry T. 1986. The Evolution of Civilizations. In *American Archaeology Past and Future*, edited by David Meltzer, Don Fowler, and Jeremy Sabloff, pp. 323–365. Smithsonian Institution Press, Washington, D.C.

Wright, Robin. 1990. Guerras e alianças nas histórias dos Baniwa do Alto Rio Negro. *Ciências Sociais Hoje* 8:217–236.

————. 1992. História indígena do noroeste da Amazônia: Hipóteses, questões e perspectivas. In *História dos Índios no Brasil*, edited by Manuela Carneiro da Cunha, pp. 253–266. Cia. Das Letras/FAPESP/SMC, São Paulo.

Yacobaccio, Hugo D. 2000. Inhumación de una cabeza aislada en la Puna argentina. *Estudios Sociales del NOA* 4(2): 59–72.

Yarrow, Thomas. 2003. Artefactual Persons: The Relational Capacities of Persons and Things in the Practice of Excavation. *Norwegian Archaeological Review* 36(1):65–73.

Yates, Tim. 1993. Frameworks for an Archaeology of the Body. In *Interpretive Archaeology*, edited by Christopher Y. Tilley, pp. 31–72. Berg, Oxford.

Young, Douglas W. 2002. *"Our Land Is Green and Black": Traditional and Modern Methods for Sustaining Peaceful Intergroup Relations among the Enga of Papua New Guinea*. Melanesian Institute Press, Goroka Papua New Guines.

Zuidema, Tom. 1989. *Reyes y Guerreros: Ensayos de Cultura Andina*. Fomciencias, Lima.

ABOUT THE CONTRIBUTORS

ELIZABETH ARKUSH received her doctorate in 2005 from the University of California, Los Angeles, where she researched late pre-Columbian forts of the southern Andes, investigating the regional patterns that emerged from conflictual and cooperative social relationships. She recently co-edited a book with Mark Allen examining multilinear trajectories of warring societies in different world regions, *The Archaeology of Warfare: Prehistories of Raiding and Conquest* (University Press of Florida 2006) and has written about the difficulties of interpreting pre-Columbian Andean warfare. Her research centers on the interplay of warfare, political power, and social identity and makes use of regional approaches, including geographic information system (GIS) analysis, to illuminate these social processes in space. She is an assistant professor at the University of Virginia.

CHARLES R. COBB, who earned his PhD from Southern Illinois University, Carbondale, recently assumed the position of director of the Institute of Archaeology and Anthropology at the University of South Carolina, where he is also a professor of anthropology. He and Dawnie Steadman have recently initiated a long-term project on prehistoric health and conflict during the Mississippian period in central Tennessee. His interest in lithic technologies during the colonial period is currently sustained by a study of Cherokee farmsteads in North Carolina. Recent publications include "Archeology and the 'Savage Slot': Displacement and Emplacement in the Premodern World" (2005), which appeared in *American Anthropologist* 107(4); and "Re-Inventing Mississippian Tradition at Etowah, Georgia" (2005), published in the *Journal of Archaeological Method and Theory* 12(3) and co-written by Adam King.

BRETTON GILES is a PhD candidate at the State University of New York, Binghamton. His dissertation concerns the development of Ohio Hopewell symbolic/mnemonic systems and how they relate to the social complexity of these communities. He is also working with Jennifer

Bauder and Marta Alfonso on a bioarchaeological analysis of the burials interred at the Helena Crossing mounds, a Middle Woodland site in Arkansas. His broader research interests include Hopewell and Mississippian societies in eastern North America, as well as a theoretical interest in phenomenology and the sociology of translation.

TAKESHI INOMATA is an associate professor in anthropology at the University of Arizona. His recent publications include "Plazas, Performers, and Spectators: Political Theaters of the Classic Maya" (2006), appearing in *Current Anthropology* 47(5); and *Archaeology of Performance: Theaters of Power, Community, and Politics* (AltaMira 2006), co-edited by Lawrence Coben. In 2005, he completed a sixteen-year period of research at the Classic Maya center of Aguateca, Guatemala, examining political change, warfare, and domestic organization. He is currently initiating a new project at Ceibal in the same region.

EDUARDO GÓES NEVES was born in São Paulo, Brazil. He received his PhD in archaeology at Indiana University. He currently belongs to the faculty of the Museu de Arqueologia e Etnologia, Universidade de São Paulo, where he teaches in the PhD program. His area of research is the Brazilian Amazon, where he coordinates two long-term interdisciplinary research projects. His works include *Arqueologia da Amazônia* (2006 Jorge Zahar Editora) and "The Political Economy of Pre-Columbian Indigenous: Landscape Transformations in Central Amazonia" in *Time and Complexity in Historical Ecology: Studies in the Neotropical Lowlands* (Columbia University Press 2006).

AXEL E. NIELSEN was educated at the University of Arizona and at the University of Córdoba, Argentina, where he is currently an assistant professor. He has been conducting research in the south Andean highlands since 1987. His work focusing on the ethnography and archaeology of caravan trade includes "Ethnoarchaeological Perspectives on Caravan Trade in the South-Central Andes" in *Ethnoarchaeology of Andean South America* (University of Michigan Press 2001), while work on pre-Columbian political systems in the south Andes produced "Plazas para los antepasados: Descentralización y poder corporativo en las formaciones políticas preincaicas de los Andes circumpuneños" (2006), appearing in *Estudios Atacameños* 31. Nielsen also has a particular interest in the role of war in social change, as can be seen in "Asentamientos, conflicto y cambio social en el altiplano de Lípez, Bolivia" (2002), an article published in *Revista Española de Antropología Americana* 32.

TIMOTHY R. PAUKETAT is a professor of anthropology at the University of Illinois in Champaign-Urbana, having previously taught at the State University of New York and the University of Oklahoma. His research focuses on the relationships between identity, politics, domestic life, materiality, and landscapes in ancient North America, particularly during the late pre-Columbian period of the Mississippi Valley with its Woodland and Mississippian mound complexes. He has directed several large-scale excavation projects around the American Indian city of Cahokia, is the author of numerous articles and book chapters, and has written or edited several recent books, including *Chiefdoms and Other Archaeological Delusions* (AltaMira 2007); *North American Archaeology* (Blackwell 2005), co-edited with D. Loren; *Ancient Cahokia and the Mississippians* (Cambridge University Press 2001); and *The Archaeology of Traditions* (University of Florida Press 2001).

Educated at Harvard, JOHN R. TOPIC is a professor of anthropology at Trent University in Peterborough, Ontario. In addition to his work on fortifications and warfare, he has done work at the large urban site of Chan Chan, with findings described in the *Latin American Antiquity* 7(4) article "From Stewards to Bureaucrats: Architecture and Information Control at Chan Chan, Peru" (2003). He has also worked in the northern highlands around Huamachuco with Theresa Lange Topic, producing "Hacia la comprensión del fenómeno Huari: Una perspectiva norteña" (2000), which appeared in *Boletín Arqueológica de la Pontificia Universidad Católica del Perú* 4. He is currently studying an oracle with Theresa Lange Topic and Alfredo Melly and has published "Catequil: The Archaeology, Ethnohistory, and Ethnography of a Major Provincial Huaca" in *Andean Archaeology I: Variations in Sociopolitical Organization* (2002 Klewer Academic/Plenum Publishers) as well as "El santuario de Namanchugo: Estructura y acción; Hacia un entendimiento de los oráculos andinos" in *Adivinación y oráculos en el mundo andino* (Instituto Francés de Estudios Andinos 2007).

THERESA LANGE TOPIC has been engaged in fieldwork in the central Andes of South America since 1970, earning a PhD from Harvard in 1977. Her early interest in coastal civilizations of Peru was followed by a focus on late prehistoric sierra cultures and their sociopolitical organization. Fieldwork in Otuzco, Huamachuco, and the Guaranda area of Ecuador followed. Her interests include warfare, fortifications, household organization, craft specialization, cosmology, and gender in prehistory, with published works including "Territorial Expansion and the Kingdom of Chimor" in *The Northern Dynasties: Kingship and Statecraft in Chimor* (1990 Dumbarton Oaks) and "The Mobility of Women in the Inka Empire and in the Spanish Colony of Peru" in *The Archaeology of Contact: Processes and Consequences* (2002 Academic Press). She is currently principal of Brescia University College, an affiliate of the University of Western Ontario in London, Ontario.

DANIELA TRIADAN received her doctoral degree from the Freie Universität Berlin in American archaeology and anthropology, European prehistory, and ethnology, and she served as a postdoctoral fellow at the Conservation Analytical Laboratory of the Smithsonian Institution. Her interests include polychrome ceramic production and distribution in Chihuahua, Mexico, and Classic Maya household organization at Aguateca, Guatemala. She is the author of *Ceramic Commodities and Common Containers: Production and Distribution of White Mountain Red Ware in the Grasshopper Region* (University of Arizona Press 1997); "Dancing Gods: Ritual, Performance, and Political Organization in the Prehistoric Southwest" in *Theatres of Power and Community: Archaeology of Performance and Politics* (AltaMira 2006), "The Political Geography and Territoriality of Fourteenth-Century Settlements in the Mogollon Highlands of East-Central Arizona" (co-written by M. Nieves Zedeño) in *Cluster Analysis: The History and Organization of Pueblo IV Period (A.D. 1275–1540) Settlement Clusters in the American Southwest* (University of Arizona Press 2004); "What Did They Do and Where? Activity Areas and Residue Analyses in Maya Archaeology" (co-written by Takeshi Inomata) in *Continuities and Changes in Maya Archaeology: Perspectives at the Millenium* (Routledge 2004).

WILLIAM H. WALKER is an associate professor of anthropology at New Mexico State University in Las Cruces. He directs New Mexico State

University's La Frontera Program, focused on the archaeology of southern New Mexico and northern Chihuahua, Mexico. In collaboration with Dr. Axel Nielsen, he has also conducted comparative studies of the later prehistory of Argentina and Bolivia. Walker's research interests include prehistoric ritual, violence, technology, and behavioral approaches to social change. His publications include "Where are the Witches of Prehistory?" (1998), appearing in the *Journal of Archaeological Method and Theory* 5(3); "Ritual Technology in an Extranatural World" in *The Anthropology of Technology* (Amerind Foundation/ University of New Mexico Press 2001); and "Stratigraphy and Practical Reason" (2002), appearing in *American Anthropologist* 104. His recent publication with Michael Schiffer, "The Materiality of Social Power: The Artifact-Acquisition Perspective" (2006) appeared in the *Journal of Archaeological Method and Theory* 13(2) and explores the role of artifact performance preferences in contexts of artifact acquisition and social change.

POLLY WIESSNER received her PhD from the University of Michigan, worked as a researcher at the Max Planck Institute for Human Ethology in Germany, and is now a professor of anthropology at the University of Utah. Her research interests include hunter-gatherers, cultural systems of sharing and exchange, ethnoarchaeology, ethology, ecology, warfare, and oral history. She has done research among the Kung bushmen of the Kalahari Desert on subsistence and social security systems, and for the past twenty years, she has conducted ethnohistorical studies among the Enga of highland New Guinea, tracing developments in warfare, ritual, and exchange. She is currently looking at the impact of the introduction of high-powered weapons on tribal fighting in Enga. Seminal publications include *Historical Vines: Tracing Enga Networks of Exchange, Ritual, and Warfare among the Enga of Papua New Guinea* (Smithsonian Institution Press 1998), co-written by Akii Tumu; and *From Inside the Women's House: The Lives and Traditions of Enga Women* (Robert Brown 1992), co-written by Alome Kyakas.

Açutuba, 148, 150, 152–154, 160
agency, 3, 7, 14, 18–19, 48, 60, 61, 220
aggregation, 11, 197, 218. *See also* settlement
    patterns
Aguada, NW Argentina, 237
Aguateca, 64, 67–69, 75–77, 80
Allen, Catherine, 27
alliances, 8, 47, 53, 142, 189, 190, 196, 205–208,
    216–217, 240, 264, 265
Amazonia, 11, 139–164
    Central, 140, 145–152
    changes with European colonization,
        140–143
ambush, 88, 91, 184
Amerind Foundation, 9, 17, 164, 243, 261
Ancestor worship, 13, 201, 203, 220, 231–234,
    236–242
Andes
    Central, 12, 17, 197–217
    South, 13, 218–242
anthropocentrism, 112, 113
anthropological history, 9, 17, 20
anthropology of war, 7
Arapaço, 155
Arawaks, 142, 159
archaeology of war, 1, 9, 61, 82
archaeological evidence of warfare, 2, 49
architecture. *See* defensive architecture
    iconographic. *See* iconography
    osteological, 41, 47, 50, 199–200
    settlements. *See* settlement patterns
    weaponry. *See* weapons
Archer, Margaret, 166
armies, 28, 39, 44, 46–48, 63, 89, 228, 230, 259
armory, 70, 226, 236, 239
Asillo, 213
artifact agency, 114–118, 222
atlatl. See spearthrower
Avatip, 114

Awatovi, 128–129
Awqaruna. See warrior
axes, 89, 101, 120, 124, 174, 153, 226, 228, 229,
    235, 242
ayllu, 232
Aymara, 24, 27, 201, 213, 242, 267
Aztecs, 63, 78, 81, 251, 259

ballgame, 78, 79, 252, 257, 259
Becan, 66
behavior, definition of, 114
beheading. *See* decapitation
Bertonio, Ludovico, 27, 213
biological approaches to war, 2
bodily practice, 60
Bonampak, 67, 69, 75, 77
booty, 28, 165
Bourdieu, Pierre, 14, 59, 192, 246
bow and arrow, 71, 88–89, 122
Brown, James, 94
Brownrigg, Leslie, 27
buffer zones, 179, 196, 205, 213, 214, 216, 239,
    240
Buopé, 155–156
Burnt Corn Pueblo, 256
Bustina Méndez, David, 200

Cahokia Mound Center, 91, 257–258, 260
Calakmul, 66
Canas, 198, 203, 212–213, 216
cannibalism, 121, 142, 161–162, 252
capitalist expansion. *See* colonial expansion
Capoche, Luis, 209
captive. *See* prisoners
Caracol, 79
caravans. *See* trade
Caribs, 142, 159
Carneiro, Robert, 145, 159, 163
Cartesian dualism, 246. *See also* practice

Casas Grandes, Chihuahua, 123, 245
Casma Valley, 32–35, 50, 252
Castalian Springs Mound Center, 94, 96
casualties in war, 168, 185
causes of war. *See* war
Cayhuamarca fortresses, 35
centralization. *See* political centralization
ceremony. *See* ritual
Cerro de La Cruz site, 45
Cerro Oreja site, 45
Cerros, 66
Cerro Sechín, 21, 30
Chacchob, 69
chacmools, 77
Chaco Canyon, New Mexico, 120–122, 131,
    135, 258, 260
Chagnon, Napoleon, 90, 141
ch'ajwa, 26, 240, 257
champions, 27, 41, 44, 45, 51, 53, 54, 134
Chamussy, Victor, 30
Chankillo, 21, 33–35, 252
Chernela, Janet, 156
Chichen Itza, 70–72, 77, 79
Chickasaw Indians, 92
chiefdoms, warfare in, 1, 144, 159, 218, 219, 260
    Amazonia, 143, 145, 160, 163
    Andes, 218–219
    Fiji, 90
    Mississippi, 88, 90, 92
Chimu, 44–48, 53, 54
Choctaw Indians, 92, 100
chullpas, 202, 203, 233, 234, 240, 251, 253, 255,
    264, 266
Chunchucmil, 69
Cieza de León, Pedro, 200, 201
Cigliano, Eduardo, 234
Circumpuna area. *See* Andes, South
circumscription, geographical and social, 145
Clastres, Pierre, 11, 139, 163
Cobo, Bernabé, 215
codes of conduct in war. *See* cultural codes
Collas, 198–216, 254, 265
Collingwood, Robin G., 19
Colonial expansion, consequences
    Amazonia, 139–144, 155, 156, 164
    Andes, 20
    North American Southwest, 132
    Papua New Guinea, 174–176, 179–180,
communication,
    theory of, 115–118
    signaling. *See* intervisibility

Cook, Captain, 60
Copan, 71
corporate mode of political action, 13, 223, 241
cosmology, 6, 93, 97–98, 102, 104, 105, 197,
    220, 237, 246, 252
    Andean, 9, 10, 23–29, 42–50, 220, 237,
        241–242
    Mississippian, 85–86, 94–97, 103
    North American Southwest, 124, 126–128
costumes, 36, 37, 39, 48, 65, 75, 97, 103, 124,
    228, 236, 239. *See also* warrior attire
Creamer, Winifred, 120, 122
Crow Creek Site, 245, 249–250
Crown, Patricia, 123
Cubeo, 156
Cuca, 69
cuirass. *See* armory
cultural codes, 10, 56, 222
    in war, 9–10, 56, 61–71, 73–74, 79, 82–83,
        180, 246–247
cultural ecology, 178. *See also* materialist
    explanation
cultural explanations of war, 3, 192
cultural logic, 14, 56, 220
culture, 3
    concept of, 58
    cross-cultural comparison, 1, 80, 144, 166
    history, 144
    (*see also* explanations of war)
Cupisnique, 31

decapitation, 34, 79, 97, 98, 100, 226, 231,
    233–236, 251
defensive architecture. *See also* pukaras
    ditches, moats, 11, 45, 46, 66, 88, 150, 152–
        153, 156–160, 245, 249, 250, 254, 255, 260
    forts, fortresses, and fortifications, 10, 31–35,
        39, 43–45, 49, 51–53, 63, 66, 67, 69, 71,
        73, 88, 90, 99, 107, 122, 158, 190, 193, 197,
        215, 225, 231, 233, 244, 249, 251–261
    hilltops, 12, 35, 45, 69, 120, 198, 199, 203,
        213, 239, 251, 252, 255, 260
    palisades, 88, 90, 91, 120, 150, 153, 160, 249,
        252
    parapets, 33, 44, 46
    walls, 30–36, 43–46, 51, 52, 66–68, 133, 193,
        199, 203, 203, 225, 231, 233, 245, 249, 254,
        255, 259, 260
Demarest, Arthur, 73, 134, 145,
demographic pressure, 2, 12, 52, 53, 145, 165
Denevan, William, 141

Dickson, Bruce, 92
disembodied heads. *See* decapitation
dismemberment, 192,
    Mississippian, 93, 98–101
ditch. *See* defensive architecture
Donatti, Patrícia, 152
Donnan, Cristopher, 41
Dos Pilas, 67
doxa, 224
drought, 73, 218, 237–238. *See also* environ-
    mental stress
Dye, David, 92–93

Earle, Timothy, 218–220, 241
Edzna, 66
egalitarian institutions, society, 12, 106, 131,
    164, 165–174, 180–181, 185–188, 220
Egan, Fred, 123, 127–129
Ek Balam, 69
El Mirador, 66, 73
El Niño events, 32, 53
Ember, Carol, 4
Ember, Melvin, 4
embodiment, 51, 84, 86, 87, 93, 94, 98, 105,
    108, 124, 161, 220, 246, 253, 261
encomienda, 212, 213, 215
Enga, 12, 165–189, 248, 252, 256
environmental,
    change, 74, 82
    stress and war, 2, 4, 120–121, 165, 197, 220
epigraphy, 56, 63, 65, 74
ethnic
    groups and boundaries, 12, 20, 73, 123, 197,
        212–214
    groups and ceramics, 208–210, 238
    identity, 39, 194–197, 214–216, 238
Etowah Mound Center, 94–97
European expansion and war. *See* colonial
    expansion
evolutionary stages. *See* neoevolutionism
exchange, *see also* trade, Tee Cycle
    and warfare, 165, 168, 177, 179–185.
    ceremonial, 166, 169, 170, 173, 176, 177, 179,
        185–187, 248, 256
    explanations of war. *See* war, causes of

Fausto, Carlos, 161
Ferguson, R. Brian, 5, 7, 141, 193–194
feuding, 8
Fewkes, Jesse Walter, 129
figurines, 65, 74–76, 78, 202, 208, 253

Fortaleza site, 158
fortified temple, 35
forts, fortifications. *See* defensive architecture
Freidel, David, 77
Fried, Morton, 170

Galindo, 43
games, 257, 259. *See also* ballgame, ritual
Geertz, Armin, 128
Gell, Alfred, 11, 110, 115–118
Geographical information systems, 12, 207
Ghezzi, Ivan, 33, 35
Gibson, Jon, 91
gladiatorial battle, 78, 251
Goldman, Irving, 156
González, Alberto Rex, 227
González Holguin, Diego, 34
Gramsci, Antonio, 58
Great Ceremonial Wars, 12, 170, 176, 181,
    183–187, 257, 263
Guamán Poma de Ayala, 23, 224–232, 235,
    241, 266

Haas, Jonathan, 109, 120, 122, 197
habitus, 59, 60, 79, 246
hallucinogenic substances. *See* transmutation,
    shamanism
Hatahara site, 149
head taker, 31, 227, 229, 235
hegemony, concept of, 58
helmet. *See* armory
Hero Twins, 78, 79
heterarchy, 223
hilltop sites. *See* defensive architecture
Hodder, Ian, 135
Hohodene Baniwa, 160
homicide, 175, 176, 248
honor, 61, 177. *See also* warrior honor
Horton, Robin, 113
Houston, Stephen, 78
Huaca de La Luna, 21, 32, 33, 37, 39, 41, 43
Huamachuco, Peru, 34
huaraca. *See* sling
human sacrifice. *See* sacrifice

iconography, 9
    Amazonian, 140, 162, 163
    Andean, 190, 201–203, 223, 226, 229, 233
    Mississippian, 87, 93–95, 97, 99, 103–104,
        106
    Maya, 56, 62, 63, 70, 72, 74, 75

Moche, 9, 22, 31, 36–42, 51
North American Southwest, 123, 130
identity. *See* ethnic identity, warrior identity
idols, 124, 125, 201
Igbo people, Nigeria, 170
inequality. *See* social stratification
Inka Pukara, 214
Inkas, 9, 12, 28, 33–34, 46–50, 54, 159, 190,
   197–202, 211–217, 218–224, 227–235,
   239–242, 265
institutional economics, 167, 171
institutions, 70, 179, 181, 220, 224, 267
   changes in, 167, 224
Intervisibility of sites. *See* settlement patterns
Iroquois, 170
Iximche', 69

Japan, war in, 74, 79–80
Jequetepeque Valley, 37
Johnson, Mathew, 254
Julien, Catherine, 211

Keeley, Lawrence, 89, 248
Kepele Cult Network, 170, 176, 183
K'iche', 78
Kincaid Mound Center, 91
kinship and war, 196
kivas, 11, chapter 4
Knight, James, 93
Kollpayoc, 228

Lago Grande, 150–153, 160, 252
Lambayeque Valley, 37, 41
landscapes, 6, 31, 62, 125, 126, 197, 200, 203,
   215, 217, 240, 247, 249, 251, 256, 259
La Paya, 235
Larson, Lewis, 91
LeBlanc, Steven, 17, 54, 120, 122, 133, 248, 250
Lekson, Stephen, 120–121, 250
Lincoln, Abraham, 248
line-of-sight systems. *See* settlement patterns
Los Amarillos, 234
Lubbub Creek, 100
Lupacas, 198, 212–213, 216, 265

mace, 11, 32, 36, 41, 51, 86, 89, 94–98, 101,
   104–106, 200, 226
Macha, 28
Macleod, Murdo, 258–259
Mae Enga. *See* Enga
Manao Indians, 142

Martínez, José Luis, 230
Marxist theory, 59
Marx, Karl, 60
materialist explanation of war, 2, 5, 161, 178,
   193, 218–221
materiality, 6, 14, 111, 117, 221, 222, 223, 238,
   245, 253–257, 259, 261
Maya, 10, 56–83, 190, 252, 253, 257, 259, 271
Mayapan, 69
Meggit, Mervyn, 12, 165, 166, 178, 179
memory, 5, 20, 78, 111, 114, 132, 222, 239, 247,
   249, 256, 259
mentalism, 221–222, 246. *See also* practice
Mercado de Peñalosa, Pedro, 201
metaphors and war, 78, 108, 224, 236, 251
migration and war, 175–178, 248–250
military-based chiefdom, 219
Miller, Andrea, 11
Mississippian culture, warfare, 10, 84, 85, 88, 90
moat. *See* defensive architecture
Moche, 9, 18, 27–28, 31–33, 36–45, 47, 49–51,
   53–54, 190, 229
Presentation scene, 38
Mohenjo-Daro, 245
Moieties, 26, 28, 29, 32, 35, 41, 42, 122–124, 266
Molino-Chilacachi, 200
Moseley, Michael, 53
Moundville Mound Center, 105–106
multinaturalism, 161
Muralla de León, 66
murals, 36, 65, 67, 69, 70–72, 75–77, 123

Naranjo, 65, 79
Natchez Indians, 91, 98
Neoevolutionism, 119, 144
Nepeña Valley, 32–35, 41, 50
New Guinea. *See* Papua New Guinea
Nicasio, 200, 214–215
no-man's-lands. *See* buffer zones
nonhuman agents. *See* religion
Norris Farms, 90
North American Southwest, 11
   iconography, 123, 130
North American Southeast, 10

O
objectification, 86
objectified culture, 59, 62, 79, 82, 222
oral history, oral tradition, 12
   Amazonia, 155–156, 160, 169
   Enga, 169–186

Mississippian, 86, 97–98
Southwestern, 114, 125–130
Orinoco basin, 11, 146
Ostra site, 21, 30, 51, 52

Pacajes, 201, 216
pacarinas, 203
Pacatnamu, 47
palaces, 67, 77
Palenque, 77
palisade, *see* defensive architecture
Papua New Guinea, 12, 114, 133, 167, 169, 181, 182
Parakaná, 161–162
peace, 7, 92, 99, 134, 169, 177, 179–181, 187, 216, 256–258, 260
peacemaking, 8, 12, 13, 177, 179–181, 187, 189, 190, 245, 257, 258, 260
Peirce, Charles S., 116, 117, 221
performance theory, 75
Phenomenological approach, 14, 260
Piedras Negras, 69
pit houses, 11
Pizarro, Pedro, 46
plaques,
copper plates, 95. *See also* iconography, Mississippian
metal plaques, Andean, 226–228
Platt, Tristan, 26, 240,
Playa Catalán site, 30
plazas, 74, 77, 122, 123, 129, 144, 149, 150, 234, 240
political
centralization, 53, 66, 144–145, 148, 160, 163, 218, 260
change, 9, 13, 81, 218, 240–241
consequences of war, 2, 12, 65, 80, 140, 145, 160, 167, 174, 185–188, 240–242
Popol Vuh, 78, 79
population and war. *See* demographic pressure
positivist approaches to war, 2, 17, 54
power, 5, 58, 178, 220, 223
Powers Fort, 248
practice,
approaches to war, 4–8, 13, 193–194, 215–217, 218–221
definition of, 4, 6, 19
practical logic, 254
theory, 4, 8, 14, 18–20, 39, 59–62, 115–118, 166, 191–197, 220–224, 245–248, 253
pragmatic semantics, 226

prestige, 33, 42, 81, 89, 92–93, 171, 241
"primitive" warfare, 3
prisoners, 28, 32, 36–43, 47, 50, 63, 65, 75, 77–80, 100, 142, 161, 190, 259
psychological approaches to war, 4
Pucará, 213–215, 258
Pueblo peoples, 11
Hopi, 124, 127–128
Tonoans, 123
Keresans, 123
Zuni, 123
pukaras, 12, 27, 199, 203–216, 226, 231, 233, 239, 251–255, 258, 264–265
Pupu, Nitze, 167

Quebrada de Humahuaca, 228–230
Quebrada del Toro, 230
Quechua, 26, 27
Quetzalcoatl, 81
Quisque, 33
Q'umarkaj, 69

raiding, raids, 35, 89, 90, 101, 102, 105, 114, 52, 184, 256
reciprocity, 23, 26, 49, 184
religion. *See also* ancestor worship, ritual
Andean sun god, *Punchao,* 227–228
and war, 9, 11, 61, 71, 78, 190
divination, 201
gods, 50
ghosts, 256
Jurupari rites, 156, 158
katsinas, 124, 132–135
mountain spirits, 201, 203, 220, 231, 235
nonhuman agents, 11, 22, 49, 60, 110, 112–113, 220, 251
shamanism, 131, 161, 236–237
Southeastern Ceremonial Complex, 86,93
spirits, 113, 201
supernatural beings, 60, 235
wak'as, 224
witches, 124–125, 130, 251
resource scarcity. *See* environmental stress
revenge, 3, 92, 108, 173, 177, 185
Rio Negro basin, 142, 153, 159, 160
ritual, 10, 27, 73, 75, 81, 82, 161, 195, 201
battles, combat, 10, 43, 63, 78, 81, 240.
See also tinku
and warfare, 50, 56, 62, 67, 69, 71, 77, 78, 80, 176, 190, 201, 239
deposits, 118, 130, 133

rites of intensification, 102
rites of passage, 102
violence, 124–125,130, 250. *See also* cannibalism
Robarchek, Carole, 192
Robarchek, Clayton, 192
Rydén, Stig, 236

sacrifice, *see also* religion
  human, 8–10, 29, 31, 32, 36–43, 47, 50, 63, 67, 71, 74, 75, 77–81, 91, 98, 100, 142, 161, 190, 201, 235, 252, 259, 260
  of buildings, 80
sacrificer. *See* head taker
Sacsahuaman, Cusco, 34
Sahlins, Marshall, 60, 110, 130, 246
samurái. *See* Japan
San Jose de Moro site, 37
San Pedro de Atacama, 237
Santa Cruz Pachacuti, 34, 50, 228
Santa Valley, 30, 31, 33, 35, 50
Schaafsma, Polly, 123
Schiffer, Michel B., 11
sedentism and war, 2
segmentary organization, 29, 168, 240
Seibal, 64
semantics, 13, 221–224, 234–238, 241–242, 266
semiology, 221
semiotics, 221–224
settlement patterns,
  as indications of conflict, 120, 122, 202
  site intervisibility, 12, 122, 205–208–214
shamanism. *See* religion
Shanks, Michael, 87
shields, 27, 36, 41, 44, 52, 71, 121, 201, 226, 230, 235
siege, 46, 67
signs. *See* semiotics
Sipan, Peru, 37, 41
site clusters, 204–217
skeletal trauma, 47, 200, 202
slaves, 161
slave trade, 142–143, 156
slingers, 30
slings, 45, 226, 231–232, 242, 253
slingstones, 30, 33, 43–46, 51, 52, 199, 200
Snead, James, 255–256
social,
  complexity, 12, 119, 163, 167, 170, 173, 187, 189, 223
  stratification, 66, 84
soldiers, 7, 28, 34, 45–48, 65, 84, 228, 230

spatiality, 245, 255, 256, 259
spear, 69, 226, 228
spearthrower, 69–71
sports and war, 78, 257
Spurling, Geoffrey, 211–213
Society for American Archaeology, 9, 17, 261
soldiers, 7, 28, 34, 45–48, 65, 84, 228, 230
Spiro Mound Center, 94
Stanish, Charles, 205
state-level warfare, 1, 53, 54, 145
stelae, 63–65, 74–77
Suhler, Charles, 77
supernatural entities. *See* religion
swords, wooden, 122, 200
symbolic capital, 60, 163
symbolism
  of objects, 93, 104–108
  of war, 190
symbols, 3, 40, 80, 97–98. *See also* semantics
  collective, 58
  and practice, 62

Tariana, 155–160
Tastil, 234
Taylor, Edward, 113
Tee Cycle, 12, 166, 168, 170, 173, 176, 177, 180, 181, 183, 185, 186, 263
Teotihuacan, 69, 70, 126
terras pretas, 143, 148
territorial expansion, 53, 150, 218
Tezcatlipoca, 81
Than Te' K'inich, 67
theater, theatrical, 75–80, 252, 260
Thorpe, Nick, 245
Tikal, 66, 70, 77
tinku, 10, 18, 26–29, 32, 35–37, 39, 41–43, 47, 49–51, 53, 240, 242, 251–252, 266
Titicaca Basin, 12, 190–217, 229
Tiwanaku, 198, 201, 202, 229, 235–237, 258, 260, 266
Tlaloc, 70
towers, 11, 35, 120, 122, 132–134, 251
trade, 2, 12, 20, 122, 141–143, 153, 167, 169–174, 182–185, 229, 236–239
tradition, 20–24, 43, 49–50, 54, 80, 84, 89, 108, 119–121, 129, 156, 224, 256
transmutation, 13, 235–236, 241. *See also* shamanism
trauma. *See* skeletal trauma
tribal warfare, 1, 89, 90, 141, 159, 174, 175, 250, 257

"tribal zones," 141, 142, 159
trophies, 28, 190
trophy heads. *See* decapitation
trumpets, 11, 226, 229–231, 235, 253
Tschopik, Marion, 208
Tukano, 155, 158
Tulum, 69
Tumu, Akii, 167
Tupa Inka, 34, 214–215
Turner, Christy, 121
Turner, Jacqueline, 121
Tupinambá, 142, 252

Uma, Umasuyu, 212–216, 265
Urco, Urcusuyu, 212–216
Uxmal, 69

Vaca de Castro, Cristobal, 215
Vandkilde, Helle, 86–87, 247
Van Dyke, Ruth, 132
Verano, John, 234
violence, 1, 2, 5–8, 13, 31, 78, 79, 86, 88, 104,
      110, 120–122, 127, 129–135, 168, 175, 190,
      192, 195, 196, 217, 220, 238, 244, 245,
      248–261, 263, 266
Virú Valley, Peru, 35
Vivante, Armando, 234

Wak'as. See religion
Walker, William, 192
Wanano, 156
wanka monoliths, 240
Wankas, ethnic group, 218–220, 241
Waorani, 192
War
   and social change, 140, 145, 159, 163, 166,
      168, 194
   causes and explanations of, 1–2, 17, 52, 55,
      56, 159–161, 165, 166, 195, 244, 249
   consequences of, 1–2 , 55, 160, 163, 166, 185
   definition of, 8, 165
   evidence of, 90, 120–121, 123, 145, 150, 245,
      248
   frequency, 169
   leaders, 7, 107, 124, 184, 218–220, 235
   motivation for, 3, 156, 165

war chiefs, 124
warfare. *See* war
Wari State, 33
warrior, 5, 7, 45, 49, 53, 80, 82, 142, 165, 183–
      186, 190, 220, 258, 259
   attire, 36, 39, 63, 64, 75, 105, 227, 228, 236.
      See also costumes
   ideal, 74, 86, 87, 99, 104, 108
   identity, 9, 10, 86, 87, 190, 192, 247
   fraternities, societies, 123, 124, 134
   honor, 56, 62, 70, 71, 79
   representations of, 11, 30, 36, 37, 39, 41, 65,
      71, 74–78, 87, 96–108, 226–229, 235, 253
   sacrifice, 9, 36, 41, 79. *See also* sacrifice
   status, prestige, 40, 41, 63, 71, 102, 104, 236,
      241
   transmutation in battle, 241
warrior king, 77, 79
warrior priest, 36–38
warrior twins 124–127, 130. *See also* cosmology
weapons, weaponry, 34, 41, 43, 48, 52, 63, 65,
      69, 71, 73, 78, 93, 107, 120, 252
   meaning of, 13, 200, 226–233, 239
Weber, Max, 19
Wilcox, David, 120
Willey, Gordon, 35
Williams, Raymond, 58
Wilson, David, 31
witchcraft, 5, 124–125, 130, 171, 251. *See also*
      religion
women,
   as captives, victims, 28, 121, 141
   roles in war, 51, 62, 65, 75, 156, 181
Wuayna Capac, 34, 212, 231

Xingu, 159

Yanomami, 141, 162, 263
Yaxchilan, 69, 79
Yaxuna, 80
Yich'aak B'alam, 64

Zac Peten, 69